STRUCTURE AND REACTIVITY

MOLECULAR STRUCTURE AND ENERGETICS

Series Editors

Joel F. Liebman
University of Maryland Baltimore County

Arthur Greenberg
New Jersey Institute of Technology

Advisory Board

Israel Agranat • *The Hebrew University of Jerusalem (Israel)* • Thomas C. Bruice • *University of California, Santa Barbara* • Marye Anne Fox • *University of Texas at Austin* • Sharon G. Lias • *National Bureau of Standards* • Alan P. Marchand • *North Texas State University* • Eiji Ōsawa • *Hokkaido University (Japan)* • Heinz D. Roth • *Rutgers University, New Brunswick* • Othmar Stelzer • *University of Wuppertal (FRG)* • Ronald D. Topsom • *La Trobe University (Australia)* • Joan Selverstone Valentine • *University of California, Los Angeles* • Deborah Van Vechten • *Sesqui Science Services* • Kenneth B. Wiberg • *Yale University*

Other Volumes in the Series

Chemical Bonding Models
Physical Measurements
Studies of Organic Molecules
Biophysical Aspects
Advances in Boron and the Boranes
Modern Models of Bonding and Delocalization
Fluorine-Containing Molecules
Mechanistic Principles of Enzyme Activity
Environmental Influences and Recognition in Enzyme Chemistry

STRUCTURE AND REACTIVITY

Edited by
Joel F. Liebman
Arthur Greenberg

Joel F. Liebman
Department of Chemistry
University of Maryland Baltimore County
Baltimore, Maryland 21228

Arthur Greenberg
Division of Chemistry
New Jersey Institute of Technology
Newark, NJ 07102

Library of Congress Cataloging-in-Publication Data

Structure and reactivity / edited by Joel F. Liebman, Arthur
  Greenberg.
    p.   cm. — (Molecular structure and energetics ; v. 7)
  Includes bibliographies and index.
  ISBN 0-89573-712-4
  1. Chemical structure. 2. Reactivity (Chemistry) I. Liebman,
Joel F. II. Greenberg, Arthur. III. Series.
QD461.M629 1986 vol. 7
[QD471]
540 s—dc19
[541.2′2]
                                        88-19225
                                           CIP

© 1988 VCH Publishers, Inc.
This work is subject to copyright
All rights are reserved, whether the whole or part of the material is concerned, specifically those of translation, reprinting, re-use of illustrations, broadcasting, reproduction by photocopying machine or similar means, and storage in data banks.

Registered names, trademarks, etc. used in this book, even when not specifically marked as such, are not to be considered unprotected by law.

Printed in the United States of America.

ISBN-0-89573-712-4   VCH Publishers
ISBN-3-527-26958-4   VCH Verlagsgesellschaft

Published jointly by:

VCH Publishers, Inc.
220 East 23rd Street
Suite 909
New York, New York 10010

VCH Verlagsgesellschaft mbH
P.O. Box 10 11 16
D-6940 Weinheim
Federal Republic of Germany

Contributors

Dieter Cremer, Göteborgs Universitet, 41124 Göteborg, Sweden

Arthur Greenberg, Chemistry Division, New Jersey Institute of Technology, Newark, New Jersey 07102

Sury Iyer, Energetics and Warheads Division, AED, U.S. Army Research, Development, and Engineering Center, Picatinny Arsenal, New Jersey 07806-5000

Cheryl L. Klein, Department of Chemistry, Xavier University of Louisiana, New Orleans, Louisiana 70125

Elfi Kraka, Theoretical Chemistry Group, Argonne National Laboratory, Argonne, Illinois 60439

Tadeusz M. Krygowski, Department of Chemistry, Warsaw University, ul Pasteura 1, 02093 Warsaw, Poland

Tsutomu Mitsuhashi, Department of Chemistry, Faculty of Science, University of Tokyo, Tokyo, Japan

Jane S. Murray, Department of Chemistry, University of New Orleans, New Orleans, Louisiana 70148

Peter Politzer, Department of Chemistry, University of New Orleans, New Orleans, Louisiana 70148

Reinhard Schulz, University of Marburg, Marburg, West Germany

Armin Schweig, University of Marburg, Marburg, West Germany

Norman Slagg, Energetics and Warheads Division, AED, U.S. Army Research, Development, and Engineering Center, Picatinny Arsenal, New Jersey 07806-5000

Edwin D. Stevens, Department of Chemistry, University of New Orleans, New Orleans, Louisiana 70148

Series Foreword

Molecular structure and energetics are two of the most ubiquitous, fundamental and, therefore, important concepts in chemistry. The concept of molecular structure arises as soon as even two atoms are said to be bound together since one naturally thinks of the binding in terms of bond length and interatomic separation. The addition of a third atom introduces the concept of bond angles. These concepts of bond length and bond angle remain useful in describing molecular phenomena in more complex species, whether it be the degree of pyramidality of a nitrogen in a hydrazine, the twisting of an olefin, the planarity of a benzene ring, or the orientation of a bioactive substance when binding to an enzyme. The concept of energetics arises as soon as one considers nuclei and electrons and their assemblages, atoms and molecules. Indeed, knowledge of some of the simplest processes, e.g., the loss of an electron or the gain of a proton, has proven useful for the understanding of atomic and molecular hydrogen, of amino acids in solution, and of the activation of aromatic hydrocarbons on airborne particulates.

Molecular structure and energetics have been studied by a variety of methods ranging from rigorous theory to precise experiment, from intuitive models to casual observation. Some theorists and experimentalists will talk about bond distances measured to an accuracy of 0.001 Å, bond angles to 0.1°, and energies to 0.1 kcal/mol and will emphasize the necessity of such precision for their understanding. Yet other theorists and experimentalists will make equally active and valid use of such seemingly ill-defined sources of information as relative yields of products, vapor pressures, and toxicity. The various chapters in this book series use as their theme "Molecular Structure and Energetics," and it has been the individual authors' choice as to the mix of theory and of experiment, of rigor and of intuition that they have wished to combine.

As editors, we have asked the authors to explain not only "what" they know but "how" they know it and explicitly encouraged a thorough blending of data and of concepts in each chapter. Many of the authors have told us that writing their chapters have provided them with a useful and enjoyable (re)education. The chapters have had much the same effect on us and we trust readers will share our enthusiasm. Each chapter stands autonomously as a combined review and tutorial of a major research area. Yet clearly there are interrelations between them and to emphasize this coherence we have tried to have a single theme in each volume. Indeed the first four volumes of this series were written in parallel, and so for these there is an even higher degree of unity. It is this underlying unity of molecular structure and energetics with all of chemistry that marks the series and our efforts.

Another underlying unity we wish to emphasize is that of the emotions and of the intellect. We thus enthusiastically thank Alan Marchand for the opportunity to write a volume for his book series, which grew first to multiple volumes, and then became the current, autonomous series for which this essay is the foreword. We also wish to emphasize the support, the counsel, the tolerance and the encouragement we have long received from our respective parents, Murray and Lucille, Murray and Bella; spouses, Deborah and Susan; parents-in-law, Jo and Van, Wilbert and Rena; and children, David and Rachel. Indeed, it is this latter unity, that of the intellect and of emotions, that provides the motivation for the dedication for this series:

"To Life, to Love, and to Learning."

Joel F. Liebman
Baltimore, Maryland

Arthur Greenberg
Newark, New Jersey

Introduction

Prediction of chemical reactivity from the ground-state structure of a molecule remains a prime goal in chemistry. Difficulties encountered include the nature of reagents and reactions media and concomitantly, the structures of transition states. Nevertheless, the refinement of experimental techniques, such as high resolution X-ray crystallography, offer highly precise measurements of atomic coordinates as well as real electron density distributions. Furthermore, calculational techniques complement experimental studies and provide a new structure of energy parameters and insights. Even with these tools, translation of the conclusion to the solution phase, where most chemistry is done, is not self-evident. This volume presents a series of essays concerned with different aspects of these approaches.

The first chapter by Politzer and Murray introduces their calculational construct of the bond deviation index, and relates this to the strain of various carbocycles and to the electrostatic potentials in these molecules. The latter provide insight into the nature of attack by various reagent classes. The following chapter, by Klein and Stevens, continues the discussion of electron density distributions and electrostatic potentials from the experimentalist's point of view, thus complementing the preceding theoretical chapter. The nature of the crystallographic technique, its advantages, limitations, and applications to reactivity are described. Cremer and Kraka reexamine the concept of molecular strain completely from the perspective of modern quantum chemistry. Again, the topology of electron density distribution provides valuable insights into ground-state strain as well as reactivity. The fourth chapter, by Greenberg, explores the distortion of the (seemingly) familiar amide linkage. Distortion of this linkage, so vital in proteins, lactam antibiotics, and polymers may transform a stable amide into a reactive one, and thence into an "aminoketone" with "anomalous" acid/base and other reactivity properties. In chapter 5 Mitsuhashi explores a crucial aspect of the relationship between solid-state structure and reactivity. Specific classes of compounds with "respectable" carbon-carbon bonds spontaneously undergo reversible heterolysis in polar solution. To what extent does solid state structure hint at this behavior and what is the role of solvation, so critical in all solution chemistry? The following chapter, by Krygowski, examines crystallographically derived structures of molecules, and explicates them in terms of the contributions of canoni reference structures. This approach provides insight potentially applicable to ground-state stability as well as chemical reactivity. The chapter by Iyer and Slagg identifies the structural features which are recognizable in the molecules of explosives, propellants, and other high energy species. Some pedagogical discussion of the prerequisites for and mechanism of explosion are included for the nonexpert. The final chapter, by Schulz and Schweig, delves into one of the most active areas of UV photoelectron spectroscopy, that is, the identification of valence shell ionization energies in transient molecular species. Aside from providing ionization potentials which are relevant toward understanding redox behavior, the approach identifies intramolecular bond interactions that are subtle, yet useful in understanding chemical and physical properties.

Contents

Series Foreword vii

Introduction ix

1. Bond Deviation Indices and Electrostatic Potentials of Some Strained Hydrocarbons and Their Derivatives 1
Peter Politzer and Jane S. Murray

   1. Introduction 1
   2. Calculated Properties 3
   3. Discussion 6
   4. Summary 22
   Acknowledgments 23
   References 23

2. Experimental Measurements of Electron Density Distributions and Electrostatic Potentials 25
Cheryl L. Klein and Edwin D. Stevens

   1. Introduction 25
   2. Method 27
   3. Examples of Experimental Density Distributions and Electrostatic Potentials 38
   4. Conclusion 61
   References 62

3. The Concept of Molecular Strain: Basic Principles, Utility, and Limitations 65
Dieter Cremer and Elfi Kraka

   1. Introduction 66
   2. The Concept of Strain 68
   3. Quantitative Assessment of Strain and Strain Energy 71
   4. Chemical Consequences of Strain 79
   5. Comparison of Strain Energies in Small Cycloalkanes 82
   6. A Molecular Orbital Approach to Strain 89
   7. An Electron Density Approach to Strain 95

CONTENTS

8. A Step Toward a Unified Description of Strain: The Laplacian of the Electron Density — 104
9. Ways of Assessing the Strain Energy from Quantum Chemical Calculations — 109
10. Calculation of the Strain Energy from in situ Bond Energies — 111
11. Quantum Chemical Evaluation of the Molecular Strain Energy Using the Westheimer Approach — 114
12. Pros and Cons of σ Aromaticity — 117
13. Limitations of the Concept of Strain — 123
14. Conclusions — 130
Acknowledgments — 132
References — 133

4. Twisted Bridgehead Bicyclic Lactams — 139
Arthur Greenberg

1. Introduction — 139
2. Structure of the Amide or Lactam Linkage — 142
3. Nonplanar Amides and Lactams — 149
4. Syntheses of Bridgehead Lactams and Related Molecules — 151
5. Spectroscopic Studies — 156
6. Calorimetric and Crystallographic Data — 168
7. Structure, Strain, and Reactivity: Conclusions — 172
8. New Antibiotics? — 172
9. Model Substrates for Proteases — 175
References — 175

5. Polar Effects on the Lability of Carbon–Carbon Bonds — 179
Tsutomu Mitsuhashi

1. Introduction — 179
2. Polar Effects on Homolytic Cleavage — 181
3. Heterolytic Cleavage — 187
4. Polar Effects on Pericyclic Cleavage — 216
5. Concluding Remarks — 226
Acknowledgments — 226
References — 227

6. Resonance Structure Contributions Derived from the Experimental Geometry of Molecules — 231
Tadeusz M. Krygowski

1. Introduction — 231
2. Factors Influencing the Observed Geometry of Chemical Species in the Crystalline State Obtained by X-Ray Diffractometry — 234

3. Stabilization Energy and Resonance (Canonical) Structure
Contributions Derived from Experimental Bond Lengths ... 236
4. Applications ... 245
5. Concluding Remarks ... 251
References ... 253

7. Molecular Aspects in Energetic Materials ... 255
Sury Iyer and Norman Slagg

1. Introduction ... 255
Glossary ... 258
2. Sensitivity of Energetic Materials ... 258
3. Detonations ... 273
4. Summary ... 285
References ... 286

8. Ultraviolet Photoelectron Spectroscopy and Matrix Isolation: A Combined Approach to the Study of Reactive Species ... 289
Reinhard Schulz and Armin Schweig

1. Methods ... 289
2. Molecules ... 295
3. Outlook ... 354
Acknowledgments ... 356
References ... 356

Addendum ... 365

General Index ... 368

CHAPTER 1

Bond Deviation Indices and Electrostatic Potentials of Some Strained Hydrocarbons and Their Derivatives

Peter Politzer and Jane S. Murray

Department of Chemistry, University of New Orleans, New Orleans, Louisiana

CONTENTS

1. Introduction 1
2. Calculated Properties 3
3. Discussion 6
4. Summary 22
Acknowledgments 23
References 23

1. INTRODUCTION

The most common conception of a strained molecule is of a system in which the bond angles deviate significantly from certain expected values. For tetracoordinate carbon, for example, these would be in the neighborhood of 109.5°, while for tricoordinate and dicoordinate carbons the anticipated values are approximately 120° and 180°, respectively. Any marked deviation from these angles is regarded as introducing strain and destabilizing the system. It should be pointed out, however, that the term "molecular strain" is sometimes used in a very broad sense to include other factors that destabilize a molecule, such as distortions of bond distances and unfavorable conformations. (A detailed discussion has been presented by Greenberg and Liebman.[1]) In this chapter, our focus will be on some of the consequences of the more common source of strain, bond angle distortion.

A particularly interesting feature resulting from strained bond angles is that the bonds themselves are not concentrated along the internuclear axes, as is usually the case, but rather follow a curved path between the nuclei; they are commonly described as "bent."[2-6] In cyclic systems, the curvature is normally to the outside of the molecule. Bent bonds have been observed repeatedly in crystallographic studies, which show that the major buildups of electronic density in the bond regions are not on the internuclear axes.[7-15]

A very good approach to studying bent bonds is by means of the bond path concept.[16,17] This is the path between two bonded nuclei that goes through the saddle point in the internuclear electronic density and follows the maximum density gradient to each nucleus. Thus it corresponds to the ridge of highest electron density that links the two nuclei. For many bonds, such as those between the carbons in propane and cyclohexane, the bond path is essentially identical with the internuclear axis; in other instances, however, such as strained bonds, there can be a significant difference between the two (Figure 1-1).[17,18]

While the bond path permits a graphic depiction of bent bonds, it is also desirable to have a quantitative means of representing the extent of the

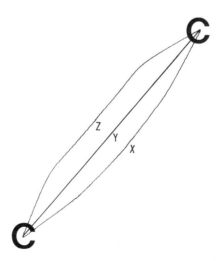

Figure 1-1. Bond path (Z), internuclear axis (Y), and reference path (X) for a C—C bond in tetrahedrane. Paths Z and X were computed at the STO-6G level.

curvature. For this purpose we have introduced the "bond deviation index," a numerical measure of the degree to which a given bond path differs from a defined reference path.[18] This index is a property of each individual bond in a molecule and provides an effective basis for characterizing strained bonds.

A second property that we have found to have a special significance for strained bonds is the molecular electrostatic potential. The C—C bonds in strained hydrocarbons and some of their derivatives show the rather unusual feature of having negative potentials near their midpoints, to the outside of the molecule. These bonds can accordingly serve as initial sites for electrophilic attack.

In this chapter we summarize and analyze these two properties—the bond deviation index and the electrostatic potential—for the bonds in a number of strained hydrocarbons and certain of their derivatives. Our discussion will bring out the important contrast between these properties, one of which focuses specifically on the electron densities in individual internuclear regions while the other reflects the net result, at any point in space, of the total charge distribution of a molecule. We shall show that it is possible by this approach to achieve a better understanding of the behavior of these molecules and greater insight into their similarities and differences.

2. CALCULATED PROPERTIES

A. Structures

Our initial step in the study of each molecule has been to calculate its optimized geometry, using the ab initio self-consistent field GAUSSIAN programs.[19] This has been done at various computational levels, in terms of basis sets ranging from minimum (STO-3G and STO-5G) to split-valence plus d orbitals (6-31G*). The choice of basis set was usually dictated by the size of the molecule. These computed geometries are in generally good agreement with the available experimentally determined ones. This means that it is meaningful to compare, at least qualitatively, the properties of molecules that have been optimized in terms of different basis sets.

In this context, it is of considerable interest to note that there appears to be no consistent general relationship, for these strained molecules, between the nature of the basis set and the accuracy of the optimized structural parameters. (It cannot be assumed that a basis set that yields an improved energy will also be better for calculating the structure or other properties.) This can be seen, for example, from the work of Wiberg and Wendoloski, who compared the effectiveness of basis sets ranging from STO-3G to 6-31G** in predicting the structures of 16 different hydrocarbons, most of them strained.[20] The same conclusion follows from a recent comparative study of the geometry of cyclopropylamine, as computed with several different basis sets.[21] It cannot even be assumed that the use of d orbitals necessarily improves the computed structure, although their inclusion is certainly generally desirable (especially when there are substituents with lone pairs).[22]

B. Bond Deviation Index

One approach to using bond paths as a basis for characterizing chemical bonds, especially those that are strained, is to establish in each instance some well-defined reference path and then to determine quantitatively the extent to which it differs from the actual one. We chose to define the reference paths in any molecule in terms of the superposed electronic densities of the corresponding free atoms, placed at the same relative positions as in the molecule. The reference paths are therefore the maximum gradient paths between the nuclei and the internuclear saddle points in the superposed free atom electronic densities. Our reference state is the same as that commonly used by crystallographers in presenting their density difference plots.

Since our reference paths are determined from free atom electronic densities, they reflect a situation with no repulsion between electrons on different atoms. Thus, to the extent that interatomic electronic repulsion is responsible for the strain associated with distorted bond angles, our reference state may be regarded as strain-free. Indeed, in molecules such as propane and cyclohexane, generally regarded as unstrained, the C—C reference paths are essentially the same as the actual bond paths.[18] In small, strained cyclic structures, on the other hand, the reference paths typically curve somewhat inward, while the bond paths curve outward (Figure 1-1).

The internuclear axis is of course another possibility for the reference path, and it might initially seem to be quite a reasonable one. However there appears to be no intrinsic physical reason for choosing it for this purpose. In particular, it seems appropriate that the reference paths for a strained cyclic molecule such as cyclopropane should be somewhat different in form from those for a straight-chain hydrocarbon. Our definition of reference path, which is molecule-dependent, allows for such differences; the internuclear axis, which can be viewed as a molecule-independent definition, does not.

There are certainly a number of different ways in which one could quantitatively measure the deviation of an actual bond path from the reference path. We chose to use the formula given in Equation 1-1, which defines the property we call the "bond deviation index" λ[18]:

$$\lambda = \frac{\left[\frac{1}{N}\sum_i^N r_i^2\right]^{1/2}}{R} \tag{1-1}$$

The quantities r_i are the distances between the actual and the reference paths at N equally spaced points. We take N to be 320, which is well above the value at which further increases in N lead to no significant changes in λ. The division by R serves to make λ a scaled dimensionless quantity and allows bonds of different lengths to be compared.

The significance of λ is of course entirely relative. The greater the extent to which the actual bond path deviates from the reference, the larger the magnitude of λ. When the two paths coincide, as for the C—C bonds in propane and cyclohexane, then λ is zero.

An important feature of the bond deviation index is that it is a property of each individual bond in a molecule, calculated rigorously and separately in each case. It reflects specifically the nature of the electronic density distribution in each internuclear region, and thus permits the characterization of the individual bonds. This is in contrast to the molecular strain energy, which is a property of the molecule as a whole. As usually estimated, strain energy may include contributions from various destabilizing factors, such as were mentioned at the beginning of this chapter. Accordingly, it should not be anticipated that there exists some straightforward relationship between these two properties, the bond deviation index and the molecular strain energy; they approach the quantification of strain in rather different ways.

C. Electrostatic Potential

Any distribution of charge creates an electrical potential $V(\mathbf{r})$ in the surrounding space. A point charge $\pm Q$ located at $\mathbf{r}$ would interact with this potential with an interaction energy of $\pm QV(\mathbf{r})$. In applying these concepts to the interpretation of molecular reactive behavior, the electrons and nuclei of the molecule are normally viewed as a static distribution of charge, and $V(\mathbf{r})$ is accordingly the molecular electrostatic potential, given rigorously by Equation 1-2:

$$V(\mathbf{r}) = \sum_A \frac{Z_A}{|\mathbf{R}_A - \mathbf{r}|} - \int \frac{\rho(\mathbf{r}')d\mathbf{r}'}{|\mathbf{r}' - \mathbf{r}|} \qquad (1\text{-}2)$$

where Z_A is the charge on nucleus A, located at $\mathbf{R}_A$; $\rho(\mathbf{r})$ is the electronic density function of the molecule, which we determine computationally from the molecular wave function.

The first term on the right-hand side of Equation 1-2 expresses the potential produced by the nuclei, while the second represents that of the electrons. Thus, negative values of $V(\mathbf{r})$ are indicative of regions in which the effects of the electrons are dominant; it is to such regions that an approaching electrophile will be attracted initially. When $\rho(\mathbf{r})$ is taken to be the ground state electronic density distribution of the molecule, $V(\mathbf{r})$ does not reflect the charge polarization that is likely to accompany the approach of some other chemical species. Hence, analyses involving $V(\mathbf{r})$ should be restricted to the early stages of interaction, to ensure separations large enough (eg, greater than the sum of the respective van der Waals radii) that such polarization is not significant.

Despite this limitation, molecular electrostatic potentials have proved to be a very effective tool for interpreting and predicting the reactive properties of molecules, especially toward electrophiles.[23-25] Treating nucleophilic interactions is somewhat more complicated,[24-26] although methods for doing so have been proposed.[26,27] (A positive potential does not necessarily indicate a tendency to react with nucleophiles. The positive charges of nuclei are highly concentrated, in contrast to the negative charges of the dispersed electrons, and create strong positive potentials that may outweigh the negative electronic contributions but yet not reflect an affinity for nucleophiles.) There have also been developed procedures for adding a polarization term to $V(\mathbf{r})$, using second-order perturbation theory.[27-29] Usually, however, it is $V(\mathbf{r})$ alone that is calculated, with the focus on regions in which it is negative, and in particular on the points at which $V(\mathbf{r})$ achieves its most negative values (the electrostatic potential minima). Finally, it is important to emphasize that the electrostatic potential, as given by Equation 1-2, is a real physical property; it can be determined experimentally, by diffraction techniques,[25] as well as computationally. At present, however, the latter approach is by far the dominant one.

3. DISCUSSION

A. General Features of Calculated Results

The strained hydrocarbons for which bond deviation indices and electrostatic potentials have been computed are shown in Figure 1-2, and the results are presented in Table 1-1. Their derivatives for which these properties have also been calculated are described later in this chapter.

Looking first at the hydrocarbons themselves, the bond deviation indices are seen to vary over a rather large range, from nearly zero in *p*-bishomocubane to 0.116 in tetrahedrane. By this criterion, therefore, the latter molecule has the most highly strained bonds of any that we have investigated. The case of cyclopropane is of particular interest because it illustrates our earlier comments concerning the absence of any direct relationship between the bond deviation index and the molecular strain energy. The value of λ is rather high for the C—C bonds in cyclopropane, 0.080, which seems quite reasonable in view of the C–C–C bond angles of 60°, certainly greatly distorted from the tetrahedral 109.5°. However, the strain energy of the molecule is commonly quoted as being only 27 kcal/mol.[1] This is only 9 kcal/mol per carbon atom or C—C bond, whereas the corresponding values for cubane, for example, in which the C–C–C bond angles are 90°, are 21 and 14

PROPERTIES OF STRAINED HYDROCARBONS   7

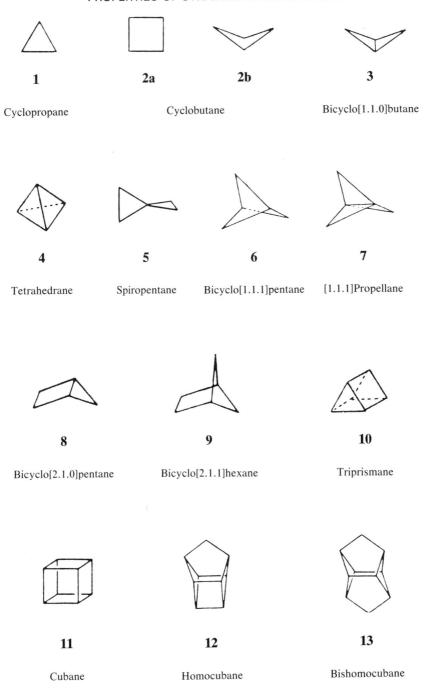

Figure 1-2. Structures of the strained hydrocarbons discussed in this chapter.

TABLE 1-1. Bond Deviation Indices and Bond Potential Minima of Some Strained Hydrocarbons[a]

Molecule	Structure	Bond deviation indices	Bond potential minima (kcal/mol)
1: Cyclopropane		0.080[b]	−13.0[c]
2A: Cyclobutane		0.021	−4.1
2B: Cyclobutane		0.023	−4.3
3: Bicyclo[1.1.0]butane[d]	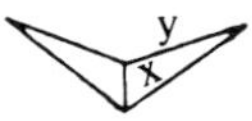	x: 0.084 y: 0.080	x: −2.3 y: −13.2
4: Tetrahedrane	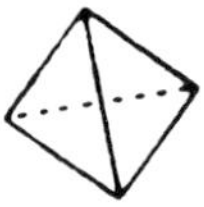	0.116[b]	−12.6
5: Spiropentane	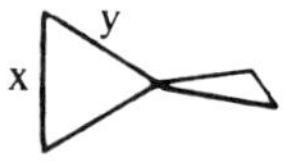	x: 0.086 y: 0.081	x: −13.2 y: −8.2
6: Bicyclo[1.1.1]pentane[d]	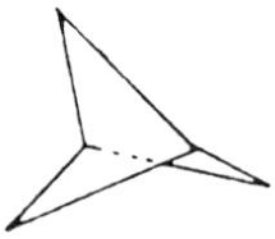	0.039	−4.9
7: [1.1.1]Propellane[d]	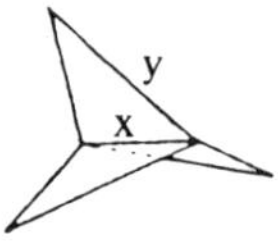	x: 0.000 y: 0.064	—
8: Bicyclo[2.1.0]pentane	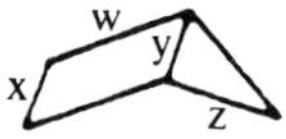	w: 0.023 x: 0.022 y: 0.081 z: 0.078	w: −4.8 x: −2.9 y: −9.5 z: −13.4
9: Bicyclo[2.1.1]hexane	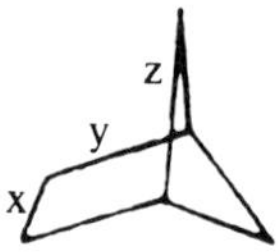	x: 0.004 y: 0.012 z: 0.028	x: −2.3 y: −4.7 z: −4.2
10: Triprismane[e]	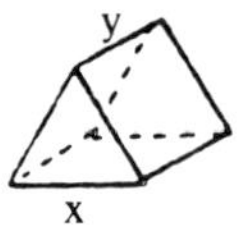	x: 0.080 y: 0.033	x: −13.2 y: −4.9

PROPERTIES OF STRAINED HYDROCARBONS

TABLE 1-1. Cont.

Molecule	Structure	Bond deviation indices	Bond potential minima (kcal/mol)
11: Cubane[f]	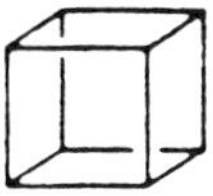	0.029	−4.6
12: Homocubane[f]		x: 0.030 y: 0.020 z: 0.009	x: −5.7 y: −2.4 z: −3.2
13: Bishomocubane[f]	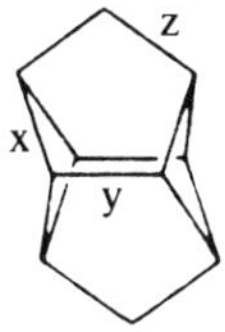	See note g.	x: −1.5 y: −1.2 z: −3.9

[a] Results for which references are not explicitly cited were calculated as part of the present work.
[b] Reference 18.
[c] Reference 31.
[d] Reference 34.
[e] Reference 33.
[f] Reference 32.
[g] As discussed in Reference 32, the C—C bond deviation indices in bishomocubane are all less than those of the analogous bonds in homocubane.

kcal/mol, respectively.[1] Thus the high bond deviation indices of cyclopropane appear to be inconsistent with its low strain energy. Dewar has pointed out, however, that cyclopropane has a very significant degree of aromatic character, which introduces a stabilizing effect and diminishes what would otherwise be a high strain energy, estimated to be 85 kcal/mol.[30] Cyclopropane is accordingly a good example of how the strain energy may reflect factors other than bond angle distortion (although it is probably somewhat unusual for such a factor to be a stabilizing one).

Proceeding now to the electrostatic potentials of the hydrocarbons in Figure 1-2, a very significant finding has been that the strained C—C bonds in these molecules are characterized by regions of negative potential located near their midpoints.[31–35] This is shown for cyclobutane in Figure 1-3. Such negative regions are not associated with more typical C—C single bonds, as in ethane and propane[36]; they constitute an important and distinctive feature

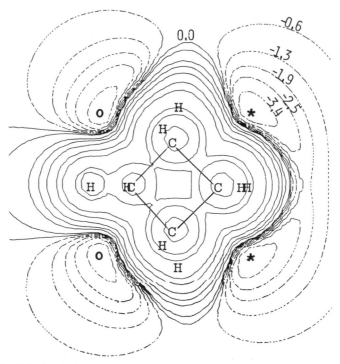

Figure 1-3. Calculated electrostatic potential (STO-5G) of the puckered form of cyclobutane (**2b**) in the plane containing the three carbons on the right-hand side of the figure. Dashed contours are negative. The magnitudes of the positive (solid) contours are 0.6, 3.1, 6.3, 12.6, 31.4, 50.2, 62.8, 313.8, 627.5, and 3137.5 kcal/mol. The positions of the most negative potentials are indicated by * and ○, with the respective values of −4.3 and −4.2 kcal/mol.

of these hydrocarbons. ([1.1.1]Propellane is a special case, in that the large and relatively strong negative potentials of the bridgehead carbons extend into the regions where the bond potentials would be expected; hence the latter are not separately observed.[34])

The presence of these negative bond potentials means that the bonds themselves can serve as initial sites for electrophilic attack. Observed examples of this include the olefin-like properties of cyclopropane[37,38] and bicyclobutane,[39,40] the tendency of the former for edge hydrogen bonding,[41] and the cation-catalyzed rearrangement reactions undergone by cubane and homocubane, in which the initial step is an interaction between the positive ion and a C—C bond.[1,42,43] (See also the discussion in Reference 32 regarding these rearrangement reactions.)

While it may be tempting to assume that these negative C—C bond potentials are directly related to the bent natures of these bonds, such an assumption is not warranted. To take an extreme example, the substitution of the electron-withdrawing —NO_2 group in cyclopropane completely eliminates the negative potentials associated with the C—C bonds, but leaves the bond paths and bond deviation indices essentially unchanged[31]; that is, the degree of bending of the bonds is virtually unaffected, even though density difference plots show that electronic charge has been removed from the internuclear regions. Even without introducing substituents, a number of instances will be seen in Table 1-1 of strained hydrocarbon C—C bonds that have relatively high λ values but weak potentials. Thus, two important points emerge. First, bent bonds do not necessarily produce negative bond potentials, and second, the degree of bending is not directly related to the quantity of electronic charge in the bond region.

B. Comparisons and Characterizations of Strained Hydrocarbons Based on Bond Deviation Indices

Many strained hydrocarbons can formally be regarded as composed of two or more smaller cyclic structures that are fused or linked together to form a larger molecule. Thus, spiropentane, **5**, can be viewed as two cyclopropanes that share one carbon, and bicyclo[2.1.0]pentane, **8**, as a cyclopropane and cyclobutane that share one C—C bond. This approach has sometimes been used in predicting and rationalizing the strain energies of larger hydrocarbons. For example, the strain energy of bicyclo[2.1.0]pentane is well approximated as the sum of the cyclopropane and cyclobutane strain energies; on the other hand, that of spiropentane is 10 kcal/mol more than the sum of two cyclopropanes.[1]

Strain energy, however, is not a rigorously defined property, and it can encompass a variety of factors that may cancel or reinforce each other in unknown ways. Thus it is not really a sufficient criterion for assessing the validity, in individual cases, of treating larger strained hydrocarbons as combinations of smaller ones. We suggest that the bond deviation index, as a direct and well-defined measure of the extent of bent bonding, provides a more fundamental basis for making this assessment. In the remainder of this section, we shall give several examples of this application of λ, using the calculated electrostatic potential to provide additional perspective. Both these properties are sensitive indicators of the distribution of electronic charge, the former for bonding regions and the latter for the entire molecule. Their values for the systems to be discussed are listed in Table 1-1.

a. Cyclobutane-**2**

If cyclobutane is to be used as a model for the four-sided faces of molecules such as bicyclo[2.1.0]pentane (**8**) and cubane (**11**), it is necessary to recognize that these faces are planar, whereas the equilibrium form of cyclobutane is nonplanar (Figure 1-2), with an experimentally determined dihedral angle of 152°.[44] Thus it must first be established how much difference in strain there is between the equilibrium and the planar cyclobutane structures. This turns out to be very little. Table 1-1 shows that the bond deviation indices and the electrostatic potentials associated with the C—C bonds have very similar values for the two forms of cyclobutane. In terms of strain considerations, therefore, the difference between **2a** and **2b** is relatively insignificant. Furthermore, since the experimentally determined energy barrier between these two conformations is only 1.4 kcal/mol,[44] it seems likely that it will not be possible to distinguish chemically between the two forms of cyclobutane.

It is interesting that **2a**, which is the less stable, is indicated as having the less strained C—C bonds. This supports earlier interpretations that view the geometry of cyclobutane as primarily a balance between ring strain, which tends to make cyclobutane planar, and the desire to avoid eclipsing neighboring methylene groups.[44]

b. Bicyclo[1.1.0]butane: 3

The bond deviation indices of bicyclobutane are consistent, on the whole, with regarding it as two fused cyclopropane rings. Bond x, between the bridgehead carbons, has a slightly higher degree of strain; this bond also shows olefin-like reactive properties, such as the addition of H_2O and I_2 to yield the corresponding cyclobutane derivatives.[39,40]

The electrostatic potentials associated with the bonds of type y are fully consistent with the analogy to cyclopropane (Table 1-1). Because of the bent nature of the molecule (our calculated dihedral angle is 118°), these negative regions overlap and produce a reinforced potential of -16.2 kcal/mol within the dihedral angle[35]; this helps to explain the observation that electrophilic attack occurs from this direction in proton addition[45] and in reaction with benzyne.[46]

Although bond x has the highest bond deviation index, it has the weakest bond potential, showing again that there is no direct correlation between the two properties. The magnitude of this potential, which is much less than that of the cyclopropane bonds, may reflect in part the positive contributions of the bridgehead hydrogens, which are coplanar with this bond path.

c. Tetrahedrane: 4

Based on its bond deviation index of 0.116, tetrahedrane is predicted to have the most highly strained C—C bonds of any in Table 1-1. Even though its bond potentials are very close to those of cyclopropane, its bonds are considerably more strained, due presumably to the rigidity imposed by its polyhedrane framework. It is not surprising, therefore, that this molecule has not yet been isolated. It has, however, been implicated as an intermediate in some reactions,[1] and its tetra-*t*-butyl derivative has been prepared.[47] The tetralithio derivative of tetrahedrane has been reported,[48] but recent ab initio calculations predict that it does not exist.[49]

d. Spiropentane: 5

The obvious model for spiropentane is two perpendicular cyclopropane rings linked through a shared carbon. The essential validity of this picture is borne out by the bond deviation indices, although these reveal a slightly increased strain in the bonds of type x (see Table 1-1). For these bonds, the electrostatic potentials are very similar to the cyclopropane values, whereas those of the y-type bonds have been modified by the presence of the other ring.

e. Bicyclo[1.1.1]pentane and Bicyclo[2.1.1]hexane: 6 and 9

One way of viewing both bicyclo[1.1.1]pentane and bicyclo[2.1.1]hexane is as cyclobutane with either a one-carbon or two-carbon bridge above the ridge of the dihedral angle. For bicyclohexane, this interpretation is supported by the bond deviation indices and the potentials of the bonds of type z, which are close to the cyclobutane values even though the analogous dihedral angle is now calculated to be 132° instead of 152° as in cyclobutane. The computed angles between bonds z and y and between x and y (respectively, 102° and 99°) are not too severely distorted from the tetrahedral 109.5°; thus the two-carbon bridge in bicyclohexane is a relatively relaxed portion of the molecule, as is confirmed by the low λ values for the x and y bonds. This allows the remainder of the molecule to retain a similarity to cyclobutane. For example, the angles between the z bonds at the secondary carbons are 84°, quite close to cyclobutane's 89°.

The situation is very different in bicyclo[1.1.1]pentane. There is a considerable degree of strain associated with the one-carbon bridge (eg, the C–C–C angles at the secondary carbons are about 74°),[20] and this presumably causes the other portion of the molecule to be significantly distorted from the geometry of cyclobutane. Thus, in addition to the bond angles at the secondary carbons now being 74° instead of the 89° of cyclobutane, the dihedral angles are 120° compared to 152° in the latter. The bond deviation indices in bicyclo[1.1.1]pentane, 0.039 (vs 0.023 in cyclobutane), reflect the resulting increased strain.

f. [1.1.1]Propellane: 7

Propellane is a particularly interesting molecule for several reasons, including the strong negative electrostatic potentials to the outside of its bridgehead carbons,[34] as well as the continuing controversy regarding the existence and nature of bonding between these carbons.[34,50–52] (In this chapter, propellane is drawn with a bond between the bridgehead positions.) The rather unique features of this molecule lead to bond deviation indices that are markedly different from any others listed in Table 1-1, indicating that no combination of the various smaller rings thus far discussed would produce a reasonable model for propellane.

g. Bicyclo[2.1.0]pentane: 8

Bicyclo[2.1.0]pentane offers an example of two fused rings of different sizes. The anticipated model for this molecule would be cyclobutane fused to cyclopropane (even though the present four-membered ring is planar), and the bond deviation indices confirm that this is indeed a reasonable representation of the system.

An interesting point that comes up here for the first time is the nature of the bond that is shared by fused rings of different sizes. Will it resemble the bonds in one or the other of the rings, or will it be somewhere intermediate between them? In the present instance, the higher degree of strain associated with cyclopropane-like bonds is found to be dominant, and it determines the nature of the shared bond. A similar situation is found in the case of triprismane.

h. Triprismane: 10

Triprismane (often simply called prismane) and cubane (**11**) may both be regarded as members of the prismane family (cubane can also be called tetraprismane). Accordingly, one might expect a similarity between the four-sided faces of the two molecules. This expectation is partially confirmed by the data in Table 1-1, which show the bond deviation indices of the y-type bonds of triprismane to be quite close in magnitude to those of cubane, as are also the electrostatic potentials. For the bonds of type x, on the other hand, both these properties have virtually the same values as for cyclopropane. This is another example, therefore, of the highly strained three-sided face determining the nature of the bond through which it is fused to a less-strained ring. [See Subsection g, above.] Thus, from the standpoint of strain, triprismane can be viewed as composed of two cyclopropanelike faces linked by three cubane-like bonds.

i. Cubane, Homocubane, and Bishomocubane: 11–13

Although it is tempting to view cubane as composed of six fused planar cyclobutanes, the bond deviation indices to show a 38% increase, probably due to the increased strain resulting from the imposition of the polyhedrane geometry. The situation is somewhat analogous to the tetrahedrane–cyclopropane relationship (in which case the increase in bond deviation indices was 46%).

Homocubane and *p*-bishomocubane may be regarded as formed from cubane by the insertion of methylene groups into cubane C—C bonds. This is accompanied by a general relaxation of the structure in the vicinity of the CH_2 group, as can be seen from our optimized geometries for these three molecules, published earlier.[32] For example, the bond angles change significantly from their cubane values toward the tetrahedral 109.5°.

This structural relaxation leads to a decrease in strain, which is reflected in the bond deviation indices. In homocubane, for instance, these diminish markedly in progressing from its cubane-like portion (bond x) to the intermediate region (bonds y) to bonds z, which are relatively little strained.

The preceding discussion has shown that the bond deviation index provides a quantitative basis for determining when and to what extent it is valid to treat a larger strained hydrocarbon as composed of fused or linked smaller ones. Indeed the analysis can be carried even further, to the level of individual bonds (eg, in the case of triprismane), since λ is specifically a bond property.

An important observation that has come out of these analyses relates to a bond that is shared by two fused rings. When strained hydrocarbons have been modeled by combinations of smaller rings in the past, the usual implicit (although somewhat contradictory) assumption was that shared bonds could be viewed as having characteristics consistent with both the rings, even when these were of different sizes. However, in the only such cases studied in this work—bicyclo[2.1.0]pentane and triprismane—we have found that the shared bond resembles those of the more strained ring.

In making these statements, we are saying in effect that the bond deviation index has a certain degree of transferability—that the C—C bonds in three-membered rings, for instance, will generally have λ in the neighborhood of 0.08 regardless of the particular molecule containing the ring, whereas for four-membered rings it will be in the range 0.02–0.03, except for bonds that are fused to three-membered rings (see above). These are of course only rough generalizations and certainly subject to exceptions (eg, tetrahedrane). Nevertheless a significant level of transferability does exist.

Although it is characteristic of the C—C bonds in strained hydrocarbons to have negative electrostatic potentials associated with them, we have pointed out several times that there is no correlation between the magnitudes of these potentials and the λ values of the bonds. This reflects the fact that the bond deviation index is determined only by the electronic density distribution in the internuclear region, whereas the electrostatic potential is a "global" property, representing the net (integrated) electrostatic effect at each point of all the charges in the entire molecule. Thus the C—C bond potential may be modified, for example, by the positive contributions of nearby hydrogens, even though these may have no effect on the strain.

C. Calculated Bond Deviation Indices and Electrostatic Potentials of Some Substituted Strained Hydrocarbons

Bond deviation indices and electrostatic potentials of some substituted cyclopropanes, bicyclobutanes, triprismanes, and cubanes are listed in Tables 1-2 and 1-3, respectively. Substituents ranging from amino and isocyanate groups to strong electron-withdrawing groups such as trifluoromethyl and nitro are included, showing a wide range of effects. The bond deviation

TABLE 1-2. Bond Deviation Indices of Some Substituted Strained Hydrocarbons

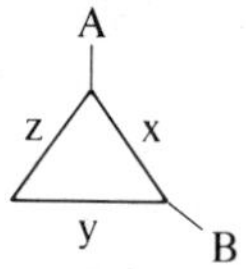

Cyclopropanes[a]

Substituents		Bond deviation indices		
A	B	x	y	z
NO$_2$	H	0.079	0.080	0.079
NO$_2$	NO$_2$	0.080	0.079	0.079
CF$_3$	H	0.080	0.079	0.079
CF$_3$	CF$_3$	0.078	0.080	0.079
NH$_2$	H	0.078	0.082	0.079

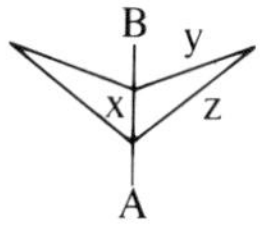

Bicyclo[1.1.0]butanes[b]

Substituents		Bond deviation indices		
A	B	x	y	z
NO$_2$	H	0.080	0.080	0.077
NO$_2$	NO$_2$	0.080	0.078	0.078
NO	H	0.078	0.077	0.077
NO	NO	0.075	0.075	0.075
NH$_2$	H	0.090	0.081	0.077
NH$_2$	NH$_2$	0.095	0.078	0.078
NCO	H	0.086	0.081	0.080
NH$_2$	NO$_2$	0.072	0.064	0.064
NCO	NO$_2$	0.082	0.076	0.078

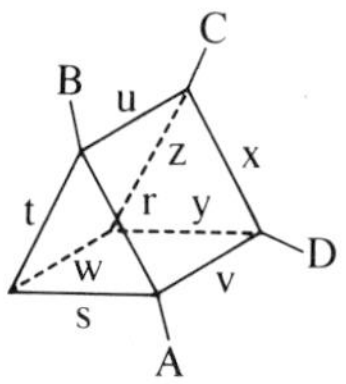

Triprismanes[c]

Substituents				Bond deviation indices								
A	B	C	D	r	s	t	u	v	w	x	y	z
NO$_2$	H	H	H	0.082	—	0.078	—	0.038	0.032	—	0.079	—
NO$_2$	NO$_2$	H	H	0.082	0.079	—	0.035	0.035	0.031	0.077	—	—
H	NO$_2$	NO$_2$	H	0.079	0.076	—	0.042	0.030	—	0.080	0.077	—

PROPERTIES OF STRAINED HYDROCARBONS

TABLE 1-2. Cont.

Substituents				Bond deviation indices								
A	B	C	D	r	s	t	u	v	w	x	y	z
H	NO₂	H	NO₂	—	0.077	0.080	0.036	0.036	0.029	0.080	—	—
NH₂	H	H	H	—	0.079	0.077	0.032	0.034	—	0.079	—	0.079
NH₂	NH₂	H	H	—	0.079	0.077	0.034	0.035	0.032	—	0.079	0.079
H	NH₂	NH₂	H	—	0.077	0.080	0.036	0.032	—	0.081	0.078	—
H	NH₂	H	NH₂	—	0.077	0.081	0.033	0.033	0.032	0.080	0.079	—

Cubanes[a]

Substituents: A	Bond deviation indices		
NO₂	o, p, q: 0.034	r, s, t, u, v: 0.028	
Cl	o, p, q: 0.035	r, s, v, w: 0.028	t, u, y: 0.029
CF₃	o: 0.031 t, u, y: 0.029	p: 0.029 v, z: 0.028	q: 0.030
NH₂	o: 0.033	p, q: 0.031	r, s, t, u, v, w: 0.029

Lithiocubanes[d]

Substituents: A	Bond deviation indices					
	u	v	w	x	y	z
—C(=O)NH₂	0.019	0.021	0.034	0.034	0.030	0.030
—C(=N)(O-C)(C) (oxazoline)	0.020	0.022	0.034	0.033	0.030	0.030
CF₃	0.018	0.022	0.035	0.036	0.030	0.030
CH₃	0.023	0.022	0.030	0.031	0.030	—
NO₂	0.024	0.021	0.038	0.041	0.029	0.029

[a] R. Bar-Adon, K. Jayasuriya, and B. A. Zilles, unpublished work.
[b] Reference 35.
[c] Reference 33, and unpublished work by K. Jayasuriya and B. A. Zilles.
[d] K. Jayasuriya, unpublished work.

indices of the C—C bonds in these substituted compounds all bear some similarity in magnitude to those of the parent unsubstituted hydrocarbons, but to varying extents. The electrostatic potentials, on the other hand, are largely dominated by the substituent group potentials; in many cases the negative bond potentials are completely eliminated. In view of the contrasting nature of these properties, which is especially striking in these substituted systems, our discussion will focus on each one separately.

In looking first at the bond deviation indices of the cyclopropane derivatives, it is seen that the λ values of the C—C bonds are virtually unaffected by the presence of nitro, trifluoromethyl, and amino groups (Table 1-2). The cyclopropane-like bonds of triprismane also show no significant change in λ upon substitution with nitro and amino groups, and the same is true in general for the bonds y and z in bicyclobutane, a molecule that can be viewed as two fused cyclopropane rings.

However, the bridgehead bond in bicyclobutane, when substituted at either or both of its positions, does not consistently fit this pattern. For example, upon introduction of one and two amino groups, the bond deviation index of the bridgehead bond increases from 0.084 to 0.090 and 0.095, respectively. For the mono- and dinitroso derivatives, on the other hand, the bond deviation index of this bond decreases to 0.078 and 0.075, respectively. It is interesting to observe that the bridgehead bond in bicyclobutane, which is the only one in the series of three-membered ring systems to show such anomalous behavior, is also the only one that is coplanar with the bonds to the substituents. One may speculate that this facilitates resonance-type interactions between the bridgehead substituents and the bond.

As an example, such interaction does appear to play a very significant role in the case in which amino and nitro groups occupy the bridgehead positions. This combination of substituents can lead to the weakening of the bridgehead bond by the following mechanism for the case of the complete disruption:

$$:NH_2 \quad \overset{O^-\;\;O^-}{N} \quad \overset{C-C}{} \quad \rightarrow \quad {}^+NH_2 \quad \overset{O^-\;\;O^-}{{}^+N} \quad C \quad C$$

It is consistent with this mechanism that our calculated structure for this system shows a significantly increased bond length compared to unsubstituted bicyclobutane (1.61 vs 1.48 Å).[35] It is not surprising that this major structural modification is accompanied by significant changes in C—C bond deviation indices throughout the molecule.

For the triprismanes, the introduction of one or two amino groups does not greatly affect the bond deviation indices of either cyclopropane-like or cubane-like bonds. With the mono- and dinitrotriprismanes, however, there is a tendency for a small increase in λ for the cubane-like bonds leading to the sites of substitution. The largest increase appears where substituents B and C are nitro groups (see Table 1-2); bond u has a λ value of 0.042 compared to 0.033 for unsubstituted triprismane.

A similar trend is seen for the mono-substituted cubanes. The amino and trifluoromethyl groups have no significant effect upon the λ values of the C—C bonds leading to the site of substitution. For nitrocubane and chlorocubane, however, increases in the bond deviation indices of bonds o, p, and q are observed (to 0.034 and 0.035, respectively, compared to 0.029 for unsubstituted cubane). Other bonds in nitrocubane and chlorocubane are unaffected by the introduction of —NO$_2$ and —Cl. This is in fact a general trend; even when a substituent has an effect on the λ values of the adjoining C—C bonds, the cubane-like bonds farther removed from the site of substitution retain λ as in the unsubstituted polyhedrane. (Bond deviation indices have also been calculated for 1,3- and 1,4-diaminocubane and 1,4-dinitrocubane. These results show the same patterns as observed for the monosubstituted cases.)

In proceeding to the 2-substituted lithiocubanes, it is important to recognize that the Li—C bond in lithiocubane differs markedly from the substituent–carbon bonds in the mono-substituted cubanes previously discussed in that it is highly polar. (This polarity leads to the rare feature of relatively strong negative electrostatic potentials associated with the carbon.[53]) Because of this distinction, it is not surprising that the bond deviation indices of the C—C bonds adjoining the lithiated carbons in the 2-substituted lithiocubanes show a different trend from that observed for the mono-substituted cubanes. The λ values of the bonds leading to the lithiated carbons are smaller than those found for unsubstituted cubane C—C bonds, reaching as low as 0.018 for bond u when group A is trifluoromethyl. The two remaining bonds leading to A (bonds w and x) show an increase in λ relative to cubane, except when A = CH$_3$. Other bonds farther removed from lithium and the substituents A show no significant change in λ. This is fully consistent with the general trend that has been observed for the substituted polyhedranes (see above).

Whereas the bond deviation indices of the substituted strained systems are similar to those of the unsubstituted parent hydrocarbons, the calculated electrostatic potentials of the substituted cyclopropanes, bicyclobutanes, triprismanes, and cubanes presented in Table 1-3 differ markedly from those of the unsubstituted compounds in two respects. First, the negative bond potentials that appear outside the C—C bonds in the latter systems now either have modified minimum values or are completely eliminated. Second, the overall electrostatic potentials of these substituted molecules are largely dominated by the potentials of the substituent groups, all of which have large negative regions associated with them; the most negative minimum values associated with the heteroatoms in these groups are listed in Table 1-3.

Looking initially at the effects of substituent groups on bond potentials, it is seen that the introduction of just *one* strong electron-withdrawing nitro, chloro, or trifluoromethyl group in all instances results in the elimination of all negative bond potentials. Substitution of a single amino group, however, has mixed effects; some negative bond potentials increase or decrease in magnitude relative to those of the parent hydrocarbon, while others become positive. Specifically for the case of bicyclobutane, the introduction of one nitroso or isocyanate group at a bridgehead position is seen to reduce, but not eliminate, negative bond potentials relative to unsubstituted bicyclobutane. However for all the cases of disubstitution given in Table 1-3, negative bond potentials are absent.

It should be emphasized at this point that this discussion is limited to the substituents and strained hydrocarbons included in Table 1-3 and is not intended to be completely general. A striking trend that is observed, how-

TABLE 1-3. Electrostatic Potential Minima (kcal/mol) of Some Substituted Strained Hydrocarbons

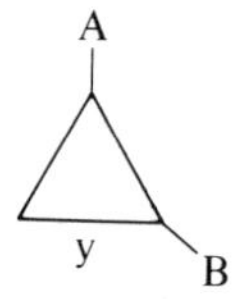

Cyclopropanes[a]

Substituents		
A	B	Electrostatic potential minima
NO_2	H	$V(NO_2) = -56.4$
NO_2	NO_2	$V(NO_2) = -44.5$
CF_3	H	$V(CF_3) = -18.6$
CF_3	CF_3	$V(CF_3) = -13.5$
NH_2	H	$V(NH_2) = -100.0$, $V(y) = -10.5$

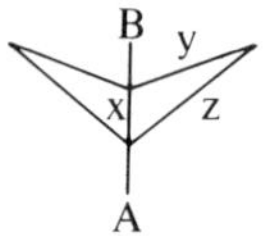

Bicyclo[1.1.0]butanes[b]

Substituents		
A	B	Electrostatic potential minima
NO_2	H	$V(NO_2) = -56.7$
NO_2	NO_2	$V(NO_2) = -46.8$
NO	H	$V(N) = -65.8$, $V(O) = -46.5$, $V(y) = -1.2$, $V(z) = -4.3$
NO	NO	$V(N) = -57.1$, $V(O) = -38.5$
NH_2	H	$V(NH_2) = -84.0$, $V(x) = -1.4$, $V(y) = -15.8$
NH_2	NH_2	$V(NH_2) = -88.0$
NCO	H	$V(N) = -2.9$, $V(O) = -50.2$, $V(y) = -2.6$, $V(z) = -3.6$
NH_2	NO_2	$V(NH_2) = -29.6$, $V(NO_2) = -68.6$
NCO	NO_2	$V(O; NCO) = -40.5$, $V(NO_2) = -53.7$

PROPERTIES OF STRAINED HYDROCARBONS

TABLE 1-3. Cont.

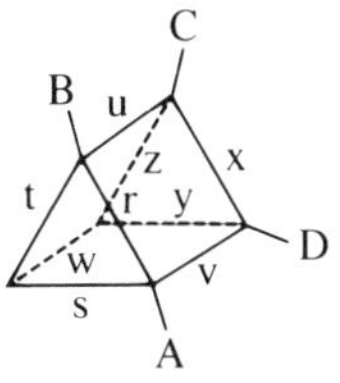

Triprismanes[c]

Substituents				
A	B	C	D	Electrostatic potential minima
NO$_2$	H	H	H	V(NO$_2$) = −55.6
H	NO$_2$	NO$_2$	H	V(NO$_2$) = −49.1
H	NO$_2$	H	NO$_2$	V(NO$_2$) = −47.5
NH$_2$	H	H	H	V(NH$_2$) = −96.6, V(s) = −18.6, V(t) = −13.4
				V(u) = −3.3, V(v) = positive, V(z) = −10.8

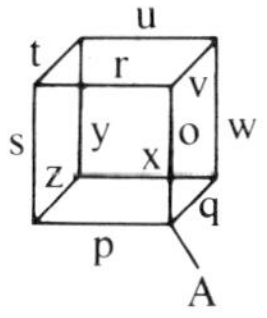

Cubanes[d]

Substituents: A	Electrostatic potential minima
NO$_2$	V(NO$_2$) = −56.7
Cl	V(Cl) = −27.3
CF$_3$	V(CF$_3$) = −20.6
NH$_2$	V(NH$_2$) = −101.0, V(o) = positive,
	V(r) = −2.0, V(s) = −3.2, V(t) = −3.6
	V(u) = −3.4, V(y) = −4.3

[a] Reference 31 and unpublished work by R. Bar-Adon, K. Jayasuriya, and B. A. Zilles.
[b] Reference 35.
[c] Reference 33 and unpublished work by K. Jayasuriya and B. A. Zilles.
[d] Unpublished work by R. Bar-Adon, K. Jayasuriya, and B. A. Zilles.

ever, is that monosubstition of a strongly electron-withdrawing group eliminates all negative bond potentials. On the other hand, the substitution of an amino group, which is a strong resonance donor as well as inductive withdrawer of electronic charge, tends to have mixed effects. We speculate that the introduction of one weakly electron-withdrawing group would tend to simply reduce the magnitude of the negative bond potentials relative to those in the parent hydrocarbon, as is observed for the specific case of mononitroso bicyclobutane.

In summary, the calculated results presented in Table 1-2 indicate that the introduction of substituent groups to strained hydrocarbons does not have a major effect on the bond deviation indices of the C—C bonds relative to those of the unsubstituted systems. This suggests that the amount of strain in C—C bonds is not highly substituent-dependent, but rather more strongly dependent on the structure of the parent hydrocarbon. The bond deviation indices of cyclopropane-like bonds are virtually unaffected by substitution (with a few exceptions). The introduction of a substituent group to a polyhedrane with cubane-like C—C bonds affects, if at all, only the bond deviation indices of those cubane-like bonds that lead to the site of substitution, leaving the λ values of bonds more than one removed from this site unchanged.

While the bond deviation indices of the individual C—C bonds in strained hydrocarbons in general remain fairly constant upon the introduction of substituents, the overall electrostatic potentials change significantly and are largely dominated by the substituent group potentials. The negative bond potentials associated with the parent hydrocarbons are completely eliminated upon substitution of a strong electron-withdrawing group. Monosubstitution of groups that are not strongly electron-withdrawing has mixed effects or merely weakens the negative bond potentials.

4. SUMMARY

In this chapter, we have analyzed and compared a number of strained hydrocarbons and some of their derivatives. Our analysis has been in terms of two key properties: the bond deviation index λ, which is a measure of the degree of bending (and strain) associated with a bond, and the electrostatic potential $V(\mathbf{r})$, a well-established tool for interpreting and predicting molecular reactive behavior. It is important to note the contrast between these two properties: λ focuses specifically on individual internuclear regions, whereas $V(\mathbf{r})$ directly reflects the charge distribution in the entire molecule. For this reason, no close relationship between λ and $V(\mathbf{r})$ should be anticipated, nor is one observed. While we defined the very interesting feature that these are consistently negative potentials near the midpoints of the C—C bonds in strained hydrocarbons (but not necessarily their derivatives), there is no correlation between the magnitudes of these potentials and the corresponding bond deviation indices. A related significant conclusion is that the degree of bond bending is not directly related to the amount of electronic charge in the bond region.

We have demonstrated that the bond deviation index is an effective quantitative means for assessing the validity, in each case, of treating a larger strained hydrocarbon as composed of fused or linked smaller ones. The limited evidence available indicates that when a bond is shared by two fused rings, it resembles those of the more strained one. An implication of these observations is that the bond deviation index possesses a certain degree of transferability.

For substituted strained hydrocarbons, the C—C bond deviation indices are generally similar in magnitude to those of the parent hydrocarbons. The electrostatic potentials, on the other hand, are usually dominated by relatively strong negative regions associated with the substituents themselves; the weaker C—C bond potentials of the original hydrocarbons are now often considerably modified or completely absent, although the bonds do remain bent. For example, strongly electron-withdrawing groups, such as —NO_2, —CF_3, and —Cl, cause the elimination of all negative bond potentials, but either increase or leave essentially unchanged the bond deviation indices. With regard to our principal interests in this chapter—bond bending and angular strain—our bond deviation indices permit the generalization that the degree of strain in C—C bonds depends much more on the parent hydrocarbon than on the natures of the substituents.

ACKNOWLEDGMENTS

We express our thanks to Dr. Ruth Bar-Adon, Dr. Keerthi Jayasuriya, and Dr. Barbara A. Zilles for calculating some of the bond deviation indices and electrostatic potentials given in this chapter. We also greatly appreciate the support of this work by the Office of Naval Research.

REFERENCES

1. Greenberg, A.; Liebman, J. F. "Strained Organic Molecules." Academic Press: New York, 1978.
2. Coulson, C. A.; Moffitt, W. E. *J. Chem. Phys.* **1947**, *15*, 151.
3. Coulson, C. A.; Moffitt, W. E. *Phil. Mag.* **1949**, *40*, 1.
4. Burnelle, L.; Coulson, C. A. *Trans. Faraday Soc.* **1956**, *53*, 403.
5. Flygare, W. H. *Science* **1963**, *140*, 1179.
6. Pebers, D. *Tetrahedron* **1963**, *19*, 1539.
7. Fritchie, C. G. *Acta Crystallogr.* **1966**, *20*, 27.
8. Hartmann, A; Hirshfeld, F. L. *Acta Crystallogr.* **1966**, *20*, 80.
9. Mathews, D. A.; Stucky, G. D. *J. Am. Chem. Soc.* **1971**, *93*, 5954.
10. Ito, T.; Sakurai, T. *Acta Crystallogr.* **1973**, *B29*, 1594.
11. Irngartinger, H. In "Electron Distributions and the Chemical Bond," Coppens, P.; and Hall, M. B., Eds.; Plenum Press: New York, 1982, p. 361.
12. Eisenstein, M.; Hirshfeld, F. L. *Acta Crystallogr.* **1983**, *B39*, 61.
13. Bianchi, R.; Pilati, T.; Siminetta, M. *Helv. Chim. Acta* **1984**, *67*, 1707.
14. Angermund, K.; Claus, K. H.; Goddard, R.; Krüger, C. *Angew. Chem. Int. Ed. Engl.* **1985**, *24*, 237.
15. Pant, A. K.; Stevens, E. D. Unpublished studies of 1,3-dinitrocubane.
16. Bader, R. F. W.; Nguyen-Dang, T. T. *Adv. Quantum Chem.* **1981**, *14*, 63.
17. Bader, R. F. W.; Tang, T.-H.; Tal, Y.; Biegler-Konig, F. W. *J. Am. Chem. Soc.* **1982**, *104*, 940, 946.
18. Politzer, P.; Abrahmsen, L.; Sjoberg, P.; Laurence, P. R. *Chem. Phys. Lett.* **1983**, *102*, 74.
19. (a) GAUSSIAN 70: Hehre, W. J.; Lathan, W. A.; Ditchfield, R.; Newton, M. D.; Pople, J. A. *Quantum Chem. Progr. Exch.* **1973**, *11*, 236. (b) GAUSSIAN 80: Binkley, J. S.;

Whiteside, R. A.; Krishnan, R.; Seeger, R.; DeFrees, D. J.; Schlegel, H. B.; Topiol, S.; Kahn, L. R.; Pople, J. A. *Quantum Chem Progr. Exch.* **1981,** *13,* 406. (c) GAUSSIAN 82: Binkley, J. S.; Frisch, M.; Raghavachari, K.; DeFrees, D.; Schlegel, H. B.; Whiteside, R.; Fluder, E.; Pople, J. A. GAUSSIAN 82, Release B. Pittsburgh: Carnegie–Mellon University, 1982.
20. Wiberg, K. B.; Wendoloski, J. J. *J. Am. Chem. Soc.* **1982,** *104,* 5679.
21. Rall, M.; Harmony, M. D.; Cassada, D. A.; Staley, S. W. *J. Am. Chem. Soc.* **1986,** *108,* 6184.
22. Pulay, P.; Fogarasi, G.; Pang, F.; Boggs, J. E. *J. Am. Chem. Soc.* **1979,** *101,* 2550.
23. Scrocco, E.; Tomasi, J. *Adv. Quantum Chem.* **1978,** *11,* 116.
24. Politzer, P.; Daiker, K. C. In "The Force Concept in Chemistry," Deb, B. M., Ed.; Van Nostrand Reinhold: New York, 1981, Chapter 6.
25. Politzer, P.; Truhlar, D. G., Eds. "Chemical Applications of Atomic and Molecular Electrostatic Potentials." Plenum Press: New York, 1981.
26. Politzer, P.; Landry, S. J.; Wärnheim, T. *J. Phys. Chem.* **1982,** *86,* 4767.
27. Francl, M. M. *J. Phys. Chem.* **1985,** *89,* 428.
28. Bartlett, R. J.; Weinstein, H. *Chem. Phys. Lett.* **1975,** *30,* 441.
29. Bertran, J.; Silla, E. Fernandez-Alonso, J. I. *Tetrahedron* **1975,** *31,* 1093.
30. Dewar, M. J. S. *Bull. Soc. Chim. Belg.* **1979,** *88,* 957.
31. Politzer, P.; Domelsmith, L. N.; Sjoberg, P.; Alster, J. *Chem. Phys. Lett.* **1982,** *92,* 366.
32. Politzer, P.; Domelsmith, L. N.; Abrahmsen, L. *J. Phys. Chem.* **1984,** *88,* 1752.
33. Politzer, P.; Jayasuriya, K.; Zilles, B. A. *J. Am. Chem. Soc.* **1985,** *107,* 121.
34. Politzer, P.; Jayasuriya, K. *J. Mol. Struct. (Theochem)* **1986,** *135,* 245.
35. Politzer, P.; Kirschenheuter, G. P.; Alster, J. *J. Am. Chem. Soc.* **1987,** *109,* 1033.
36. Politzer, P.; Daiker, K. C. *Chem. Phys. Lett.* **1975,** *34,* 294.
37. Cromwell, N. H.; Graff, M. A. *J. Org. Chem.* **1952,** *17,* 414.
38. Charton, M. In "Chemistry of the Alkenes," Vol. 3, Zabicky, J. Ed.; Wiley–Interscience: New York, 1970, Chapter 10.
39. Wiberg, K. B. *Rec. Chem. Prog.* **1965,** *26,* 143.
40. Wiberg, K. B.; Lampuran, G. M.; Ciula, R. P.; Connor, D. S.; Schertler, P.; Lavanish, J. *Tetrahedron* **1965,** *21,* 2749.
41. Legon, A. C.; Aldrich, P. D.; Flygare, W. H. *J. Am. Chem. Soc.* **1982,** *104,* 1486.
42. (a) Paquette, L. A. *Acc. Chem. Res.* **1971,** *4,* 280. (b) Paquette, L. A.; Boggs, R. A.; Ward, J. S. *J. Am. Chem. Soc.* **1975,** *97,* 1118.
43. (a) Cassar, L.; Eaton, P. E.; Halpern, J. *J. Am. Chem. Soc.* **1970,** *92,* 3515, 6366. (b) Eaton, P. E.; Cassar, L.; Hudson, R. A.; Hwang, D. R. *J. Org. Chem.* **1976,** *41,* 1445.
44. Cremer, D. *J. Am. Chem. Soc.* **1977,** *99,* 1307.
45. Dauben, W. G.; Smith, J. H.; Saltiel, J. *J. Org. Chem.* **1969,** *34,* 261.
46. Pomerantz, M.; Wilke, R. N.; Gruber, G. W.; Roy, U. *J. Am. Chem. Soc.* **1972,** *94,* 2752.
47. Irngartinger, H.; Goldmann, A.; Jahn, R.; Nixdorf, M.; Rodewald, H.; Maier, G.; Malsch, K.-D.; Emrich, R. *Angew. Chem.* **1984,** *96,* 967; *Angew. Chem. Int. Ed. Engl.* **1984,** *23,* 993.
48. Rauscher, G.; Clark, T.; Poppinger, D.; Schleyer, P. v. R. *Angew. Chem. Int. Ed. Engl.* **1978,** *17,* 276.
49. (a) Ritchie, J. P. *J. Am. Chem. Soc.* **1983,** *105,* 2083. (b) Disch, R. L.; Schulman, J. M.; Ritchie, J. P. *J. Am. Chem. Soc.* **1984,** *106,* 6246.
50. Newton, M.D.; Schulman, J. M. *J. Am. Chem. Soc.* **1972,** *94,* 773.
51. (a) Wiberg, K. B.; Walker, F. H. *J. Am. Chem. Soc.* **1982,** *104,* 5239. (b) Wiberg, K. B. *Acc. Chem. Res.* **1984,** *17,* 379.
52. Jackson, J. E.; Allen, L. C. *J. Am. Chem. Soc.* **1984,** *106,* 591
53. Jayasuriya, K.; Alster, J.; Politzer, P. *J. Org. Chem.* **1987,** *52,* 2306.

CHAPTER 2

Experimental Measurements of Electron Density Distributions and Electrostatic Potentials

Cheryl L. Klein

Department of Chemistry, Xavier University of Louisiana, New Orleans, Louisiana

Edwin D. Stevens

Department of Chemistry, University of New Orleans, New Orleans, Louisiana

CONTENTS

1. Introduction 25
2. Method 27
3. Examples of Experimental Density Distributions and Electrostatic Potentials 38
4. Conclusion 61
References 62

1. INTRODUCTION

The role of single-crystal X-ray diffraction in structural studies is well recognized. Much of our understanding of the details of molecular structure,

structure–activity relationships, intermolecular interactions, and even chemical bonding has been drawn from the many structures that have been determined by X-ray diffraction.

Inherent in the X-ray experiment is the potential for providing direct information on the electronic structure of molecules. Since X-rays are scattered by the electrons in a crystal, an experimental electron density distribution can be obtained directly for accurate high-resolution X-ray intensity measurements.[1,2]

Until recently, this property could be obtained only from theoretical calculations.[3-5] The experimental distribution of electrons determined by X-ray scattering reveals not only the atomic positions, but also the redistribution of electrons as covalent bonds and lone pairs are formed, which provides fundamental information on chemical bonding. In the addition to electron distribution itself, other related properties of the molecular electronic structure such as net atomic charges, dipole moments, electrostatic potentials, and intermolecular interaction energies may be calculated from the X-ray intensities.[6-9]

Over the past 20 years, experimental techniques and methods of analysis have been developed to the point at which experimental charge density studies are possible on most materials that can be crystallized. When compared with large basis set ab initio calculations on small molecules, theoretical and experimental results agree to within the estimated error in the experimental density. Smaller basis set calculations and other less rigorous methods are easily discriminated against using experimental charge densities.

Since the difficulty of experimental electron density studies increases only moderately with the size of the molecule (linearly for data collection and as the second power of the size for least-squares refinement), compared with the fourth-power dependence of theoretical calculations, experimental studies offer a significant advantage for molecules of moderate size (50–200 atoms). Over the past several years, experimental electron density distributions of drugs and other molecules of biological interest, and of transition metal complexes, have been measured experimentally. In these cases, obtaining the same information by rigorous theoretical methods would be extremely difficult.

One of the goals of measuring electron density distributions is to obtain information about the molecule's reactivity either directly from the electron densities or from other one-electron properties calculated from the electron densities (ie, net atomic charges, dipole moments, electric field gradients, electrostatic potentials, intermolecular interaction energies).

Many drugs are believed to exhibit their pharmacological activity by binding to specific receptors. While the stereochemical requirements of drug–receptor binding have received much attention, it is reasonable to assume that a complementarity must also exist between the charge distribution of drug and receptor. Thus, the X-ray experiment has the potential of providing information on both the stereochemical and charge distribution requirements of drug–receptor binding (Figure 2-1).

The strain energy in many molecules, which can be tied to the reactivity of those molecules, can be observed as a function of the electron deformation density. The X-ray experiment provides information on the strain in small-ring systems by confirming the existence of and the extent of bond "bending" in those systems.

A dominant feature of the electron density distribution of transition metal complexes is the large asphericity observed around the metal atom due to the preferential occupancy of the d orbitals in the ligand field (Figure 2-2). With multipole refinement methods, as discussed in the next section, these features can be analyzed to yield d-orbital occupancies. In such an analysis, however, the contribution of metal–ligand covalent bonding is neglected. A model that specifically includes such interactions is required if experimental charge densities are to be used to study the role of transition metals in chemical catalysis.

In this chapter, we present an overview of the current experimental methods used in charge density studies, with many examples of experimental studies on drugs and other molecules of biological interest, transition metal complexes, and some other organic molecules including some highly reactive strained systems. We also include some hydrogen bond systems and describe predictions of properties and reactivities that have been made.

2. METHOD

A. Deformation Density

A periodic function such as the electron density distribution in a crystal can be represented by a Fourier series:

$$\rho(\mathbf{r}) = \frac{1}{V} \sum_{\mathbf{H}} F_{\mathbf{H}} \exp(2\pi i \mathbf{H} \cdot \mathbf{r}) \tag{2-1}$$

where V is the volume of the crystallographic unit cell, the $\mathbf{H}(= \mathbf{k}/2\pi)$ are vectors in reciprocal space, and the Fourier series coefficients $F_\mathbf{H}$, are X-ray structure factors. Measurement of the intensity of scattered X-rays provides experimental values of the structure factors, since $I_\mathbf{H} \propto |F_\mathbf{H}|^2$. Thus, given sufficiently accurate intensity measurements, the full three-dimensional electron density distribution of a crystalline solid may be obtained.

Two fundamental problems are associated with using Equation 2-1 to calculate an experimental electron density distribution. First, the X-ray structure factors are in general complex numbers, $F_\mathbf{H} = A_\mathbf{H} + iB_\mathbf{H}$, and thus have both an amplitude, $|F_\mathbf{H}| = (A_\mathbf{H}^2 + B_\mathbf{H}^2)^{1/2}$, and a phase, $\alpha_\mathbf{H} = \tan^{-1}(B_\mathbf{H}/A_\mathbf{H})$. From X-ray intensity measurements it is possible to obtain the ampli-

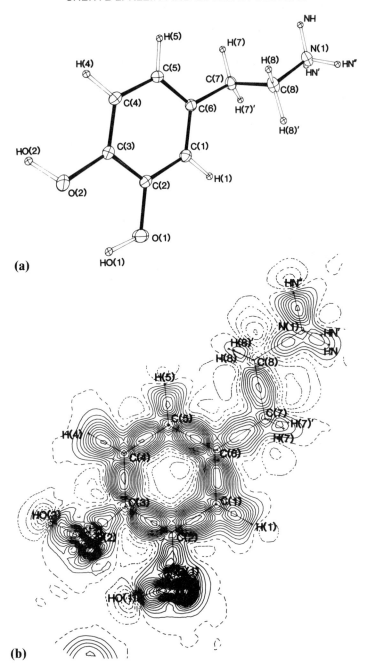

Figure 2-1. (*a*) Molecular structure of the neurotransmitter dopamine hydrochloride as determined by single-crystal X-ray diffraction at 90 K. (*b*) Contour map of the experimental electron deformation density of dopamine plotted in planes passing through all nonhydrogen atom positions. The map clearly shows a buildup of density in all covalent bonds and next to the oxygen atoms due to the nonbonding "lone pair" electron density. Contours are drawn at intervals of 0.05 e/Å^3; the zero and negative contour lines are dashed.

EXPERIMENTAL MEASUREMENTS OF EDD AND ESP

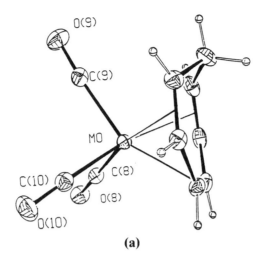

(a)

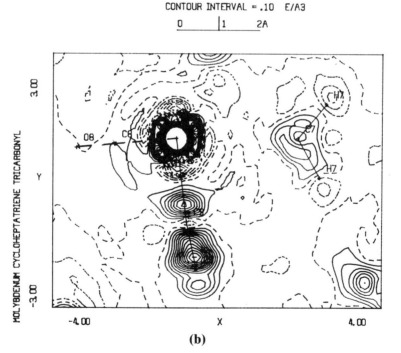

(b)

Figure 2-2. (a) Molecular structure of molybdenum cycloheptatriene tricarbonyl at 90 K. (b) Experimental electron density plotted in a plane containing the Mo atom and the C≡O bond and bisecting the cycloheptatriene ring. The map shows sharp peaks near the metal atom due to occupancy of the t_{2g} orbitals by d electrons and a depletion of density in the directions of the ligands. Peaks are also observed for the C≡O bond and for the lone pairs on both the carbon and oxygen atoms. Contours are drawn at intervals of 0.10 e/Å^3.

tude, but not the phase. Second, the summation in Equation 2-1 should be in principle over an infinite number of terms, but in practice only a finite number of measurements, determined by the X-ray wavelength, are possible in any experiment.

The first problem can be overcome by using the phase calculated for a model of the electron density. For centrosymmetric structures (crystals that possess inversion symmetry), the phase angle is restricted to one of two values, 0 and π, and a knowledge of the atom positions is sufficient to calculate the correct phase angle. For acentric structures, which lack inversion symmetry, the phase angle may have any value between 0 and 2π, and a more detailed model of the electron distribution must be used to calculate proper phases for the structure factors (Figure 2-3).

The problem of limited resolution will result in serious series termination errors in the electron distribution even when measurements are made to relatively high resolution. To overcome this problem, a difference density may be calculated by subtracting a reference density:

$$\Delta\rho(\mathbf{r}) = \rho_{obs} - \rho_{ref} = \frac{1}{V} \sum_{\mathbf{H}} (F_{\mathbf{H},obs} - F_{\mathbf{H},ref}) \exp(2\pi i \mathbf{H} \cdot \mathbf{r}) \qquad (2\text{-}2)$$

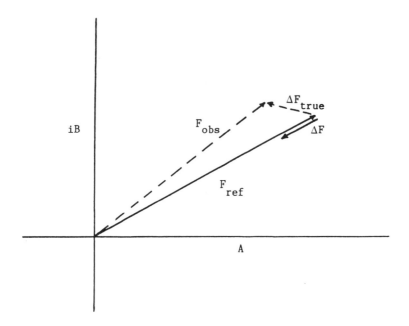

Figure 2-3. Graph showing the error that can result in the calculation of ΔF for an acentric structure if the proper phase of F_{obs} is not used in the calculation of the electron deformation density maps. In the absence of an estimate of the phase of F_{obs} from a multipole refinement, the phase of F_{obs} is assumed to be the same as that of F_{calc}.

The reference density is most often taken as the superposition of the spherically averaged densities of isolated atoms calculated from Hartree–Fock atomic wave functions and placed at the nuclear positions with the observed thermal motions. This reference density is called the "promolecule" or IAM (independent atom model). The resulting difference is termed the "deformation" density, to distinguish it from the usual low-resolution difference densities of X-ray structural studies, and because it displays the deformations that occur in the atomic electron distributions as the result of chemical bonding. Series termination is greatly reduced because the very high-order terms that are omitted correspond to the scattering from the core densities, where the actual and reference densities closely coincide. In some cases, some series termination effects may still exist for sharp features of the valence density such as lone pair peaks on oxygen atoms and especially nonbonding d-electron density on transition metal atoms.

Subtle changes in the electron distribution that result from chemical bonding are more clearly visible in the deformation density than in the total electron density distribution, which is dominated by large, spherical contributions centered at the atomic positions. Features of the deformation density may be interpreted easily with a qualitative valence bond model of chemical bonding, showing peaks due to both covalent bonding and "lone pair" nonbonding electron distributions.

The appearance of the deformation density is, of course, sensitive to the nature of the reference density subtracted. For example, the use of non-spherical atomic reference densities may be helpful in interpreting the details of covalent bond formation. In almost all experimental studies, however, spherical atom reference densities are subtracted, both for convenience in calculation and because it represents a well-defined reference for comparison between studies. It should be noted that an alternative method exists for analysis of chemical bonding in terms of the total density.[10]

To subtract the reference density, accurate positions and thermal motion parameters of each of the atoms are required. The values obtained for these parameters from conventional least-squares analysis of the X-ray data cannot be used, however, since the least-squares method seeks to find parameters that minimize the quantity $\Sigma(F_{obs} - F_{ref})^2$. The resulting parameters will thus be biased by the aspherical features of the valence electron distribution. An extreme example is the position of a hydrogen atom in a C—H bond as determined by least-squares refinement of X-ray data. The refined position of the hydrogen (center of the electron density on hydrogen) is shifted by about 0.1 Å from the position of the nucleus toward the carbon atom.

Unbiased parameters can be obtained by a separate neutron diffraction experiment, which yields the positions and thermal parameters of the atomic nuclei. Electron difference density maps calculated using X-ray measurements for F_{obs} and parameters from neutron diffraction measurements to calculate F_{ref} are termed "X–N" deformation densities. Alternatively, unbiased parameters can be obtained by a refinement including only X-ray data collected at high scattering angles, where the scattering is due largely to the core electrons. Difference maps calculated using parameters from refinement of high-order measurements are termed "X—X(high order)" deformation densities.

Since associated with each X-ray intensity measurement is some statistical uncertainty in addition to possible systematic errors, the deformation density calculated by Fourier series summation will contain some random noise.[11] The average error level in the deformation density can be estimated by $\sigma(\rho) = (\sqrt{2}/V)(\Sigma\sigma^2(F))^{1/2}$ and typically has values in the range 0.04–0.06 e/Å^3. Ironically, as more observations are included to increase the resolution and reduce series termination errors, the average error in the deformation density also increases. For molecules of biological interest, the increased size of the structure often leads to a large number of weak reflections, yielding higher noise levels in the deformation density. To reduce the noise, experimental densities are now more commonly fit using a multipole refinement model, which filters out much of the experimental noise.

In addition to the contribution from errors in the X-ray observations, errors in the parameters used in the reference density also contribute to the uncertainty in the deformation density.[11,12] As a result, the error increases rapidly near the atomic nuclei, where the magnitude of the reference density is high. Thus, the result of experimental deformation density measurements should not be considered reliable within about 0.3 Å of the atomic nuclei.

B. Multipole Refinement Models

As an alternative to obtaining the deformation density by Fourier series summation, as described above, the X-ray data may be analyzed using an expanded least-squares refinement model. Rather than using the scattering factors of neutral, spherical atoms (the model used in conventional refinement of X-ray data), the refinement model can be expanded to include parameters that describe the distortions of the atomic electron distribution as a result of chemical bonding and the crystalline environment. The most popular of these models involve the description of the distorted atom ("pseudoatom") density in terms of atom-centered multipole expansions, either in the form:

$$\rho_{\text{pseudoatom}}(\mathbf{r}) = \rho_{\text{atom}}(\mathbf{r}) + \sum_{l,m} P_{lm} R_l \cos^n \theta_m \qquad (2\text{-}3)$$

due to Hirshfeld,[13] where θ_m are the angles with respect to a set of vectors $\mathbf{r}_m$, defined for each atom, or in the form[14,15]:

$$\rho_{\text{pseudoatom}}(\mathbf{r}) = \rho_{\text{core}}(\mathbf{r}) + \sum_{lm} P_{lm} R_l y_{lm}(\theta, \phi) \qquad (2\text{-}4)$$

where the P_{lm} are refinable parameters, the R_l are radial functions, and the $y_{lm}(\theta, \phi)$ are spherical harmonic angular functions in real form. If the angular functions are defined in terms of local coordinate systems on each atom, then molecular symmetry may be imposed by selecting only the symmetry-allowed P_{lm} parameters to be refined.

The multipole refinement model offers several advantages in the analysis of high-resolution X-ray data. The electron density distribution is described in terms of a small number of analytical functions rather than a large grid of points. All the X-ray data (and neutron data, if available) can be included in the refinement, properly weighted in accord with the uncertainty in the measurement. If the multipole functions describing the density are plotted out, either directly, or by a Fourier summation:

$$\Delta\rho(\mathbf{r}) = \frac{1}{V} \sum_{\mathbf{H}} (F_{\text{multipole}} - F_{\text{promolecule}}) \exp(2\pi i \mathbf{H}\cdot\mathbf{r}) \quad (2\text{-}5)$$

the random noise of the experiment is effectively filtered out. Such densities are properly designated as "model" deformation densities rather than experimental deformation densities.

The multipole parameters are frequently defined in terms of local coordinate systems on each atom, making it easy to impose noncrystallographic symmetry constraints on the refinement. For example, the density in an aromatic ring is commonly constrained to have a mirror plane of symmetry in the plane of the ring. The validity of any such constraints, as well as the success of the least-squares fit in general, can be evaluated by examination of the residual density:

$$\Delta\rho_r(\mathbf{r}) = \frac{1}{V} \sum_{\mathbf{H}} (F_{\text{obs}} - F_{\text{multipole}}) \exp(2\pi i \mathbf{H}\cdot\mathbf{r}) \quad (2\text{-}6)$$

which should be featureless except for random experimental noise.

For acentric structures, the uncertainty in the phase of the observed X-ray structure factor leads to a lowering of peak heights and loss of resolution in the features of the experimental deformation density.[16] With a multipole refinement model, however, the correct amplitude and phase of ΔF can be obtained, since the phases of both structure factors in the expression for the model density (Equation 2-5) are known. The multipole refinement method is thus essential if quantitative information on the electron distribution is to be obtained from acentric structures (Figure 2-4).

Since atoms in a crystal are in constant motion, the various experimental densities described above represent a time average of the electron density over the atomic displacements due to thermal motion. Parameters describing thermal motion are included in the refinement model. Assuming that thermal motion is adequately described in the model, plotting out the multipole functions directly without the thermal motion parameters should yield a "static"

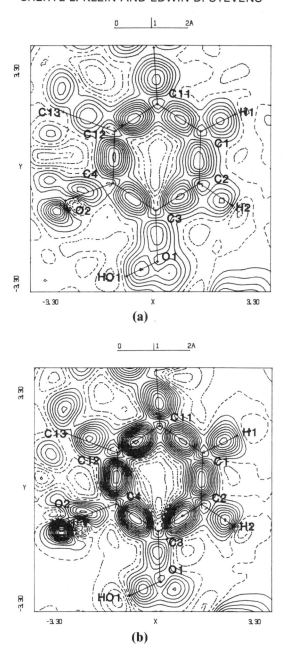

Figure 2-4. (a) Experimental model deformation density map in the aromatic ring plane of naloxone assuming that the phases of $F_{\text{multipole}}$ are the same as those of $F_{\text{promolecule}}$. (b) Experimental model deformation density maps in (a) using the correct phases of $F_{\text{multipole}}$ to calculate the ΔF's. The effect of correcting for the difference in phase angle is to increase the heights of the deformation density features. Contours are drawn at intervals of 0.05 e/Å^3 in both figures. (Reproduced with permission from Reference 20.)

electron density distribution. However, the inherent convolution of thermal motion with the electron density distribution greatly limits the information available in the experimental data on sharp features of the valence density. The reliability of static electron distributions is therefore still open to debate. Our personal preference is to stay closer to the experiment by reporting dynamic deformation densities in most cases.

C. Other Properties

Given the electron density distribution of a molecule, any other one-electron property of the molecule can be calculated.[17] These include both "outer moments" (expectation values of $\mathbf{r}^n$, where n is positive), such as the dipole moment and the quadrupole moment, and "inner moments" (where n is negative), such as the electric field and the electric field gradient. For any operator $\hat{O}$, the corresponding property will be given by integrating over the electron distribution,

$$\langle \hat{O} \rangle = \int \hat{O} \rho(\mathbf{r}) \, d\mathbf{r} \qquad (2\text{-}7)$$

If the electron distribution has been fit with a multipole model, this integral may be easily evaluated for many operators.

If the Fourier transform of the operator can be evaluated analytically:

$$\Omega_\mathbf{H} = \int \hat{O} \exp(-2\pi i \mathbf{H} \cdot \mathbf{r}) \, d\mathbf{r} \qquad (2\text{-}8)$$

then Equation 2-7 can be evaluated by substituting Equation 2-1 for $\rho(\mathbf{r})$, yielding an expression for the desired quantity as a Fourier summation:

$$\langle \hat{O} \rangle = \frac{1}{V} \sum_\mathbf{H} \Omega_\mathbf{H} F_\mathbf{H} \exp(2\pi i \mathbf{H} \cdot \mathbf{r}) \qquad (2\text{-}9)$$

thus allowing an evaluation of the property directly from the experimental structure factors. Equation 2-9 may still suffer from series termination errors, however, depending on the convergence behavior of $\Omega_\mathbf{H}$.

A property of the electron density distribution that has found widespread use in chemistry is the distribution of net atomic charges. The value of the net atomic charge on an atom in a molecule or crystal is model dependent, however, because atomic charge is a defined property and depends on how the continuous three-dimensional electron distribution is partitioned among the atoms. Net atomic charges are most frequently obtained by Mulliken population analysis of theoretical wave functions. This method involves an arbitrary division of orbital products between atoms, and is well known to be basis-set dependent.

Net atomic charges obtained from X-ray measurements are similarly dependent on partitioning methods.[9] If the operator in Equation 2-7 is a step function defining a box around an atom, with a value of one inside the box and zero outside, then Ω_H can be easily obtained, and evaluation of Equation 2-9 will yield the total charge inside the box. To account for the differing sizes of atoms, the crystal unit cell may be divided into a large number of small boxes, and the charge in each box assigned to an atom according to its position and the atomic radii, yielding a generalized Wigner–Seitz cell about each atom.[18] In practice, experimental errors in the X-ray structure factors contribute surprisingly little to the uncertainty in the resulting charge. The values of the respective net atomic charges are found to be strongly dependent on the position of the boundary between cells, however.

When the data have been fit with a multipole refinement model, the leading term in the expansion of the atomic density (the "monopole") yields a value for the net atomic charge. In this case, the partitioning is between a set of continuous, spherical density functions centered on each atom. As with the basis-set dependence of theoretical charges, the experimental monopole charges depend on the form of the density function and its radial dependence. A further ambiguity arises from the tendency of higher order terms in the multipole expansion to have a significant contribution to the density features on neighboring atoms. One way of reducing the influence of nonlocal contributions is to take net atomic charges from a restricted refinement in which only the monopole term and its radial dependence are refined.[9] Alternatively, locality can be imposed by applying the "stockholder" concept, in which each point in the model density is weighted by a factor given by the contribution of the unperturbed atom to the density of the promolecule.[19]

To further reduce the effects of partitioning errors on experimental net atomic charges, we have introduced the concept of a "local area charge,"[20] given by the sum of the net atomic charge on an atom and the charges on all other atoms to which it is bonded. The purpose of defining local area charges is to give, in particular areas of the molecule, a measure of the charge that is less sensitive to the ambiguity of partitioning. When calculated for two similar opiate molecules, naloxone and morphine (see Section 3), the local area charges are found to be more consistent in chemically similar areas than net atomic charges derived from monopole refinements.[20,21]

D. Experimental Electrostatic Potentials

A property of the electron density distribution that has received considerable attention is the molecular electrostatic potential.[22] It is given by:

$$V(\mathbf{r}) = \sum_A \frac{Z_A}{|R_A - \mathbf{r}|} - \int \frac{\rho(\mathbf{r}') \, d\mathbf{r}'}{|\mathbf{r} - \mathbf{r}'|} \qquad (2\text{-}10)$$

where the first term represents the contribution from the nuclear charges Z_A and the second term the contribution from the molecular electron distribution $\rho(\mathbf{r})$. Equation 2-10 gives the potential to a point positive charge at a point $\mathbf{r}$ due to the total (unperturbed) charge distribution.

The electrostatic potential as given by Equation 2-10 is well defined for a molecule and does not suffer from problems with partitioning. It has therefore been promoted as a more reliable measure of the electrostatic character than net atomic charges. In addition, it has proved useful as a predictor of chemical reactivity and sites of chemical attack. For example, negative potentials near nitrogen and oxygen atoms correlate well with sites of protonation and electrophilic attack. Areas of positive electrostatic potential correlate less reliably with sites of nucleophilic attack, however. At any rate, the use of electrostatic potentials is becoming increasingly common in the study of intermolecular interactions, especially in the study of biomolecular systems.

The electrostatic potential is most often obtained from a theoretical calculation. However, it may also be obtained from high-resolution X-ray measurements. As described in the preceding section, the potential can be calculated either by numerical integration of the experimental electron density or by a modified Fourier series summation[23]:

$$V(\mathbf{r}) = \frac{1}{4\pi V} \sum F'_\mathbf{H} \frac{\exp(2\pi i \mathbf{H} \cdot \mathbf{r})}{(\sin \theta/\lambda)^2} \qquad (2\text{-}11)$$

where $F'_\mathbf{H}$ is the structure factor for the total charge density including both the electron and nuclear charge contributions. If the density has been fit using a multipole model, the modified structure factor can be easily calculated by:

$$F'_\mathbf{H} = \sum_A (Z_A - f_{\text{multipole},A}) T_A(\mathbf{H}) \exp(-2\pi i \mathbf{H} \cdot \mathbf{r}) \qquad (2\text{-}12)$$

where $f_{\text{multipole},A}$ is the atomic scattering factor corresponding to pseudoatom A, and $T_A(\mathbf{H})$ is the atomic thermal motion factor.

If the electrostatic potential is evaluated using a Fourier series summation, as in Equation 2-11, care must be taken to avoid series termination errors. However, because of the additional factor of $(\sin \theta/\lambda)^{-2}$, which decreases as $\mathbf{H}$ increases, the series will converge faster than the series for $\rho(\mathbf{r})$ (Equation 2-1).

Equation 2-11 yields the electrostatic potential within the crystal lattice. To evaluate the electrostatic potential of an isolated molecule requires a partitioning of the electron density in the crystal into molecular fragments. Since little electron density is found in intermolecular regions, this should result in fewer problems than the partitioning into atoms. If the volume of the crystallographic unit cell is artificially increased, the Fourier series summation method may still be used to calculate the potential of effectively isolated molecules.

Whether calculated by Fourier series or by direct integration of a partitioned density, the experimental electrostatic potential of the "isolated" molecule will reflect the influence that the surrounding atoms of the lattice have on the molecular electron density distribution. Compared with the theoretical calculations on isolated molecules, experimental "isolated" molecular electrostatic potentials appear generally more negative in areas outside the van der Waals surface, which may reflect the influence of the crystal environment. However, a detailed comparison between theoretical and experimental potentials that includes the effects of basis-set size, series termination, and thermal motion has yet to be made.

3. EXAMPLES OF EXPERIMENTAL DENSITY DISTRIBUTIONS AND ELECTROSTATIC POTENTIALS

The survey of recent experimental results presented here serves to illustrate the variety of systems that may be studied using these techniques. Several reviews are available,[1-5] and we therefore choose not to attempt an exhaustive review, but rather to concentrate on recent work and on studies that contain interesting interpretations of the results.

In general, our interest is in the distribution of valence electrons and any predictions of properties and reactivity that can be made from a knowledge of this distribution. The core electron distribution is of little interest but adds an unavoidable contribution to the X-ray scattering experiment. As a result, the most reliable measurements are those from crystals containing only light elements. The higher ratio of core to valence electrons in structures with heavy elements leads to more experimental difficulties. However, as the accuracy of experiments is improved, along with the methods of analysis, studies on compounds with increasingly heavy elements will become feasible. Currently, reasonable results may be routinely obtained on molecules containing elements in the periodic table up through the second-row transition metals.

A. Organic Molecules

Experimental deformation densities typically show peaks in covalent bonds and additional peaks associated with nonbonding or "lone pair" electron distributions. Covalent bond peaks increase in height for a series of bonds that are formally single, double, or triple. A good correlation has been reported between bond peak heights of 9-t-butylanthracene and theoretical overlap populations calculated using the extended Hückel method.[1] Double bonds and bonds participating in delocalized π-bonding show bond peaks that are elongated at the bond midpoint perpendicular to the plane of the molecule, reflecting a combination of σ- and π-covalent interactions. Likewise, cumulenes show alternating bond peak elongations, while triple bond peaks are again cylindrically symmetric, reflecting the combination of one σ interaction and two equal π interactions.

Given the appearance of the deformation densities in organic and other small molecules, it is very tempting to interpret the results in terms of a naive valence bond model. Overlap of atomic valence bond orbitals results in a buildup of density in the covalent bonds, while doubly occupied nonbonding orbitals yield lone pair density peaks. Atomic hybridization is often readily apparent from the deformation density. Indeed, the success of the simple valence bond model in predicting structure and other properties of organic molecules is certainly due to the underlying basis of bonding in the actual distribution of electrons.

One must keep in mind that such an interpretation of the deformation density has a number of limitations. The features of the deformation density represent only a fraction of the total density in the bond and lone pair regions. In addition, in some cases very little density is observed in the bond peak (or even a deficiency of density leading to a hole) despite the formation of a reasonably strong covalent bond. For example, the experimental deformation density of a molecule cleverly designed to contain C—N, C—O, N—N, N—O, and O—O single bonds has been reported by Dunitz and Seiler.[24] A deficiency is observed at the center of the O—O bond, and only weak accumulations are observed in the N—N and C—O bonds. Although some have asserted that such results cast doubt on the reliability of experimental results, it is now well understood that the lack of density is a result of the reference density being subtracted. For atoms with more than one electron per valence orbital (ie, those in the right half of the periodic table), if a spherically averaged atomic density is used for the reference, the reference density subtracted will have a greater density in the bond region than the density resulting from a covalent bond formed by the overlap of two singly occupied atomic orbitals. In these cases, subtracting hybridized atomic reference densities will result in bond peaks in all the bonds. Alternatively, the bonding can be analyzed in terms of the total density.

The experimental electron density also provides information on strain in small-ring systems. As one might anticipate, covalent bond peaks in the deformation density lie outside the direct line joining the atom positions, providing confirmation for the concept of "bent" bonding. An excellent example is provided by a recent low-temperature X-ray study of 1,3-dinitrocubane[25] (Figure 2-5). Some other recent examples include X-ray deformation density studies of molecules containing cyclopropane,[26] bicyclobutane,[27] tetrahedrane,[28] cyclobutadiene,[29] and [Dewar]anthracene[1] ring systems. The degree of bond bending can be estimated by the location of the bond maxima. Alternatively, as a more quantitative measure of the bending, a bond deviation index λ can be defined as the extent to which the actual bond path (ridge in the total density) differs from the corresponding refer-

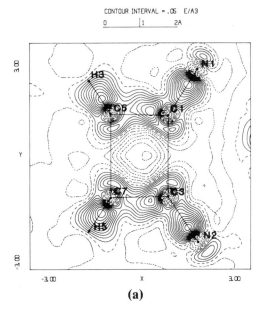

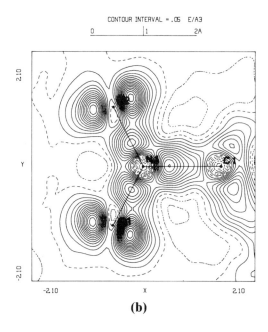

Figure 2-5. (a) Experimental electron deformation density of 1,3-dinitrocubane at 90 K plotted in a plane passing diagonally through the cube and containing C—C bonds on opposite edges. The maximum density in the bond peaks is found to be well outside the straight line connecting the carbon atom centers. (b) Experimental electron density distribution plotted in the plane of one of the nitro groups in 1,3-dinitrocubane.

B. Biologically Active Molecules

In an effort to use the experimental electron densities obtained by X-ray diffraction to predict biological and pharmacological reactivities within classes of compounds, many studies of the charge densities of molecules of these types have been done. Calculations of the electron density distribution and other one-electron properties (eg, net atomic charges, dipole moments, and electrostatic potentials) can give an accurate map of the electronic character of biologically interesting molecules and of the charge distribution and structural requirements of the receptor sites.

a. Opiates

The opiates form a class of compounds that are known to bind to a highly specific receptor and produce an analgesic effect. These molecules are of moderate size (50–110 atoms), hence are not easily studied by theoretical methods. Experimental difficulties are also often encountered, since these molecules are moderately large. Often the molecules crystallize in acentric space groups (see Subsection B of Section 2) and do not scatter well to very high resolution.

In studies of drug–receptor interactions, it is reasonable to expect that in addition to shape, the distribution of charge in a drug molecule will be of importance in receptor recognition and binding. The charge density is expected to determine long-range interactions, while steric fit and specific interactions such as hydrogen bonds are expected to be of equal or greater importance at shorter distances. To investigate the feasibility and potential usefulness of accurate electron density determinations, a series of opiates was chosen for study. Beginning with molecules that have a high degree of rigidity, such as morphine and naloxone, leaves little doubt as to the active conformation of the molecules. The conformation of the rigid opiates can be readily described as "T-shaped," with the structures of the rigid skeletons being essentially identical. Figure 2-6a shows the model deformation density

ence path (either a straight line or the bond path in the promolecule).[30] In 1,3-dinitrocubane the average bond deviation index is estimated to be 0.041 compared to a theoretical value of 0.029 for cubane itself. See Chapter 1 by Politzer and Murray for a further discussion of the bond deviation index.

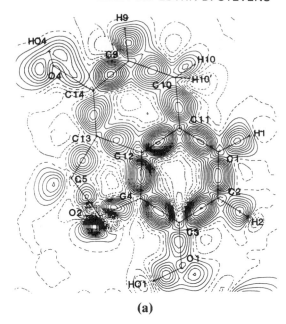

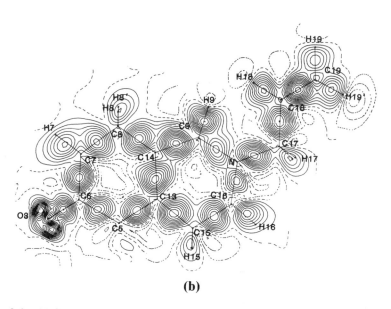

Figure 2-6. (a) Composite experimental electron deformation density map of naloxone hydrochloride at 90 K calculated in a series of planes passing through all nonhydrogen atoms that define one face of the "T." (b) Composite model deformation density map calculated in planes passing through the atoms that describe the perpendicular face of the "T" and the N-allyl group. (Reproduced with permission from Reference 20.)

map calculated through all of the atoms that describe one face of the "T" in naloxone hydrochloride.[20] At physiological pH, the nitrogen will be protonated, and the hydrochloride salt thus corresponds to the presumed active form of the molecule. Figure 2-6b shows the model deformation density map calculated through all the atoms that describe the perpendicular face of the "T" and the N-allyl group. The deformation densities in the hydroxyl groups in naloxone (O_1–HO_1 and O_4–HOH) show peaks of electron density corresponding to nonbonded lone pairs. The lone pairs have merged together into a single elongated peak with the maximum density in the plane plotted. The furan oxygen atom, O_2, also shows the same elongated peak above and below the plane with maximum in the plane, although the lone pair density on O_2 appears to be much sharper than on the hydroxyl oxygen atom. The deformation density in the lone pairs on the ketone oxygen, O_3, are well resolved into two distinctive peaks of electron density. The C_6=O_3 bond peak shows additional buildup of electron density consistent with the "double-bond" character. Elsewhere in the molecule there appears to be additional buildup of electron density in the allyl group double bond and in the aromatic ring, as expected. It should be indicated at this point that because naloxone hydrochloride crystallizes in an acentric space group, the phases of the structure factors had to be calculated from the multipole deformation refinement (Subsection B of Section 2). This had the effect of increasing the size of the features in the deformation density by a factor between 1.1 and 1.8.

The net atomic charges on each atom in the two faces of naloxone hydrochloride were calculated from the monopole populations (P_v) obtained from a monopole refinement of the data (Subsection C of Section 2) and can be found in Figures 2-7a and 2-7b. These net atomic charges do not appear to give consistent information concerning areas of the molecule that are electron rich or electron deficient except for the oxygen atoms and the nitrogen atom, which appear to be electron rich (all have negative net atomic charges). However, calculation of local area charges results in the establishment of trends of charge through certain areas of the molecule (Figures 2-7c and 2-7d). The fused atoms in the aromatic ring tend to be more negative than the peripheral atoms. Additionally, the charge on the molecule appears to become more negative in the region around the hydroxide group bonded to C_{14} and in the region around the nitrogen atom. In fact, this region of naloxone hydrochloride has the most negative local area charges in the molecule. Interestingly, the —OH group bonded to the aromatic ring (as well as the ketone oxygen) does not appear to be significantly charged. Again, one must refrain from assigning too much physical significance to the individual atomic charges. However, it is useful to point out that very similar (although not identical) trends have been observed in the local area charges calculated from the X-ray data on morphine hydrate free base[21] and methadone hydrochloride, a flexible opiate.[31]

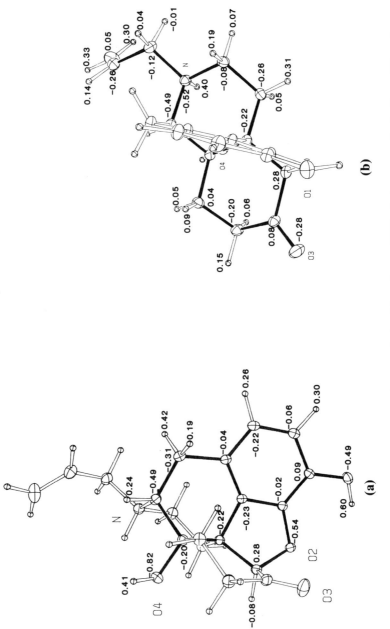

Figure 2-7. (a) and (b) Experimental net atomic charges calculated for naloxone HCl in the two perpendicular faces of the naloxone skeleton. (c) and (d) "Local area charges" for naloxone HCl calculated from the experimental net atomic charges. (Reproduced with permission from Reference 20.)

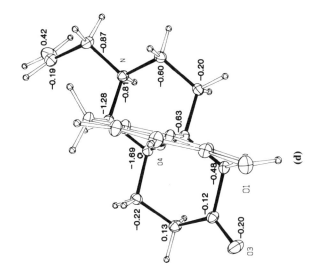

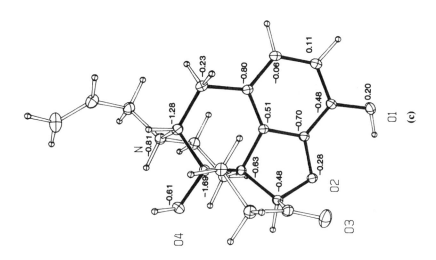

Figure 2-7. Cont.

b. Chemical Carcinogens

The experimental electron density distributions of several chemical carcinogens have been determined with the goal of using the experimental information to help elucidate the role of electronic factors in directing the steps of metabolic activation and reaction with cellular targets.

Polyaromatic hydrocarbons are important environmental carcinogens that undergo metabolic activation by mixed-function oxidase enzymes.[32] In several cases, the ultimate carcinogenic metabolite has been identified as a diol epoxide. The experimental electron density distribution for a model of these metabolites, *anti*-3,4-dihydroxy-1,2,3,4-tetrahydronaphthalene 1,2-oxide (NDE), has been determined from X-ray intensity measurements at 100 K.[33] Although not carcinogenic itself, NDE has been identified as an in vivo metabolite of naphthalene. Of greatest interest is the electron distribution of the epoxide ring (Figure 2-8), since this is the functional group believed to be responsible for the reactivity of the diol epoxides of carcinogenic hydrocarbons. The bond density peaks lie outside the straight lines connecting the

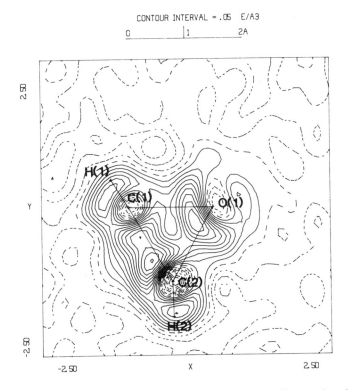

Figure 2-8. Model deformation density map calculated in the epoxide ring plane in *anti*-3,4-dihydroxy-1,2,3,4-tetrahydronaphthalene 1,2-oxide, a model for the ultimate carcinogenic metabolite of benzo[*a*]pyrene. (Reproduced with permission from Reference 33.)

atom centers, indicating "bent" bonding as expected from a three-membered ring. Large peaks observed directly above and below the oxygen atom suggest largely unhybridized s and p lone pairs on the oxygen (a pure s lone pair, being spherical, will show few features in a deformation density, since most of the density will be subtracted by the promolecule). The C_1—C_2 bond peak is polarized toward C_2, indicating that C_2 is more electronegative than C_1. Aside from steric considerations, one expects on the basis of the electron distribution that chemical attack by a nucleophile would occur at C_1. Reaction of epoxide metabolites of carcinogens such as benzo[a]pyrene and benz[a]anthracene with DNA has been shown to result in adducts in which the carcinogen is covalently bound to DNA bases, predominantly guanosine. The site of reaction, which can be considered as nucleophilic attack by the nitrogen of a DNA base, is observed to be the carbon atom corresponding to C_1 in NDE.[34]

In another study, the experimental electron density distribution of 7,12-dimethylbenz(a)anthracene has been determined from medium-resolution data collected at 180 K.[35] Studies of the parent hydrocarbon should give some indication of the preferred sites of reaction in the first step of metabolic activation, which involves attack by electrophilic oxygen. The experimental electron density (Figure 2-9) shows a buildup of density in the fused aromatic

Figure 2-9. Composite electron deformation density through all atoms that define the carbon skeleton of 7,12-dimethylbenz[a]anthracene, a potent chemical carcinogen. Contours are drawn at intervals of 0.05 e/Å³. (Reproduced with permission from Reference 35.)

C—C bonds and some excess density in the bay region. The K region, an area known to be chemically reactive, does not appear to have any additional density.

Nitrosamines are a class of chemical carcinogens significant because of their occurrence in the diet. In many cases, they require metabolic activation. Because the intermediates are generally unstable and have not been isolated, relatively little is known experimentally about the mechanistic details of nitrosamine carcinogenesis. The experimental electron density distribution of N-nitrosodiphenylamine has been determined from X-ray data collected at 100 K.[36] In the deformation density calculated in the plane of the nitroso group, large lone pair peaks are observed on the oxygen and nitroso nitrogen atoms. The N—N and N—O bond peaks fall off the bond axes, apparently a result of repulsion by the large lone pair densities.

c. Neurotransmitters

The experimental electron density distributions of two important neurotransmitters have been determined with the goal of using the information to help elucidate the electronic charge density and its role in the biological processes.

Dopamine, a decarboxylation product of DOPA, is a precursor for the hormones norepinephrine and epinephrine and has its own biological function. Excess dopamine in the central nervous system results in depression, schizophrenia, and paranoia, while insufficient levels result in the rigidity and tremors associated with Parkinsonism.

A high-resolution data set was collected for dopamine hydrochloride at 90 K.[37] A composite model electron density distribution (EDD) map for a molecule of dopamine HCl was shown in Figure 2-1b. Both hydroxyl oxygen atoms show peaks of electron density corresponding to nonbonded lone pairs. The lone pairs have merged together into a single elongated peak with the maximum in the plane plotted. These hydroxyl groups are the most electron-rich regions of the protonated dopamine molecule, which is consistent with the negative net atomic charges on the oxygen atoms of -0.68e and -0.66e for O_1 and O_2, respectively.

γ-Aminobutyric acid (GABA) is also a neurotransmitter in the mammalian central nervous system. The experimental electron density distribution for the zwitterionic form of GABA was determined by Craven and Weber at 122 K.[38] They also reported net atomic charges for the molecule calculated from the refinement results of the X-ray data and found that the hydrogen atoms that are involved in hydrogen bonding schemes are significantly more electropositive than those that are not, including a hydrogen atom that is involved in an intramolecular "methylene bridge."

d. Phenothiazines

Phenothiazine and many of its derivatives belong to a class of compounds that are generally used to treat the symptoms of psychosis. Many of these compounds have been used as tranquilizers, antidepressants, and antischizophrenics. It has been suggested that many of these antipsychotic drugs function as competitive antagonists of dopamine and that the phenothiazines block postsynaptic dopamine receptors in the brain and peripheral nervous system.

A study of the experimental electron density distribution using medium resolution data for chlorpromazine hydrochloride[39] (a potent tranquilizer) and levomepromazine sulfoxide,[40] the sulfoxide metabolite of levomepromazine (which is used in Europe as a neuroleptic with pronounced sedative properties) shows interesting trends of charge throughout the molecules.

As in the opiates, the net atomic charges do not give consistent information concerning areas of the molecules that are electron rich or electron deficient. However, the "local area charges" for both chlorpromazine hydrochloride and levomepromazine sulfoxide show that the substituted aromatic rings are more negative than the unsubstituted rings (Figure 2-10). Additionally, the charge in the region around the amino nitrogen atoms appears to become more negative despite the protonation of the nitrogen in chlorpromazine hydrochloride.

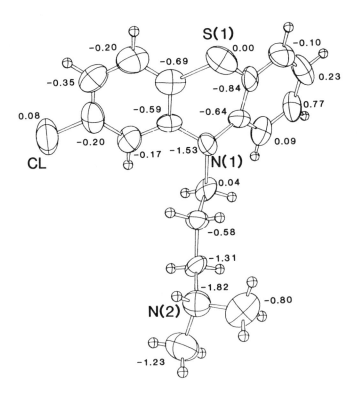

Figure 2-10. Experimental "local area charges" obtained from the electron distribution of chlorpromazine hydrochloride at 90 K, showing charge trends throughout the molecule.

e. Nucleotides, Bases, and Sugars

Bases and sugars make up the components of nucleotides and nucleosides, which are the building blocks that comprise DNA and RNA in biological systems. Several experimental electron density studies of these components have been completed. A study of β-DL-arabinose[41] at 75 K using both X-ray and neutron diffraction data served as the basis of a comparison between anomeric and exoanomeric effects obtained experimentally and effects obtained in ab initio studies. 9-Methyladenine[42] was studied at 126 K to obtain an understanding of the electronic structure and hydrogen-bonding behavior of the adenine moiety in nucleic acids and nucleotides. These 9-methyladenine data and cytosine monohydrate data[43] were refined using Hirshfeld's refinement model (Equation 2-3), and the experimental results were compared to ab initio deformation densities of isolated purines and pyrimidines.[44] The largest discrepancies occurred for atoms that participated in hydrogen bonds in the crystal.

Combined X-ray and neutron diffraction data at 85 K for putrescine diphosphate[45] were used to study the experimental charge density in the dihydrogen phosphate anion and a variety of N—H···O and O—H···O hydrogen bonds involving ammonium groups on the putrescine molecule and phosphate oxygen atoms. Medium-resolution data for the zwitterionic form of 2'-deoxycytidine-5'-monophosphate monohydrate were collected by Pearlman and Kim[46] to obtain experimental net atomic charges.

f. Amino Acids and Miscellaneous

Experimental electron density data collected on the α-glycine crystal[47] were used by Lau, Bader, Hermansson, and Berkovitch-Yellin to calculate the electron charge density illustrated in terms of the Laplacian of the electron density $(-\nabla^2\rho)$.[48] High-resolution X-ray data of α-glycylglycine were collected at room temperature[49] and at 82 K,[50] to study the electronic structure of the peptide bond. In both studies, the peptide bond shows partial double-bond character. In addition, experimental net atomic charges were calculated using least-squares methods for the atoms in the dipeptide.

Imidazole, a constituent of purine bases of nucleotides and nucleic acids, and of histamine and histidine residues of proteins, was studied at 103 and 292 K.[51] The function of the imidazole ring in biological systems depends on the ease of its ionization and possibility for the gain and loss of protons at the two nitrogen atoms. Stewart, who calculated the molecular electrostatic potential using the experimental X-ray data, found a negative potential adjacent to the unprotonated nitrogen atom and a very diffuse positive potential adjacent to the hydrogen atom bonded to the other nitrogen in imidazole.[52]

C. Hydrogen Bonds and Hydrogen-Bonded Systems

In addition to the effects of covalent bonding, distortions due to weaker forces such as intermolecular interactions in the solid can be observed in some cases. One of the strongest and most interesting among intermolecular interactions is hydrogen bonding. Hydrogen bonds vary over a considerable range from an O···O distance of approximately 2.8 Å for a typical "long" hydrogen bond to about 2.4 Å for a strong "short" hydrogen bond. Experimental densities[49,53,54] are in agreement with the theoretical description of long hydrogen bonds as largely electrostatic interactions.[55,56] However, experimental studies of systems with short hydrogen bonds suggest a significant covalent contribution.[57] The investigation of the EDD in a hydrogen bond of intermediate length (2.5 Å) in α-oxalic acid dihydrate (Figure 2-11) between the oxalic acid and water molecules shows the perturbation of the molecular density of the water by hydrogen bonding.[6] In addition, a short hydrogen bond (≈2.4 Å) in sodium hydrogen diacetate shows lone pair peaks that are further from the oxygen nucleus and more diffuse than those commonly observed in oxygen lone pairs involved in hydrogen bonds,[58] indicating a degree of covalent character of the short hydrogen bond. Further evidence of the covalent character of the short hydrogen bond is observed in a deformation density map (where the H atom density is not subtracted in the difference density) in which a continuous ridge of electric density extends along the hydrogen bond from one oxygen to the other. Normal length hydrogen bonds that are predominantly electrostatic have been observed in numerous chemical systems including recent studies of $K_2C_2O_4 \cdot H_2O$,[59] $LiNO_2 \cdot H_2O$,[60,61] $LiNO_3 \cdot 3H_2O$,[62] and $LiOH \cdot H_2O$.[63] In $LiNO_3 \cdot 3H_2O$, which contains bifurcated hydrogen bonds, EDD maps indicative of hydrogen bonds that are not unusual and typically electrostatic have been obtained.

Many pharmacologically interesting molecules crystallize as hydrohalide salts with a nitrogen atom protonated and the halide ion involved in an extensive hydrogen bond network. In dopamine hydrochloride,[37] for example, the chloride ion is involved in five unique hydrogen bonds (to five different but symmetry-equivalent dopamine molecules). None of the hydrogen bonds shows any indications of covalent character (Figure 2-12). Likewise, the chloride ion in naloxone hydrochloride[20] is involved in a network of four hydrogen bonds that only appear to be electrostatic. In both dopamine hydrochloride and naloxone hydrochloride, no significant distortions of the chloride ion from spherical symmetry is observed.

There have been some attempts to use properties derived from the electron density distributions to make predictions about the presence or absence of hydrogen bonds. For example, Craven and Weber reported net atomic charges for γ-aminobutyric acid[38] calculated from X-ray refinement results and found that the hydrogen atoms that are involved in hydrogen-bonding schemes are significantly more electropositive than those that are not. Stewart calculated molecular electrostatic potentials (MEP) in 9-methyladenine, and he found that hydrogen-bonding regions have a slightly negative MEP[52] with a potential of approximately −4 eV/unit charge.

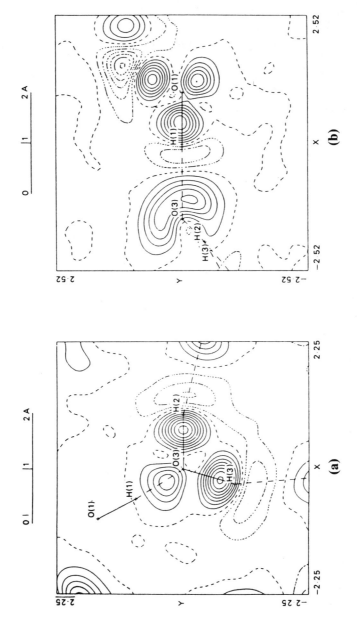

Figure 2-11. Experimental deformation density in the hydrogen bond in α-oxalic acid dihydrate between the oxalic acid and water molecules, showing the perturbation of the molecular density of the water due to the formation of a strong hydrogen bond. (*a*) Model density calculated in the plane of the water molecule. (*b*) Model density calculated in a plane perpendicular to the water molecule and containing the short hydrogen bond. (Reproduced with permission from Reference 6.)

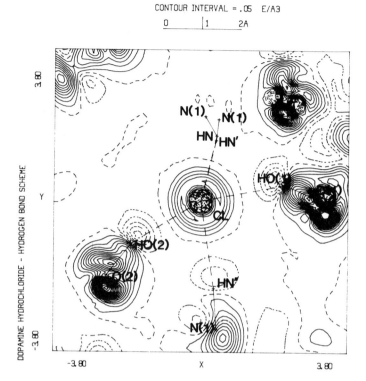

Figure 2-12. Experimental deformation density in the hydrogen bonds to the chloride ion in dopamine hydrochloride.

D. Transition Metal Complexes

For molecules containing only light atoms, careful X-ray measurements of electron distributions are in very good agreement with the results of theoretical calculations. In most such cases, however, the observed electron distribution could have been predicted, at least qualitatively, from simple valence bond models. For transition metal complexes, the nature of the bonding is more complex, and neither experimental nor theoretical methods have thus far been able to achieve the level of accuracy possible for light atom structures.

The larger size of transition metal complexes and the greater number of core electrons often forces some compromise in theoretical calculations, either in basis-set size or with respect to neglect of core electrons. The use of a truncated basis set is known to have a pronounced effect on the features of the resulting electron deformation density. In addition, since theoretical calculations are based on an energy minimization procedure, the results may be expected to be most reliable in the high-energy (core) regions of the molecule.

On the other hand, systematic errors in the X-ray measurements due to crystal effects such as absorption and extinction become more significant when transition metals or other heavy atoms are present. The results of such errors will be most pronounced near the nuclear positions, where the total electron density is high. The experimental density will be most accurate, on the other hand, in the bonding regions of the molecule. Thus, although the scattering from the valence electrons becomes a smaller fraction of the total scattering and other experimental errors increase with heavier elements, the X-ray experiment has the potential to provide extremely detailed information useful in the understanding of the electronic structure of transition metal complexes and the mechanisms of metal–metal and metal–ligand bonding.

An early X-ray study by Iwata and Saito of the electron density distribution of [Co(NH$_3$)$_6$][Co(CN)$_6$] revealed highly aspherical features around the metal atoms.[63] Although these results were originally received with some skepticism, similar sharp features near the metal atoms have been observed in numerous other complexes in which the transition metal has a low-spin electron configuration.[64] Analysis of these features yields d-orbital occupancies, the spin configuration of the metal, and other properties such as the electric field gradient at the nucleus, which can be compared with values obtained by other spectroscopic measurements.

a. d-Electron Distributions

A dominant feature of the electron deformation distribution of most transition metal complexes is the presence of sharp features near the metal due to unequal occupancies of the d orbitals. The ligand field generated by electronegative atoms cordinating to the transition metal splits the energy of the d orbitals. Generally, the d orbitals with lobes pointing in the direction of the ligands will be raised in energy, while those pointing in directions between the ligands will be lowered in energy, relative to a spherical ligand field. The splitting of the d-orbital energy levels will result in a preferential occupancy of the lower energy levels by the d electrons. Since the reference state for the density of the transition metal is spherical, with equal occupancy assigned to each orbital, the deformation density will show holes (negative density) in directions corresponding to d orbitals with less than the average occupancy, and peaks in the directions of orbitals with more than the average occupancy. An exception to this behavior will occur for a d^5 configuration when the ligand field splitting is not large enough to result in spin pairing.

In an octahedral field, the d orbitals are split into a lower t$_{2g}$ set (d$_{xy}$, d$_{xz}$, d$_{yz}$) and a higher e$_g$ set (d$_{z^2}$, d$_{x^2-y^2}$). Excess occupancy of the t$_{2g}$ orbitals (compared to a spherical distribution) will result in eight peaks in the deformation density directed in the $\langle 111 \rangle$ directions into the faces of the octahedron, while depletion of the e$_g$ orbitals will lead to six holes in the $\langle 100 \rangle$ directions pointing toward the ligands. For example, in iron pyrite (FeS$_2$), the iron atom is located in a slightly distorted octahedral environment of sulfur atoms[65] (Figure 2-13). The deformation density shows eight large peaks located in the faces of the coordination octahedron corresponding to a low-spin d^6 configuration.

Similar results have been obtained in many studies of complexes with transition metals in octahedral, pseudo-octahedral, or square-planar environments. Some recent examples include $KNiF_3$,[66] $KFeF_3$,[67] $CoTiO_3$,[68] $Cu(H_2O)_4(SO_4)\cdot H_2O$,[69] $Fe_2(SO_4)_3$,[70] $K_2[PtCl_6]$,[71] Co(tetraphenylporphyrin),[72] Ni(tetramethylporphyrin),[73] Fe(phthalocyanine),[74] and $K_2[PtCl_4]$.[75] For high-spin d^5 complexes the d orbitals are equally populated, resulting in relatively little asphericity in the deformation density near the metal atom, as observed in $MnTiO_3$,[76] $KMnF_3$,[66] and Fe(tetraphenylporphyrin)(OMe).[77]

In a tetrahedral ligand field, the d-orbital energy levels are inverted with respect to an octahedral field. The t_2 set (d_{xy}, d_{xz}, d_{yz}) is at a higher energy than the e set (d_{z^2}, $d_{x^2-y^2}$), and the preferential occupancy of the lower e set will result in six peaks in the deformation density pointing towards the centers of the tetrahedron edges. This has been observed, for example, in $CoAl_2O_4$[78] and $KFeS_2$[79] (Figure 2-14). In contrast, excess density is observed in directions corresponding to the t_2 orbitals in the experimental deformation density of the CrO_4^{2-} ion in K_2CrO_4[80] and in the theoretical deformation density of the isoelectronic MnO_4^- ion.[81] In these cases, the d-electron distribution is the result of strong metal–oxygen covalent bonding rather than ligand field splitting.

b. Metal–Metal and Metal–Ligand Bonds

The nature of the covalent interactions involved in metal–metal and metal–ligand bonding is of considerable interest particularly with regard to chemical catalysis. Transition metal complexes with two or more metal atoms in a cluster are often catalytic or may be used as models for the metal surfaces of heterogeneous catalysts. The nature of the metal–ligand bond is also of interest because such bonds must be broken or formed in catalytic reactions at the metal center.

Given the success of experimental electron density measurements on light atom structures, it was of considerable surprise that the results of an early experimental study of the electron density distribution in $[Cr(CH_3COO)_2\cdot H_2O]_2$, a complex with a formal metal–metal quadruple bond, showed little density at the center of the metal–metal bond.[82] It is now recognized that although formally a quadruple bond, the metal–metal bond in this complex is relatively long and weak due to the presence of ligands in the axial positions. Subsequent studies of the analogous molybdenum complex[83] and of a chromium complex with a much shorter metal–metal bond distance[84] do show substantial density buildup in the metal–metal bond.

Studies of several complexes containing metal–metal bonds with lower formal bond order, including $(\mu\text{-}C_2H_2)[(C_5H_5)Ni]_2$,[85] $[(C_5H_5)Fe(CO)_2]_2$,[86] $Co_2(CO)_8$,[87] $Co_3(CH)(CO)_9$,[88] $Ru_3H_3(CO)_9CCl$,[89] $[(CH_3C_5H_4)Mo(CO)_3]_2$,[90] and $(\mu\text{-}CH_2)(\mu\text{-}CO)[(C_5H_5)Fe(CO)_2]_2$,[91] have shown little electron density

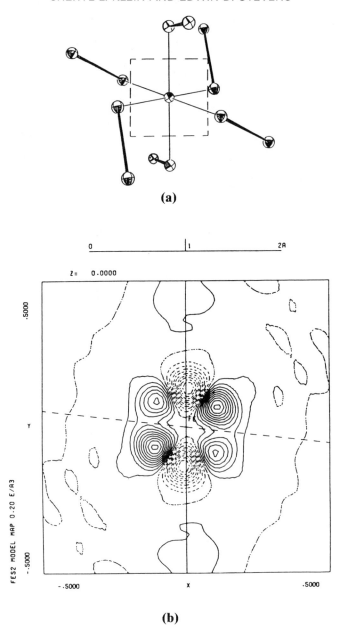

Figure 2-13. (a) Coordination sphere of the Fe(II) ion in the structure of iron pyrite. Each cation is surrounded by six S_2^{2-} anions with the nearest sulfur atoms forming a trigonally distorted coordination octahedron. The experimental deformation density is plotted in a plane passing through the vertical Fe—S bonds and bisecting the horizontal S–Fe–S angles as indicated by the dashed box. (b) Multipole model deformation density near the iron atom. Contours are plotted at intervals of 0.20 e/Å^3; the zero and negative contours are dashed. (Reproduced with permission from Reference 65.)

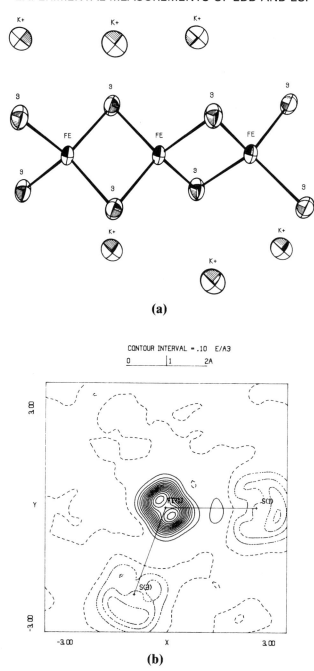

Figure 2-14. (a) Structure of KFeS$_2$ consisting of infinite chains of edge-sharing FeS$_4$ tetrahedra. (b) Multipole model deformation density near the iron atom showing the excess density in the directions of the tetrahedra edges. Contours are plotted at intervals of 0.20 e/Å^3; the zero and negative contours are dashed. (Reproduced with permission from Reference 79.)

accumulation in the metal–metal bonds. This is especially apparent in the case of $Mn_2(CO)_{10}$.[92] Since there are no bridging ligands, there must be a reasonably strong bond between the manganese atoms, yet no significant density is observed in the metal–metal bond. As pointed out by Bénard,[93] the lack of density is in part due to the reference state subtracted, as discussed earlier for bonds between light elements (Subsection A of Section 3). In going from left to right across the periodic table with transition elements, more electrons are subtracted per d orbital compared with the density expected for a covalent metal–metal bond formed by the overlap of two d orbitals, each occupied by one electron. In addition, the d orbitals are sharp functions that peak close to the metal centers, resulting in little overlap with d orbitals on other metal centers. The metal s and p orbitals, on the other hand, are very diffuse and thus also cannot produce sharp accumulations of bonding density in the metal–metal bond region.

Bonding between transition metal atoms and ligands is typically characterized in the electron deformation density by a hole near the transition metal due to the depletion of the d orbitals pointing toward the ligand mentioned above and by a peak near the ligand due to the ligand lone pair electrons. The general features of the deformation density of the coordinated ligand do not appear to be qualitatively different from the density of the uncoordinated ligand. In $Cr(CO)_6$, for example,[94] the presence of a peak in the CO bond and lone pair peaks on both the carbon and oxygen atoms do not reveal the change in bonding within the ligand and between the ligand and the metal as a result of coordination. If, however, the density calculated for a free carbonyl is subtracted, a σ buildup is observed between the chromium and the carbon atoms, as well as depletion of π density in the ligand.[94,95] This is direct experimental verification of the Dewar–Chatt–Duncanson model[96] of σ donation and π back-donation in metal–ligand interaction.

c. d-Orbital Populations

Since the deformation densities of many transition metal complexes show the d electrons to be essentially nonbonding, the electronic configuration can be analyzed, at least to a first approximation, in the spirit of crystal field theory, which neglects the covalent interactions between metal atoms and ligands. If it is further assumed that the atomic wave function for the metal atom in the crystal field can be expressed as a single Stater determinant including for the d electrons a minimal basis set of d orbitals, a simple relationship between the electron density and the d-orbital occupancies may be derived.[97-99]

If n_m is the occupancy of the mth orbital, then the density due to the d electrons, with the assumptions above, will be given by

$$\rho_{d\ electrons} = \sum_{m}^{5} n_m\ d_m^2 \quad (2\text{-}13)$$

where the d_m are the atomic d orbitals. Generally, the d orbitals will be taken as the product of a Slater-type radial function, $R_2(r) = Nr^2 e^{-\zeta r}$, and a second-order spherical harmonic angular function in real form y_{2m}, yielding:

$$\rho_{\text{d-electrons}} = (R_2(r))^2 \sum_m n_m (y2m)^2 \quad (2\text{-}14)$$

The product of spherical harmonic functions may be expressed as a linear combination of spherical harmonic functions[100]:

$$(y_{lm})^2 = \sum_{l'm'} C_{ll'mm'} y_{l'm'} \quad (2\text{-}15)$$

where the $C_{ll'mm'}$ are related to the well-known Clebsch–Gordon coefficients,[101] and include terms due to the differences in normalization between the wave functions and the density functions. The summation in Equation 2-15 over l' is from 0 to $2l$.

Thus, the electron density distribution corresponding to the d electrons will be given by a summation over spherical harmonic functions to the hexadecapole ($l = 4$) level:

$$\rho_{\text{d electrons}} = (R_2(r))^2 \sum_m n_m \left(\sum_{l'm'} C_{ll'mm'} y_{l'm'} \right) \quad (2\text{-}16)$$

A typical multipole refinement model for a first-row transition metal (Subsection B of Section 2) might have the following form:

$$\rho_{\text{metal}} = \rho_{\text{core}} + P_{4s}\rho_{4s} + R_4(r) \sum_{lm} P_{lm} Y_{lm} \quad (2\text{-}17)$$

where ρ_{core} is the density of the K, L, and M shell core, P_{4s} is the population of the 4s orbital, and the last term represents the 3d density. If the terms in Equation 2-17 are equated to the terms in Equation 2-16 with the same angular functions, a set of linear equations is obtained that relates the observed multipole populations P_{lm} to the orbital occupancies n_m and the $C_{ll'mm'}$ coefficients. Solving this set of equations, the apparent orbital occupancies can be obtained as a function of the X-ray multipole populations.

For example, in cobalt tetraphenylporphyrin (CoTPP) the metal atom is located in a tetra-coordinate square-planar environment at a crystallographic $\bar{4}$ site.[72] The crystal field may be considered to be a distorted octahedron in which the absence of axial ligands further splits the t_{2g} and e_g orbitals leaving only the d_{xz} and d_{yz} orbitals degenerate. In this case, comparison of Equations 2-16 and 2-17 yields[72,102]:

$$P_{00} = \left(\frac{1}{4\pi}\right)^{1/2} (n_1 + n_2 + n_3 + n_4)$$

$$P_{20} = \left(\frac{5}{196\pi}\right)^{1/2} (-2n_1 + 2n_2 - 2n_3 + n_4)$$

$$P_{40} = \left(\frac{1}{196\pi}\right)^{1/2} (-n_1 + 6n_2 + n_3 - 4n_4)$$

$$P_{44+} = \left(\frac{5}{28\pi}\right)^{1/2} (-n_1 + n_3) \tag{2-18}$$

where n_1, n_2, n_3, and n_4 represent the occupancies of the $d_{x^2-y^2}$, d_{z^2}, d_{xy}, and $d_{xz} + d_{yz}$ orbitals, respectively. This set of four equations and four unknowns is easily solved to yield expressions for n_1, n_2, n_3, and n_4 in terms of the experimental values of P_{00}, P_{20}, P_{40}, and P_{44+}. The occupancies obtained from refinement of low-temperature (100 K) X-ray data are given in Table 2-1. The assignment of the unpaired spin to the d_{z^2} orbital for the low-spin configuration is based on electron paramagnetic resonances results.[103]

The number of d-orbital occupancies to be determined will be five or fewer depending on the local symmetry of the crystal field, and the number of allowed multipole functions will depend on the crystallographic site symmetry. In cases of low site symmetry, the number of allowed multipoles may be larger than the number of orbital occupancies to be determined, but many of these terms cannot arise as products of d orbitals. In cases of low symmetry, the orientation of set of d orbitals with respect to the ligands may become a variable as well.

Experimental data from a number of transition metal complexes have been analyzed by the procedure described above, and in all cases, reasonable results were obtained. Negative orbital occupancies, for example, are not obtained, nor are occupancies of more than two electrons per orbital, which

TABLE 2-1. Experimental and Theoretical d-Orbital Occupancies for Cobalt(II) Tetraphenylporphyrin (CoTPP) and Cobalt(II) Phythalocyanine (CoPc)[a]

Orbital	Neutron[104]		Theoretical[103]		X-ray[105]
	(Spin) CoPc	(Charge)[b] CoPc	(Charge) CoPc	(Charge) CoTPP	(Charge) CoTPP
$b_1(d_{x^2-y^2})$	−0.21(10)	0.21	0.24	0.26	0.83(15)
$a_1(d_{z^2})$	0.79(12)	0.79	1.02	1.06	0.92(15)
$b_2(d_{xy})$	0.40(10)	1.60	1.78	1.76	1.64(15)
$e(d_{xz}, d_{yz})$	0.34(20)	3.66	3.96	3.93	3.62(30)
4s	−0.14(16)	0.14			

[a] Numbers in parentheses represent standard deviations derived from least-squares fits to the X-ray and polarized neutron data.
[b] Charge occupancies are derived from the spin occupancies by assuming maximum multiplicity (Hund's rule).

would violate the Pauli exclusion principle. Although the orbital occupancies are expected to have only the integer values 0, 1, or 2, noninteger values may arise because of the effects of covalent bonding or because more than one configuration is required to describe the electronic ground state of a complex.

The observed occupancies of CoTPP are readily interpreted in terms of a low-spin configuration with a contribution from metal–ligand bonding. The apparent occupancy of the $d_{x^2-y^2}$ is likely a result of σ donation by the lone pair electrons of the nitrogen ligands, while the apparent depletion of the d_{xz} and d_{yz} orbitals arises from back-bonding into the π system of the ligand. The X-ray results on CoTPP have been confirmed by an independent polarized neutron diffraction study of the spin density distribution in cobalt phthalocyanine,[104,105] a compound with crystal structure and coordination environment very similar to the cobalt atom (Table 2-1).

Analysis of the d-electron density in terms of the apparent orbital occupancies as discussed here provides a simple description of the electronic structure of the complex. It may permit an assignment of the ground state of the complex in ambiguous cases. It is, however, limited by the initial assumptions of the model. As the accuracy of experiments increases, as may be possible with synchrotron sources,[106] and as complementary data such as spin density measurements become available,[107,108] more detailed modeling of the electronic structure may become possible. Two-center terms might be included to account for covalency, in which case quantum mechanical constraints on the model may be desirable.[109]

4. CONCLUSION

As is evident from the work described here and elsewhere, the electron density distribution is an extremely detailed three-dimensional function of a molecular system that may be obtained from careful high-resolution X-ray diffraction measurements. Only large basis set ab initio theoretical calculations are capable of predicting this function with comparable accuracy. Much information on the electronic structure of molecules and crystals can be obtained by direct inspection of the experimental electron distribution. In addition, other one-electron properties may be calculated directly from the experimental electron density.

A more fundamental connection also exists between the electron distribution and the energy of the system. According to the Hohenberg–Kohn[111] theorem, the energy and wave function for a nondegenerate ground state are unique functionals of the electron distribution. Thus, knowledge of the electron density distribution should be sufficient to give all properties of the system. The exact nature of the functional that relates these properties to the electron density distribution is, however, still unknown.

Current results show that reliable electron distributions can be obtained experimentally on light atom structures with a moderately large number of atoms. Reasonable results can also be obtained from structures including elements up through the second row in the periodic table. In the future, the accuracy of experiments will certainly increase, allowing the study of effects such as electron correlation in small molecules, structures with heavier elements, and weaker effects such as intermolecular interactions. Methods of analysis may also be expected to improve, allowing, for example, the incorporation of spin-density measurements, and the results of other experimental probes.

Although analysis of the electron distribution itself is common, far less attention has been given to other properties such as net atomic charges, dipole moments, and electric field gradients, which may be obtained from the electron distribution. Of these properties, the molecular electrostatic potential is perhaps of most interest because it provides a possible bridge between the charge distribution and chemical reactivity. However, electrostatic potentials have been calculated from experimental data for only a handful of cases.

REFERENCES

1. Angermund, K.; Claus, K. H.; Goddard, R.; Krüger, C. *Angew. Chem. Int. Ed. Engl.* **1985**, *24*, 237–247.
2. Coppens, P. *J. Chem. Educ.* **1984**, *61*, 761–765.
3. Coppens, P.; Stevens, E. D. *Adv. Quantum Chem.* **1977**, *10*, 1–35.
4. Becker, P. "Electron and Magnetization Densities in Molecules and Crystals." Plenum Press: New York, 1980.
5. Coppens, P.; Hall, M. "Electron Distributions and the Chemical Bond." Plenum Press: New York, 1982.
6. Stevens, E. D.; Coppens, P. *Acta. Crystallogr.* **1980**, *B36*, 1864–1876.
7. Stewart, R. F. *J. Chem. Phys.* **1972**, *57*, 1664–1668.
8. Moss, G.; Feil, D. *Acta Crystallogr.* **1981**, *A37*, 414–421.
9. Coppens, P.; Guru Row, T. N.; Leung, P.; Stevens, E. D.; Becker, P. J.; Yang, W. *Acta Crystallogr.* **1979**, *A35*, 63–72.
10. Bader, R. F. W.; Anderson, S. G.; Duke, A. J. *J. Am. Chem. Soc.* **1979**, *101*, 1389–1395.
11. Rees, B. *Acta Crystallogr.* **1976**, *A32*, 483–488.
12. Rees, B. *Acta Crystallogr.* **1978**, *A34*, 254–256.
13. Hirshfeld, F. L. *Acta Crystallogr.* **1971**, *B27*, 769–781.
14. Hansen, N. K.; Coppens, P. *Acta Crystallogr.* **1978**, *A34*, 909–921.
15. Stewart, R. F. *Acta Crystallogr.* **1976**, *A32*, 565–574.
16. Savariault, J.-M.; Lehmann, M. S. *J. Am. Chem. Soc.* **1980**, *102*, 1298–1303.
17. Stewart, R. F. *Chem. Phys. Lett.* **1979**, *65*, 335–342.
18. Staudenmann, J. L.; Coppens, P.; Muller, J. *Solid State Commun.* **1976**, 29–33.
19. Hirshfeld, F. L. *Isr. J. Chem.* **1977**, *16*, 198–201.
20. Klein, C. L.; Majeste, R. J.; Stevens, E. D. *J. Am. Chem. Soc.* **1987**, *109*, 6675–6681.
21. Pant, A. K.; Stevens, E. D.; Klein, C. L. *J. Phys. Chem.* **1988**, submitted for publication.
22. Politzer, P.; Truhlar, D. G., Eds. "Chemical Applications of Atomic and Molecular Electrostatic Potentials." Plenum Press: New York, 1981.

23. Spackman, M. A.; Stewart, R. F. In "Chemical Applications of Atomic and Molecular Electrostatic Potentials," Politzer, P.; and Truhlar, D. G., Eds.; Plenum Press: New York, 1981, pp. 407–426.
24. Dunitz, J. D.; Seiler, P. *J. Am. Chem. Soc.* **1983**, *105*, 7056–7058.
25. Pant, A. K.; Stevens, E. D. *J. Am. Chem. Soc.* **1988,** submitted for publication.
26. Bianchi, R.; Pilati, T.; Simonetta, M. *Helv. Chem. Acta* **1984**, *67*, 1707–1712.
27. Eisenstein, M.; Hirshfeld, F. L. *Acta Crystallogr.* **1983**, *B39*, 61–75.
28. Irngartinger, H.; Goldman, A.; Jahn, R.; Nixdorf, M.; Rodewald, R.; Mair, G.; Malsch, K. D.; Emrich, R. *Angew. Chem. Int. Ed. Engl.* **1984**, 993–994. Irngartinger, H.; Jahn, R. **1986,** personal communication.
29. Irngartinger, H. In "Electron Distributions and the Chemical Bond," Coppens, P.; and Hall, M. B., Eds.; Plenum Press: New York, 1982, pp. 361–379.
30. Politzer, P.; Domelsmith, L. N.; Abrahmsen, L. *J. Phys. Chem.* **1984**, 1752–1758.
31. Majeste, R. J.; Stevens, E. D.; Klein, C. L. Unpublished results.
32. Harvey, R. G. *Acc. Chem. Res.* **1981**, *14*, 218–226.
33. Klein, C. L.; Stevens, E. D. *Acta Crystallogr.* **1988**, *B44*, 50–55.
34. Jeffrey, A. M.; Jennete, K. W.; Blobstein, S. H.; Weinstein, I. B.; Beland, P.; Harvey, R. G.; Kasai, H.; Miura, I.; Nakanishi, K. *J. Am. Chem. Soc.* **1976**, *98*, 5714–5715.
35. Klein, C. L.; Stevens, E. D.; Zacharias, D. E.; Glusker, J. P. *Carcinogenesis* **1987**, *8*, 5–18.
36. Foss, L. I. Ph.D. thesis, University of New Orleans, 1986, New Orleans, LA.
37. Klein, C. L.; Williams, S. Unpublished results.
38. Craven, B. M.; Weber, H. P. *Acta Crystallogr.* **1983**, *B39*, 743–748.
39. Klein, C. L.; Stevens, E. D. *J. Pharm. Sci. J. Heterocyclic Chem.* 1988 submitted.
40. Data were collected by Hough, E.; Hjorth, M.; Dahl, S. G. (*Acta Crystallogr.* **1982**, *B38*, 2424–2428) and reanalyzed by Klein, C. L.; Wilson, R. (unpublished results).
41. Longchambon, F.; Gillier-Pandraud, H.; Wiest, R.; Rees, B.; Mitschler, A.; Feld, R.; Lehmann, M.; Becker, P. *Acta Crystallogr.* **1985**, *B41*, 47–56.
42. Craven, B. M.; Benci, P. *Acta Crystallogr.* **1981**, *B37*, 1584–1591.
43. Hirshfeld, F. L. *Isr. J. Chem.* **1977**, *16*, 226–229.
44. Eisenstein, M. *Chem. Scripta* **1986**, *26*, 481.
45. Takusagawa, F.; Koetzle, T. F. *Acta Crystallogr.* **1979**, *B35*, 867–877.
46. Pearlman, D. A.; Kim, S. H. *Biopolymers* **1985**, *24*, 327–357.
47. Legros, J.-P.; Kvick, A. *Acta Crystallogr.* **1980**, *B36*, 3052–3059.
48. Lau, C.; Bader, R. F. W.; Hermansson, K.; Berkovitch-Yellin, Z. *Chem. Scripta* **1986**, *26*, 476.
49. Griffen, J. F.; Coppens, P. *J. Am. Chem. Soc.* **1975**, *97*, 3496–3505.
50. Kvick, A.; Koetzle, T. F.; Stevens, E. D. *J. Chem. Phys.* **1979**, *71*, 173–179.
51. Epstein, J.; Ruble, J. R.; Craven, B. M. *Acta Crystallogr.* **1982**, *B38*, 140–149.
52. Stewart, R. F. *God. Jugosl. cent. kristalogr.* **1982**, *17*, 1–24.
53. Stevens, E. D. *Acta Crystallogr.* **1978**, *B34*, 544–551.
54. Kvick, A.; Thomas, R.; Koetzle, T. F. *Acta Crystallogr.* **1976**, *B32*, 224–231.
55. Yamabe, S.; Morokuma, K. *J. Am. Chem. Soc.* **1975**, *97*, 4458–4465.
56. Dreyfus, M.; Pullman, A. *Theor. Chim. Acta* **1970**, *19*, 20–37.
57. Stevens, E. D.; Lehmann, M. S.; Coppens, P. *J. Am. Chem. Soc.* **1977**, *99*, 2829–2831.
58. Jovanovski, G.; Thomas, J. O.; Olovsson, I. *Acta Crystallogr.* **1987**, *B43*, 85–92.
59. Hermansson, K.; Thomas, J. O. *Acta Crystallogr.* **1983**, *C39*, 930–936.
60. Ohba, S.; Kikkawa, T.; Saito, Y. *Acta Crystallogr.* **1985**, *C41*, 10–13.
61. Hermansson, K.; Thomas, J. O.; Olovsson, I. *Acta Crystallogr.* **1984**, *C40*, 335–340.
62. Hermansson, K.; Thomas, J. O. *Acta Crystallogr.* **1982**, *B38*, 2555–2563.
63. Iwata, M.; Saito, Y. *Acta Crystallogr.* **1973**, *B29*, 822–832.
64. Coppens, P. *Coord. Chem. Rev.* **1985**, *65*, 285–307.
65. Stevens, E. D.; DeLucia, M. L.; Coppens, P. *Inorg. Chem.* **1980**, *19*, 813–820.
66. Kijima, N.; Tanaka, K.; Marumo, F. *Acta Crystallogr.* **1983**, *B39*, 557–561.

67. Miyata, N.; Tanaka, K.; Marumo, F. *Acta Crystallogr.* **1983,** *B39,* 561–564.
68. Kidoh, K.; Tanaka, K.; Marumo, F.; Takei, H. *Acta Crystallogr.* **1984,** *B40,* 92–96.
69. Varghese, J. N.; Maslen, E. N. *Acta Crystallogr.* **1985,** *B41,* 184–190.
70. Christidis, P. C.; Rentzeperis, P. J.; Kirfel, A.; Will, G. *Z. Kristallogr.* **1983,** *164,* 219–236.
71. Ohba, S.; Saito, Y. *Acta Crystallogr.* **1984,** *C40,* 1639–1641.
72. Stevens, E. *J. Am. Chem. Soc.* **1981,** *103,* 5087–5095.
73. Kutzler, F. W.; Swepston, P. N.; Berkovitch-Yellin, Z.; Ellis, D. E.; Ibers, J. A. *J. Am. Chem. Soc.* **1983,** *105,* 2996–3004.
74. Coppens, P.; Li, L. *J. Chem. Phys.* **1984,** *81,* 1983–1993.
75. Ohba, S.; Sata, S.; Saito, Y. *Acta Crystallogr.* **1983,** *B39,* 49–53.
76. Kidoh, K.; Tanaka, K.; Marumo, F.; Takei, H. *Acta Crystallogr.* **1984,** *B40,* 329–332.
77. Lecomte, C.; Chadwick, D. L.; Coppens, P.; Stevens, E. D. *Inorg. Chem.* **1983,** *22,* 2982–2992.
78. Toriumi, K.; Ozima, M.; Akaogi, M.; Saito, Y. *Acta Crystallogr.* **1978,** *B34,* 1093–1096.
79. Pant, A. K.; Stevens, E. D. *Phys. Rev. B* **1988,** *37,* 1109–1120.
80. Toriumi, K.; Saito, Y. *Acta Crystallogr.* **1978,** *B34,* 3149–3156.
81. Johansen, H. *Acta Crystallogr.* **1976,** *A32,* 353–355.
82. Bénard, M.; Coppens, P.; DeLucia, M. L.; Stevens, E. D. *Inorg. Chem.* **1980,** *19,* 1924–1930.
83. Hino, K.; Saito, Y.; Bénard, M. *Acta Crystallogr.* **1981,** *B37,* 2164–2170.
84. Mitschler, A.; Rees, B.; Wiest, R.; Bénard, M. *J. Am. Chem. Soc.* **1982,** *104,* 7501–7509.
85. Wang, Y.; Coppens, P. *Inorg. Chem.* **1976,** *15,* 1122–1127.
86. Mitschler, A.; Rees, B.; Lehmann, M. S., *J. Am. Chem. Soc.* **1978,** *100,* 3390–3397.
87. Leung, P. C.; Coppens, P. *Acta Crystallogr.* **1983,** *B39,* 535–542.
88. Leung, P. C.; Thesis, State University of New York at Buffalo, 1982.
89. Zhu, N.-J.; Coppens, P. Unpublished results reported in Reference 64.
90. Pant, A. K.; Stevens, E. D. *Inorg. Chem.* **1988,** submitted for publication.
91. Li, Y.-J.; Stevens, E. D. *Inorg. Chem.* **1988,** submitted for publication.
92. Martin, M.; Rees, B.; Mitschler, A. *Acta Crystallogr.* **1982,** *B38,* 6–15.
93. Bénard, M. In "Electron Distributions and the Chemical Bond," Coppens, P.; and Hall, M. B., Eds.; Plenum Press: New York, 1982, pp. 221–253.
94. Rees, B.; Mitschler, A. *J. Am. Chem. Soc.* **1976,** *98,* 7918–7924.
95. Sherwood, D. E.; Hall, M. B. *Inorg. Chem.* **1983,** *22,* 93–100.
96. Dewar, J. J. S. *Bull. Soc. Chim. Fr.* **1951,** *18,* C71–C79. Chatt, J.; Duncanson, J. *J. Chem. Soc.* **1953,** 2939–2947.
97. Stevens, E. D.; Coppens, P. *Acta Crystallogr.* **1979,** *A35,* 536–539.
98. Varghese, J. N.; Mason, R. *Proc. R. Soc. London, Ser. A* **1980,** *372,* 1–7.
99. Holladay, A.; Leung, P.; Coppens, P. *Acta Crystallogr.* **1983,** *A39,* 377–387.
100. Rose, M. E. "Elementary Theory of Angular Momentum." Wiley: New York, 1957.
101. Condon, E. U.; Shortley, G. H. "Theory of Atomic Spectra." Cambridge University Press: Cambridge, 1957.
102. Stevens, E. D. In "Electric Distributions and the Chemical Bond," Coppens, P.; and Hall, M. B., Eds.; Plenum Press: New York, 1982, pp. 331–349.
103. Lin, W. C. *Inorg. Chem.* **1976,** *15,* 1114–1118. Lin, W. C. In "The Porphyrins," Vol. IV; Dolphin, D., Ed.; Academic Press: New York, 1979, pp. 355–377.
104. Williams, G. A.; Figgis, B. N.; Mason, R. *J. Chem. Soc. Dalton Trans.* **1981,** 1837–1845.
105. Coppens, P.; Holladay, A.; Stevens, E. D. *J. Am. Chem. Soc.* **1982,** *104,* 3546–3547.
106. Nielsen, F. S.; Lee, P.; Coppens, P. *Acta Crystallogr.* **1986,** *B42,* 359–364.
107. Figgis, B. N.; Reynolds, P. A.; Mason, R. *J. Am. Chem. Soc.* **1983,** *105,* 440–441.
108. Coppens, P.; Koritsanszky, T.; Becker, P. *Chem. Scripta* **1986,** *26,* 463–467.
109. Mussa, L.; Goldberg, M.; Frishberg, C. A.; Boehme, R.; LaPlaca, S. *Phys. Rev. Lett.* **1985,** *55,* 622–625.
110. Massa, L. *Chem. Scripta* **1986,** *26,* 469–472.
111. Hohenberg, P.; Kohn, W. *Phys. Rev. B* **1964,** *136,* 864–871.

CHAPTER 3

The Concept of Molecular Strain: Basic Principles, Utility, and Limitations

Dieter Cremer
Universität Köln, Köln, West Germany

Elfi Kraka
Theoretical Chemistry Group, Argonne National Laboratory, Argonne, Illinois

CONTENTS

1. Introduction.. 66
2. The Concept of Strain 68
3. Quantitative Assessment of Strain and Strain Energy........ 71
4. Chemical Consequences of Strain 79
5. Comparison of the Strain Energies in Small Cycloalkanes.... 82
6. Molecular Orbital Approach to Strain 89
7. Electron Density Approach to Strain 95
8. A Step Toward a Unified Description of Strain: The Laplacian of the Electron Density........................ 104
9. Ways of Assessing the Strain Energy from Quantum Chemical Calculations 109
10. Calculation of the Strain Energy from in situ Bond Energies. 111
11. Quantum Chemical Evaluation of the Molecular Strain Energy Using the Westheimer Approach 114

12. Pros and Cons of σ Aromaticity 117
13. Limitations of the Concept of Strain 123
14. Conclusions.. 130
Acknowledgments... 132
References 133

> To know a lot does not teach reason.
> Heraclitus

1. INTRODUCTION

The continuous improvement of both experimental and theoretical methods in the past decades has swamped chemistry with a tremendous amount of data about molecules and their properties. A systematization and rationalization of the myriad of established facts is of paramount importance for progress in chemistry. The human mind can best comprehend and interpret molecular properties in terms of conceptual models. Of course, models always differ from the object being modeled. If we are using models of molecules, we relinquish accuracy but gain generality, simplicity, and feasibility. Such a compromise is acceptable provided one is always aware that models are just temporary aids, always to be revised, and eventually to be discarded as better models are developed. Also one has to bear in mind that a model may be suitable for one purpose but inadequate for other needs. Models always have to be adjusted to reflect and to cope with improved insights into nature.

A useful model for the description of molecular properties should fulfill three basic requirements. First, it should be simple and easy to memorize. Second, it should possess a sound physical basis and provide a consistent description of its objects. Finally, it should be flexible and applicable over as wide a range as possible. Insofar as these requirements are partially contradictory, each and every model is open to criticism. For example, increased simplicity of a model usually entails more insight and clarity, but also, a decrease in flexibility and applicability. A rigorous theoretical approach very often discloses inconsistencies in a model and, as a consequence, can lead to its revision.

Many models in chemistry have been inspired by classical mechanics. They describe molecules with the properties of macroscopic quantities such as balls, sticks, and springs. A typical example is the concept of molecular strain introduced into chemistry a century ago by Adolf von Baeyer to explain the relative stabilities of cycloalkanes.[1] Despite its simplicity, this model has been of astonishing value for chemistry with regard to both the rationalization of known molecular properties and the prediction of unknown properties. For example, in nearly all modern textbooks the essence of von Baeyer's strain concept is described and then used to rationalize the stability of small cycloalkanes and other strained compounds. Without exception, chemical textbooks impart to the reader the impression that the relative stabilities of ring compounds are well understood in the light of the concept of molecular strain.

This view has been challenged recently by various authors[2-9] who have pointed out that the structure and stability properties of cyclopropane and possibly other three-membered rings evade classical descriptions of chemical bonding and, thereby, also classical strain theory. Analyzing the chemical and physical properties of small cycloalkanes, Dewar[2] came to the conclusion that cyclopropane possesses a sextet of delocalized σ electrons and, therefore, is isoconjugate with benzene, an idea that can already be found in an early PPP-type description of cyclopropane by Brown and Krishna.[3] In the same way that benzene may be considered a π-aromatic system, cyclopropane may be considered to be σ-aromatic.[10] Delocalization of σ electrons adds to the stability of cyclopropane, thus compensating for part of its ring strain.

Cremer and Kraka[4,5] added support to the idea of σ-delocalization by analyzing the properties of the electron density distribution of three-membered rings. They estimated the extra stablization of cyclopropane due to σ-delocalization. This work was extended by Cremer and Gauss,[6] who compared cyclopropane and cyclobutane on the basis of ab initio calculations. These authors confirmed the idea of σ-electron delocalization and assigned energies to the various effects acting in cyclopropane.

Schleyer[11] reviewed the question of the stability of small cycloalkanes and, on the basis of his analysis, challenged the idea of σ-aromaticity. He argued that the energetics of small cycloalkanes can safely be explained within an extended concept of strain.

These controversial views indicate that contrary to what is written in chemical textbooks, the use of the concept of strain to rationalize the energetics of small cycloalkanes is questionable. Furthermore, the controversy that has recently flared up concerning strain in cyclopropane and other three-membered rings probably will not be settled in the near future. A critical review of the notion of strain is needed at this time to properly evaluate the various arguments. We attempt to present such a review in this chapter, not by listing in an encyclopedic way the many investigations that have been devoted to the subject, but by searching for the roots of the notion of strain and exploring its facets in modern chemistry. In particular, we will investigate the extent to which von Baeyer's classical strain concept can be confirmed within the realm of quantum chemistry. Is there a way of placing the concept of strain on a sound quantum mechanical basis, hopefully improving the self-consistency, flexibility, and applicability of the model but at the same time retaining its simplicity?

We will deal with this question by reviewing briefly the origin of the concept of strain and the various types of strain discussed in chemistry (Section 2). Then, we will discuss ways of quantitatively assessing molecular strain and its energetic consequences (Section 3), also pointing out the chemical consequences of strain (Section 4). In Section 5, we will focus on the strain energies determined for small cycloalkanes, giving special emphasis to the puzzling similarity of the strain energies of cyclopropane and cyclobutane. The sections that follow will be devoted to a quantum chemical approach to molecular strain, first on the basis of the molecular orbitals (MO) (Section 6), then within the realm of electron density theory (Section 7) and, finally, by considering the Laplace distribution of electrons in strained molecules (Section 8). Ways of assessing the strain energy from quantum chemical calculations will be put forward in Sections 9, 10, and 11, either on the basis of in situ bond energies (Section 10) or by a dissection of the molecular strain energy according to Westheimer (Section 11). In Section 12, we will focus on cyclopropane and will display the results of quantum chemical calculations under the heading of pros and cons of σ-aromaticity. Then we will establish the limitations of the concept of molecular strain (Section 13), by concentrating on the relationship between three-membered rings and π complexes. Section 14 contains concluding remarks. We hope that our account will sharpen the reader's eye with respect to the utility and the limitations of the concept of molecular strain.

2. THE CONCEPT OF STRAIN

A. Strain in Classical Mechanics

If forces act on an elastic body, the body is deformed. It becomes *strained*. Quantitatively, this deformation, called *strain*, is given by the relative displacements $\Delta x/x$ of the parts of the elastic body.[12] Thus, strain is a dimensionless quantity. One distinguishes dilatation (compression) strain, shearing strain, and torsional strain.

There are forces in a strained body that act to restore its original form. The restoring force per unit area is called *stress*. According to Hooke, the stress set up within an elastic body is proportional to the strain to which the body is subjected:

$$\text{stress} = k \times \text{strain} \quad \text{(Hooke's law)}$$

where k, the modulus of elasticity, is a proportionality constant that possesses the same dimension as the stress, namely force per unit area.

The potential energy per unit volume stored up in the body is called the strain energy function.[12]

B. Strain in Chemistry

The notion of "strain" was introduced into chemistry by Adolf von Baeyer in 1885,[1] at a time when the tetravalency of the carbon atom, the tetrahedral arrangement of carbon bonds, and the connection between molecular structure and bonding had been established by Kekulé, Couper, van't Hoff, Le Bel, and others. Von Baeyer postulated that the valencies (bonds) of a carbon atom may deviate from the tetrahedral directions. These deviations lead to *strain* that increases with the magnitude of the deviation.[13] According to von Baeyer, strain should be largest in small cycloalkanes such as cyclopropane and cyclobutane that possess CCC bond angles that deviate by 49° and 19°, respectively, from a tetrahedral angle of 109.5°.

Von Baeyer drew the connection to the notion of strain in classical mechanics by considering the carbon bonds as elastic springs or sticks, deformation of which can be described by Hooke's law. Thus, the "elastic body" considered in chemical strain theory is the chemical bond. *Obviously, the concept of strain is intimately connected with the concept of the chemical bond.* Any weakness, oversimplification, or inconsistency in the latter inevitably shows up in the concept of molecular strain. Clearly, both chemical bonding and molecular strain are model-bound quantities. In other words, *the strain energy of a molecule is (contrary to the strain energy of an elastic body) nonobservable* (as are the bond properties bond energy, bond polarity, etc). This has to be borne in mind when applying the concept of strain in chemistry.

C. Types of Molecular Strain

Von Baeyer defined what is termed "bond angle strain" (Table 3-1). Its equivalent in classical mechanics is the shearing strain. Since von Baeyer's work, the strain concept has been extended in chemistry and other strain types have been taken from classical mechanics.

For example, the stretching or dilatation (compression) of a bond from an idealized value leads to stretching strain. Torsion of a bond causes torsional strain. The latter type of strain was first identified by Pitzer when investigating the conformational behavior of ethane.[14] In the equilibrium form, all CH bonds of ethane are staggered (torsion angle $\tau = 60°$). CH bond eclipsing ($\tau = 0°$) leads to an increase in the energy of ethane, the reason for which is considered to be torsional strain. This type of strain has also been termed "bond opposition strain"[15] or "Pitzer strain."

The shape of a molecule is determined by internal coordinates such as bond lengths, bond angles, and torsional angles. As an alternative to bond angles and torsional angles, one can use nonbonded distances when describing molecular shape. Similarly, as in the case of internal coordinates, ideal-

TABLE 3-1. Types of Strain Used in Chemistry

Strain	Description and alternative names	Ref.
1. Baeyer strain	Bond angle strain	1
2. Pitzer strain	Torsional strain	14
	Bond opposition strain	
	Bond eclipsing strain	
3. Stretching strain	Bond length strain	19
4. Dunitz–Schomaker strain	Nonbonded strain	18
	Compression of van der Waals radii (van der Waals strain)	16
	Transannular strain or Stoll pressure (better: Stoll strain)	17
5. Steric strain	Sum of 1–4	19
6. I strain	Internal strain comprising 1–4	15
7. F strain	Front strain leading to a retardation of chemical reactions	15
8. B strain	Back strain leading to an acceleration of chemical reactions	15
9. Electrostatic strain	Actually electrostatic stress caused by more than one charge within a molecule	20
10. Superstrain	Difference between the total strain and the sum of the strain of the fused rings in polycyclic molecules	22

ized values can be defined for nonbonded distances utilizing the van der Waals radii of the atoms. If the molecular structure entails nonbonded distances that are smaller than the sum of the van der Waals radii, the molecule is compressed, an idea first proposed by F. Kehrmann in 1889.[16] In this case one speaks of steric strain, nonbonded strain, transannular strain, or Stoll pressure (in the case of ring compounds).[17] Dunitz and Schomaker[18] investigated nonbonded C,C repulsion in cyclobutane and showed it to be an important factor influencing the stability of this molecule. Therefore, nonbonded interactions are often listed under "Dunitz–Schomaker strain" (Table 3-1).

The four types of strain are usually subsumed under the term "steric strain."[19] Brown has coined the term I- (internal) strain[15] for the total molecular strain in ring systems and uses this when discussing strain in connection with molecular reactivity. Other types of strain have been introduced in connection with the concepts of steric hindrance and steric assistance: for example, F- (front) strain, leading to retardation of chemical reactions, or B- (back) strain, causing acceleration of chemical reactions.[15] Other types of strain are listed in Table 3-1.[20–22]

D. Colloquial Use of the Term "Strain"

A description of the concept of strain is most often found in textbooks of organic chemistry.[23] In all cases, the strain in cycloalkanes is discussed in terms of Baeyer strain, which is considered to dominate the relative stabilities of the ring compounds. Therefore, it is not surprising that "strain" and "Baeyer strain" often are used synonymously. For example, in encyclopedic books on chemistry a description of bond angle strain is given under the key word "strain theory."[24]

There are only a few general textbooks,[23b,23d,25] that present a clear distinction between the various types of strain. More information can be obtained from books that deal with such special topics as conformational analysis, stereochemistry, and force field calculations.[26-29] Elaborate discussions of the various types of strain can be found in a number of review articles that have appeared over the years.[30-35]

Some authors use "stress" and "strain" interchangeably; that is, they use terms like "steric pressure" or "steric forces" as substitutes for the notion of strain. Of course, this may sometimes happen because chemists are in general more interested in the energetic consequences of strain than in strain itself. Also, the terms "stress" and "strain" are differently used in other languages.[13]

3. QUANTITATIVE ASSESSMENT OF STRAIN AND STRAIN ENERGY

In a deformed elastic body, potential energy is stored in the form of strain energy. Similarly, a strained molecule possesses a strain energy (SE) that increases its potential energy in relation to a hypothetical strain-free molecule. Chemists are interested in the SE of a molecule in order to rationalize its stability and reactivity. Thus strain and strain energy provide a basis that links the structure, stability, and reactivity of molecules (Scheme I).

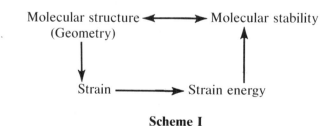

Scheme I

In contrast to classical mechanics, which defines the SE of an elastic body as a macroscopic property, the SE of a molecule is determined by the individual SEs of all the "elastic" subunits making up the molecule, namely atoms, atomic groups, and bonds. In a classical sense these can be considered as elastic balls, connected by elastic sticks or springs. To obtain the individual parts of the total molecular SE, each elastic subunit of a molecule must be defined in a hypothetical strain-free state.

There exist countless ways of defining the strain-free subunits. Each of these ways establishes a specific model of molecular strain. The value of these models can be determined only by examining the extent to which chemical properties of molecules can be rationalized. A model of strain can be considered useful if the structure and stability of a large set of molecules can be rationalized in a consistent way.

There are both theoretical and experimental ways of defining strain-free subunits and their SEs. We will briefly describe those used for the evaluation of strain and SE in saturated hydrocarbons.

A. Establishment of "Strain-free" Reference States

In classical mechanics strain is determined by the relative displacements of the parts of an elastic body, that is, by the deformation of the body.[12] Von Baeyer introduced strain into chemistry by considering tetravalent carbon as an elastic body that can be deformed by forcing the four valencies out of the tetrahedral directions.[1] Such deformations become necessary when forming a strained hydrocarbon molecule from unstrained CH_n ($n = 0, 1, 2, 3$) subunits. (It is stressed that these subunits are just model building blocks and should not be confused with atomic carbon, methylidyne, carbene, or the methyl radical.) Angle strain is relieved as soon as tetravalent carbon is extricated from the strained molecule. For sp^3-hybridized carbon, the ideal CCC, CCH, and HCH angles in hydrocarbons should be 109.5°. Other choices of ideal angles have been advocated on the grounds that the preferred angles in small hydrocarbons differ from 109.5°.[37-40] However, these choices were made with the intention of reproducing experimental ΔH_f° values or other molecular properties using the molecular mechanics approach. Thus, the molecules possessing these angles are strain-free in an operational but not absolute sense.

Von Baeyer's idea of an elastic chemical "body" has to be extended when assessing stretching and torsional strain of a molecule. In this case, unstrained reference molecules must be defined. For hydrocarbons, molecules such as methane, ethane, propane, n-butane, or isobutane are considered to possess unstrained CH and CC bonds.[37-40] Deviation of measured CC and CH bond lengths from the reference bond length is indicative of stretching strain. Similarly, ideal HCCH, CCCH, and CCCC torsional angles (60°) are taken from the staggered forms of ethane, propane, and n-butane and are used to detect torsional strain.[37-40]

The van der Waals radii of C (1.85 Å) and H (1.20 Å) have been determined from the intermolecular distances in crystalline hydrocarbons.[29,41] If the nonbonded distances of a molecule fall short of the sum of the corresponding van der Waals radii, the molecule will be considered to suffer from nonbonded strain.

Chemists are primarily interested in the energetic consequences of strain, that is, the strain energy. Therefore, the choice of the reference molecule is often guided by the intention to determine the SE of a molecule. For this purpose, molecules have been selected that lead to ideal CC or CH bond energies or, alternatively, to ideal group increments $\Delta H_f^\circ(CH_n)$ with $n = 0$, 1, 2, 3. It would be beyond the scope of this chapter to list all the possible choices suggested in the literature during the past decades. (Some of these are listed in References 37, 38, and 42–45.) Instead, we have summarized some basically different possibilities of selecting appropriate reference groups CH_n in Table 3-2. For example, one can select CH_4 as a reference molecule and subject it to angle deformations typical of strained hydrocarbons. The corresponding increase in the energy of CH_4 can be used to determine the bond angle SE.[46,47]

Hydrocarbons are made up from the four subunits CH_n, $n = 0, 1, 2$, and 3, namely $\rangle C \langle$, $\rangle CH—$, $\rangle CH_2$, and $—CH_3$. Accordingly, one needs the ΔH_f° value of just four appropriate reference molecules, each containing at least one of these groups. Solving four equations with four unknowns, the $\Delta H_f^\circ(CH_n)$ ($n = 0, 1, 2, 3$) values lead to the definition of strainless CH_n subunits to be used when assessing strain and SE in hydrocarbons. This approach has been employed in several cases. Investigators have used as reference molecules the smallest alkanes that possess the respective subunit,[48] or hydrocarbons that consist uniquely of one and the same subunit ('diagonal' reference states,[45] see Table 3-2).

Alternatively, one can set up an overdetermined system of linear equations for the four unknowns by utilizing known ΔH_f° values of a series of

TABLE 3-2. Possible Reference Molecules for Analyzing Strain and Calculating Strain Energies in Hydrocarbons

	Reference molecules	Remark	Source	Ref.
1.	CH_4	HCH angles are distorted to values of CCC angles in strained molecules; distortion energies are calculated	Wiberg et al. Schleyer	46 47
2.	CH_3CH_3, $(CH_3)_2CH_2$, $(CH_3)_3CH$, $(CH_3)_4C$	Definition of homodesmotic references	George et al.	48
3.	Ethane, cyclohexane, cubane, adamantane, diamond	Definition of "diagonal" reference states	Liebman and Van Vechten	45
4.	Acyclic and cyclic hydrocarbons	Definition of averaged groups	Franklin Benson et al. Boyd Schleyer et al. Allinger et al.	42 43, 44 36 38, 39 29a, 37

acyclic and cyclic hydrocarbons. This approach has been employed by several authors.[37-39,42-44] In case of substituted hydrocarbons, especially when electronegativity differences become pronounced, non-next-nearest-neighbor interactions (unanticipated within the Benson scheme[44b]) may play a dominant role; hence strain energies may depend strongly on uncertainties in the decision of the model group increment as pointed out by Liebman and co-workers[134] for fluorocycloalkanes.

In the case of compounds with nonclassical structures (e.g., π complexes, H-bridged structures, structures with penta- or hexacoordinate C), the definition of appropriate reference groups or reference molecules becomes problematic, if not meaningless. This, of course, has to do with the fact that von Baeyer's idea of a bond as an elastic spring is no longer useful, for example, in the case of two-electron, multicenter bonds.

B. Definition of the Molecular Strain Energy (SE)

Within the concept of strain, the SE is used to rationalize the thermodynamic stability of a strained molecule. Once the strain-free subunits of the molecule have been defined, the SE can be calculated using the Westheimer equation[19]:

$$\Delta E = \Delta E_r + \Delta E_\alpha + \Delta E_\tau + \Delta E_{nb} \tag{3-1}$$

where ΔE_r and ΔE_α are the SEs arising from total bond length and bond angle strain, respectively. They are calculated for all bond lengths r_i and bond angles α_j using Hooke's law:

$$\Delta E_r = \sum_i \frac{k_{ri}}{2} (r_i - r_i^\circ)^2 \tag{3-2}$$

$$\Delta E_\alpha = \sum_j \frac{k_{\alpha j}}{2} (\alpha_j - \alpha_j^\circ)^2 \tag{3-3}$$

The torsional SE is assessed by:

$$\Delta E_\tau = \sum_k \frac{V_{3k}}{2} (1 + \cos 3\tau_k) \tag{3-4}$$

for saturated hydrocarbons (with torsional angles τ_k) and more elaborate Fourier series for other compounds.[29]

The expression for nonbonded interactions is generally of the form of a Buckingham potential[29]:

$$\Delta E_{nb} = \sum_{mn} a_{mn} \exp\{-b_{mn} I_{mn}\} + c_{mn} I_{mn}^{-6} \tag{3-5a}$$

or a Lennard–Jones potential[29]:

$$\Delta E_{nb} = \sum_{mn} a'_{mn} I_{mn}^{-12} + b'_{mn} I_{mn}^{-6} \qquad (3\text{-}5b)$$

where constants a_{mn}, b_{mn}, c_{mn}, or a'_{mn}, b'_{mn} depend on the pair of atoms m and n separated by a distance I_{mn}. Both functions work comparably well, and the preference for one or the other may be considered as a matter of convenience.[29]

Contrary to spectroscopic force fields, in which only the $3N - 6$ degrees of freedom of a molecule are considered, the summations in Equations 3-2 through 3-5 lead to considerably more terms including all possible internal coordinates. For example, for each bonded tetravalent carbon there are four bond lengths and six bond angles in addition to all possible torsional angles and nonbonded distances. For all internal coordinates, the constants of Equations 3-2 through 3-5 (k_{ri}, r_i°, $k_{\alpha j}$, α_j°, V_{3k}, τ_k; a_{mn}, b_{mn}, c_{mn}, or a'_{mn}, b'_{mn}) must be known for the strain-free reference states in order to determine SE by Equation 3-1. One could consider determining these constants from the structural spectroscopic properties of the reference compounds listed in Table 3-2. Indeed, complete sets of these constants are known, and one might consider the calculation of SE to be straightforward. However, two basic problems impede the immediate use of Equation 3-1.

1. The bending force constants of the strain-free CH_n subunits should be determined in the absence of any 1,3-nonbonded repulsion. Any $k_{\alpha j}$ taken from small hydrocarbons contains effects from nonbonded repulsions. Accordingly, ΔE_α and ΔE_{nb} are interdependent and must be adjusted to avoid counting the corresponding energetic contributions to the SE twice. The same holds for all other terms in Equation 3-1. Therefore, the choice of the correct constants for the determination of SE becomes a difficult enterprise. This is why Equation 3-1 and more sophisticated equations derived from it are used to calculate the *steric energy* rather than the SE of a molecule.[19] The steric energy obtained from molecular mechanics calculations is not (!) equal to SE. Even for a strainless molecule like ethane, the steric energy is nonzero because Equation 3-1 is used to reproduce or to predict the heat of formation ΔH_f° of a molecule rather than its SE. The value of ΔH_f° is obtained by adding appropriate enthalpy increments for bonds and atomic groups to the steric energy. Different force fields distribute energy contributions needed for the calculation of ΔH_f° differently between steric energy and enthalpy increments. Therefore, the steric energy is of little significance and differs in magnitude considerably from force field to force field.[29] As a consequence, the SE can be obtained from ΔH_f° only by introducing another set of enthalpy increments applicable for the calculation of ΔH_f° of strainless reference molecules.[29,36,37,39]

For reasons of simplicity, we will not follow the molecular mechanics procedure to obtain SE. Instead, we will stick to the original assumption that the steric energy is equal to the SE. But even with this assumption, another serious problem remains when attempting to calculate SE from Equation 3-1.

2. The observed geometrical data r_i, α_j, τ_k, and I_{mn} of a molecule are affected not only by strain but also by other stabilizing or destabilizing effects such as π conjugation, hyper- or homoconjugation, aromaticity, antiaromaticity, and steric attraction. Hence, Equation 3-1 can be solved only if each energy term is corrected for all other effects also active in the molecule and influencing either r_i, α_j, τ_k, or I_{mn}. This means that a new set of geometrical parameters must be inserted in Equations 3-2 through 3-5 to yield energies $\Delta E'$, which add up to the actual SE:

$$SE = \Delta E'_r + \Delta E'_\alpha + \Delta E'_\tau + E'_{nb} \qquad (3\text{-}6)$$

To solve Equation 3-6 it is necessary to specify exactly the geometrical consequences of other energetic effects such as aromaticity or antiaromaticity, quantities that in turn are defined only within a specific model.

For the reasons just enumerated, an application of Equation 3-1 or 3-6 to calculate the SE has been attempted only in simple cases using various approximations. More often, thermochemical data, either from experiment or theory, have been used to define molecular SEs in one of the following ways (compare with Figure 3-1).[43-45,48-57]

1. *From heats of atomization (ΔH_a) and bond enthalpies (BE).* This approach implies a definition of appropriate bond enthalpies taken from strain-free molecules. Various procedures have been suggested to obtain normal, averaged, or intrinsic BEs.[49-51]

2. *From heats of formation (ΔH_f°) and group enthalpies [ΔH_f° (group)].*[38,42-45] Different ways of deriving $\Delta H_f^\circ(CH_n)$ are discussed in Subsection A of Section 3 and are listed in Table 3-2.

3. *From in situ bond energies (be) of the molecule in question and appropriate reference molecules.* The in situ (or instantaneous) bond energy[52] is the energy needed to homolytically break the bond in question while maintaining in the fission products all molecular features such as hybridization of atoms and bond lengths. The sum of the in situ bond energies is (per definition) equal to the atomization energy of a molecule. In situ bond energies can be calculated from overlap[53] or overlap populations, from shared electron numbers,[54] from resonance integral contributions to semiempirical energies,[55] or from the total electron density distribution $\rho(\mathbf{r})$ in the bond regions of a molecule.[56]

4. *From in situ atomic (or atomic group) energies (ae) of the molecule in question and of appropriate reference molecules.* In analogy to the in situ bond energy, the in situ atomic energy is defined as the energy of an atom within its molecular environment. The sum of the in situ atomic energies is equal to the molecular energy. However, this approach requires a clear

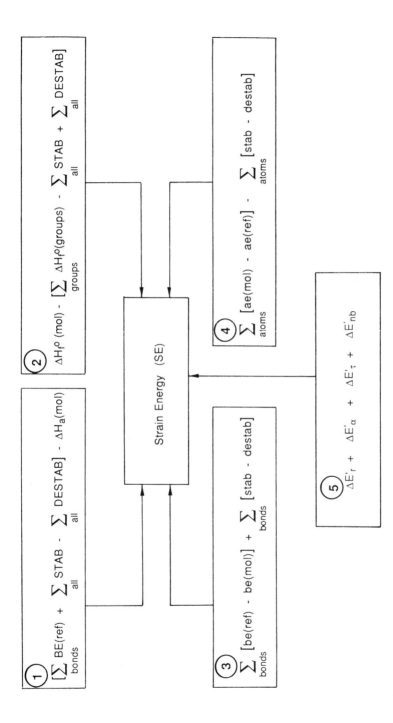

Figure 3-1. Methods of defining the strain energy (SE) of a molecule: mol = strained molecule, ref = reference, group = reference group. BE = bond enthalpy, ΔH_a = heat of atomization, ΔH_f° = heat of formation, be = in situ bond energy, ae = in situ atomic energy, STAB (DESTAB) = stabilization (destabilization) energy per bond or per atom (not arising from strain), $\Delta E_r'$, $\Delta E_\alpha'$, $\Delta E_\tau'$, $\Delta E_{nb}'$ = strain energies arising from bond length, bond angle, torsional strain, and nonbonded strain. For explanation of primed ΔE terms, see text.

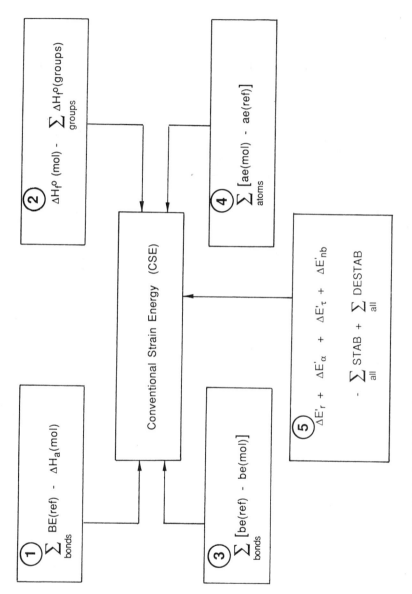

Figure 3-2. Methods of defining the conventional strain energy (CSE) of a molecule. Compare with Figure 3-1.

definition both of what is meant by an atom (or atomic group) in a molecule and of a method for evaluating the energies of atoms (or atomic groups) in their molecular environment. This can be done by the virial partitioning method of Bader,[57] which is based on a quantum mechanical dissection of $\rho(\mathbf{r})$.

When the SE is derived from enthalpies, it is advisable to speak of a strain enthalpy, SH. For a similar reason one could distinguish between SE at finite temperature and at 0 K, with and without zero-point energy corrections. However, compared to the uncertainties in determining the energetic consequences of strain, differences between SE(0), SE(T), SH, and so on are relatively small, justifying the somewhat inaccurate use of the term "strain energy."

Methods 1 through 4 are based on *model quantities* such as the bond enthalpy, group enthalpies, and in situ bond energies or atomic energies. In methods 1 and 2, model quantities are compared with observed (or calculated) enthalpies to obtain SE, while methods 3 and 4 are based on a comparison of two sets of theoretical energies defined within a given model for bonds or atoms.

In all methods, the energetic consequences of stabilizing and destabilizing effects (apart from strain) have to be determined to obtain the proper SE. For methods 1 and 2, this can be done in a cumulative way, but for methods 3 and 4 appropriate corrections have to be evaluated for each bond or each atom. Since this may be problematic for the reasons mentioned above, the term "conventional strain energy" (CSE) has been introduced.[49]

The CSE contains all stabilization and destabilization energies. This makes its evaluation far easier than that of SE, as is shown in Figure 3-2. If, however, the CSE is to be evaluated by means of Equation 3-6, stabilization or destabilization energies must be added to the SE (Figure 3-2).

Although CSE and SE values are similar in saturated hydrocarbons, it is important to note that *in general, a small CSE does not necessarily imply that the molecule is unstrained*. Stabilization and destabilization effects may just cancel each other out. Also, if two CSEs are similar, the corresponding molecules will not necessarily be equally strained. We will come back to this point when discussing the CSEs of cyclopropane and cyclobutane.

4. CHEMICAL CONSEQUENCES OF STRAIN

The molecular properties of strained molecules differ distinctly from those of unstrained molecules. This has been amply demonstrated in the case of cycloalkanes with regard to molecular geometry, thermodynamic stability (as discussed), and various spectroscopic properties.[58–60] The main achievement of the concept of strain has been the rationalization of known properties and the prediction of yet unknown properties as well.

Another objective of the concept of strain has been the prediction of molecular reactivity. A highly strained molecule should try to rearrange, decompose, or react with another molecule to adopt a less strained form. Thus, it is reasonable to expect that an increase in strain would lead to an increase in molecular reactivity and that a strained molecule would react faster than its unstrained counterpart. However, an analysis of this expectation in terms of barriers to reaction casts some doubts on whether or not strain and reactivity are related in a simple way.

Figure 3-3 shows qualitative reaction profiles of two possible extremes. One and the same reaction is considered for both the strain-free reference molecule M1 and the strained molecule M2. In Figure 3-3a, the same transition state is traversed by both M1 and M2. This implies that the activation energy $\Delta E_2^\ddagger$ ($\Delta G_2^\ddagger$) is smaller than $\Delta E_1^\ddagger$ ($\Delta G_1^\ddagger$) by the amount SE; hence the reaction involving M2 should be faster. In Figure 3-3b, the transition state traversed by M2 is energetically higher than that traversed by M1, namely by the amount SE [ie, $\Delta E_1^\ddagger = \Delta E_2^\ddagger$ ($\Delta G_1^\ddagger = \Delta G_2^\ddagger$)]. Thus, M1 and M2 should react equally fast. For both cases, the difference in the reaction energies $\Delta_R E_2 - \Delta_R E_1$ ($\Delta_R G_2 - \Delta_R G_1$) is assumed to be equal to SE.

Clearly, actual reactions will mostly fall between these two extremes, thus leading to $k_2 \geq k_1$. This inequality may be a useful relationship when rationalizing kinetic data. However, a caveat is appropriate when using these qualitative relations between the strain energy (thermodynamical stability) of a molecule and its kinetic stability; no quantitative connection between these molecular properties can be expected (see also Section 13). This will be possible only if the TS energy is known—that is, only if the predictive value of the concept of strain actually is no longer needed.

Even qualitative predictions concerning the reactivity of a strained molecule will no longer be possible if the rate determining step does not lead to strain relief.[61] Also, a strained molecule may not react at all if (a) the strain-relieving reaction is endothermic or (b) the barrier to reaction is too high (e.g., since the reaction in question is symmetry forbidden). In the latter case the paradoxical situation is encountered that a highly strained (thermodynamically labile) molecule is kinetically stable.[62] One example of this situation is the "perfluoroalkyl (R_f) effect."[62c] The R_f effect is a combination of stabilizing effects that perfluoroalkyl groups confer upon highly strained hydrocarbon rings. Although perfluoroalkylation generally thermodynamically destabilizes strained organic rings by enhancing nonbonded repulsion, perfluoroalkylated compounds are more resistant to both catalyzed and unimolecular destruction than their parent compounds. As such, their striking thermal stability has been identified as being purely kinetic in nature.[62d]

Finally, it must be noted that rate constants depend on the free energy of activation, $\Delta G^\ddagger = \Delta H^\ddagger - T\Delta S^\ddagger$, rather than simply on $\Delta E^\ddagger$. Also, measure-

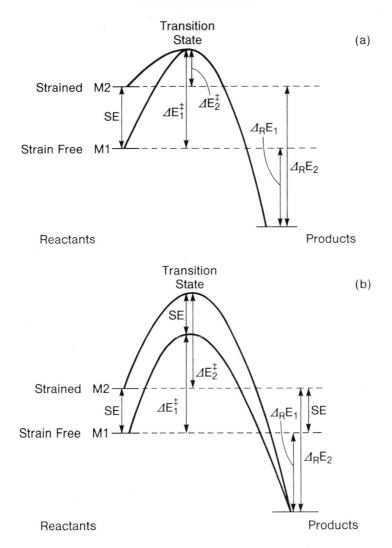

Figure 3-3. Schematic representation of the energy profiles of the reaction of a strained molecule M2 and its strain-free counterpart M1; $\Delta E^{\ddagger}$ and $\Delta_R E$ denote activation and reaction energies, respectively. (a) M1 and M2 traverse the same transition state (i.e., the whole SE of M2 is relieved when the transition state is reached. (b) M1 and M2 traverse different transition states separated by the energy SE (i.e., the SE of M2 is set free after traversing the transition state.

ments of reaction energetics lead to $\Delta_R G = \Delta_R H - T\Delta_R S$. Hence, changes in entropy may disguise effects resulting from strain.[27b,63]

Despite all the difficulties mentioned above, the concept of strain has been successfully applied when discussing reactions of strained molecules. Many examples can be found in the literature in which rate constants of strained molecules are rationalized utilizing the relation between strain and reactivity.[33,64,65] (The interested reader may also refer to the forthcoming chapter "Intramolecularity: Proximity and Strain,"[66] in which the relationship between rate constants and strain energy will be discussed in more detail.)

Also, the kinetic data for the formation of ring compounds have been successfully rationalized by considering the strain of the molecules formed.[7,136]

5. COMPARISON OF THE STRAIN ENERGIES IN SMALL CYCLOALKANES

A. Evaluation of Strain Energies from Thermochemical and ab initio Data

Table 3-3 summarizes heats of formation (ΔH_f°), heats of atomization (ΔH_a),[49] and various sets of CSEs reported in the literature.[37,38,42,44,51] CSE values that have been derived from either averaged BEs or group increments (Table 3-2) vary by less than 1 kcal/mol, a variance that results from the choice of the "strain-free" reference molecules. It is noteworthy that Schleyer[38] and Allinger[37] consider cyclohexane to be slightly strained, contrary to the general belief that this molecule is strain-free. However, these authors make use of a definition of strain in an operational sense, not necessarily in terms of absolute strain. As a consequence, all CSEs of cycloalkanes given by Schleyer and Allinger are somewhat higher than those of other authors (see Table 3-3).

If group increments are determined from the enthalpies of single molecules rather than by averaging over the enthalpies of many molecules, CSE values will differ considerably from the average CSEs given in Table 3-3. As an example, CSEs derived from homodesmotic reaction enthalpies suggested by George and co-workers[48] are listed in Table 3-4. The homodesmotic CH_2 group increment is taken from ΔH_f° of propane by subtracting from the latter the ΔH_f° value of ethane (i.e., twice the enthalpy of the CH_3 group increment). This can be expressed in terms of a formal reaction, the homodesmotic reaction[48]:

$$(CH_2)_n \rightarrow n[CH_3CH_2CH_3 - C_2H_6] \qquad 1$$

the reaction enthalpy of which is equal to the CSE. To distinguish the CSE thus obtained from those based on averaged group increments, we will speak of homodesmotic SEs (HSE).

In the same way, the "diagonal" reference states (cyclohexane in the case

TABLE 3-3. Heats of Formation (ΔH_f°), Heats of Atomization (ΔH_a), and Conventional Strain Energies (CSE) of Cycloalkanes[a]

	Cox and Pilcher (1970)		Conventional strain energies					
	$\Delta H_f^\circ(T)$	$\Delta H_a(T)$	Franklin (1949)	Cox and Pilcher (1970)	Schleyer et al. (1970)	Allinger et al. (1971)	Benson et al. (1976)	Leroy (1985)	Average
Cycloalkane	Ref.: 49	49	42	49	38	37	44	51	
C_3H_6	12.74	812.57	27.2	27.5	28.1	—[b]	27.6	27.6	27.6
C_4H_8	6.38	1093.62	26.1	26.5	26.9	27.2	26.2	26.9	26.6
C_5H_{10}	−18.46	1393.94	5.7	6.2	7.2	7.5	6.3	6.0	6.5
C_6H_{12}	−29.43	1680.10	−0.6	0	1.3	1.7	0	0.2	0.4
C_7H_{14}	−28.34	1953.91	6.1	6.2	7.6	8.0	6.4	6.4	6.8
$\Delta H_f^\circ(CH_2)$		—	−4.9	−4.9	−5.2	−5.2[b]	−4.9	−4.9[c]	−4.9

[a] All enthalpies in kcal/mole. Some authors have used slightly different ΔH_f° values.
[b] The CSE of cyclopropane could not be calculated within the molecular mechanics approach using the same parameters as for other cycloalkanes.[37]
[c] From averaged bond enthalpies.

TABLE 3-4. Experimental and Theoretical Strain Energies

(a) Homodesmotic and diagonal strain energies obtained at 298 and 0 K with and without ZPE corrections

Cycloalkane $(CH_2)_n$	$\Delta H_f^\circ(298)$		$\Delta H_f^\circ(0)^b$		ZPE^b	ZPE correction		$E(THEO),^c$
	HSE	DSE	HSE	DSE		HSE	DSE	HSE = DSE
n = 3	26.5	27.5	25.5	26.7	49.6	3.4	2.2	28.9
4	25.1	26.0	24.4	26.0	67.1	3.6	2.0	28.0
5	4.5	6.1	4.3	6.3	85.5	2.8	0.8	7.1
6	0.4	0	−2.4	0	103.6	2.4	0	0
CH_2 group	−4.6	−4.9	−2.9	−3.3				

(b) Strain energies obtained with various basis sets at the HF level^a

n	STO-3G	4-31G	6-31G(d)	6-31G(d, p)
3	46.8^d	30.4	28.8	28.0
4	29.5	27.0	26.6	27.3
5	6.4		6.8	
6	0.8		0.9	

^{a} All enthalpies and energies in kilocalories per mole. The ΔH_f° values of Table 3-3 have been used to calculate CSEs.
^{b} ZPE values and vibrational corrections are obtained from frequencies given in Reference 102.
^{c} For the definition of E(THEO) see text and Reference 67.
^{d} Value from Lathan et al.[72]

of the CH_2 group), suggested by Van Vechten and Liebman[45] lead to the formal reaction:

$$a\ (CH_2)_n \rightarrow b\ (CH_2)_6 \quad (a \cdot n = 6 \cdot b) \qquad 2$$

that can be used to determine CSE = $-\Delta_R H/a$. In this case we will speak of diagonal SEs (DSE).

Reactions **1** and **2** are useful for a theoretical calculation of CSEs. Table 3-4 gives experimental and theoretical HSEs and DSEs for some cycloalkanes. The HSEs are 1–2 kcal/mol smaller than the CSE values obtained from an averaged $\Delta H_f^\circ(CH_2)$. *Obviously, the CH_2 group in propane is slightly strained*. DSE values, on the other hand, are similar to those given in Table 3-3, which means that a CH_2 group in cyclohexane comes closer to the ideal strain-free CH_2 group than that in propane.

Correcting ΔH_f° values from 298 to 0 K leads to a slight decrease of CSEs. This decrease is compensated when zero-point energy (ZPE) corrections are taken into account and the CSEs are calculated for the motionless molecules at 0 K. These values are listed in Table 3-4 under the heading E(THEO).[67] Two interesting observations can be made.

1. *The energies E(THEO) for the homodesmotic and the diagonal reference state CH_2 are identical (i.e., HSE and DSE are the same for all cycloalkanes).*
2. In the case of small cycloalkanes, the CSEs at 0 K for the motionless molecules are 1–3 kcal/mol larger than the CSE values normally used in chemistry (Table 3-3).

These trends can be explained in the following way. The SEs derived from E(THEO) represent the *electronic consequences* of strain. These will be changed by molecular vibrations where the changes depend on the reference state chosen. The ZPE per CH_2 group is higher for propane-ethane (17.7 kcal/mol) than that for cyclohexane (17.3 kcal/mol). It seems that this has to do with the fact that in the following homodesmotic reaction:

$$(CH_2)_6 \rightarrow 6 \; (CH_3CH_2CH_3 - C_2H_6) \qquad 3$$

there are 6 degrees of vibrational freedom fewer (6 degrees of translational/rotational freedom more) on the side of the cyclohexane molecule [total number: $(3 \times 18) - 6 + 6 = 48$ on both sides]. That is, per CH_2 group there are formally just 8 degrees of vibrational freedom on the left but 9 on the right-hand side of reaction **3**. Since translational/rotational degrees of freedom do not contribute at 0 K, the ZPE correction per CH_2 group is higher in the case of a homodesmotic definition of CH_2 than in the case of a diagonal definition (based on cyclohexane).

Since ZPE values have to be added to energies E(THEO),[67] the *electronic* SE derived for motionless molecules will be reduced and HSE and DSE values are no longer identical. When the temperature is raised from 0 to 298 K, the difference between HSE and DSE is essentially maintained since $\Delta H_f^\circ(CH_2$, homodesmotic) and $\Delta H_f^\circ(CH_2$, diagonal) change only slightly. This is schematically shown in Figure 3-4.

The definition of HSE and DSE is advantageous when determining the CSE by theoretical means. Calculation of the reaction energies $\Delta_R E$ of reactions **1** and **2** can be performed at the Hartree–Fock (HF) level of theory, since correlation effects play a minor role if number and types of bonds are conserved in a reaction.[48,68] In addition, one can take advantage of the fact that HSE = DSE for theoretical energies E(THEO) and confine the evaluation of CSE to just one reaction.[69] In the following, we will use homodesmotic reaction energies to obtain the CSE = HSE of a cycloalkane. In Table 3-4, HF values of CSEs are given for the STO-3G, 4-31G, 6-31G(d), and 6-31G(d, p) basis set developed in the Pople group.[71] Apart from the STO-3G value for cyclopropane, useful HF CSE values can already be obtained with minimal or split-valence basis sets. The failure of the STO-3G basis in the case of the three-membered ring is well known and has to do with an inadequate description of the strained CC bonds in cyclopropane.[72]

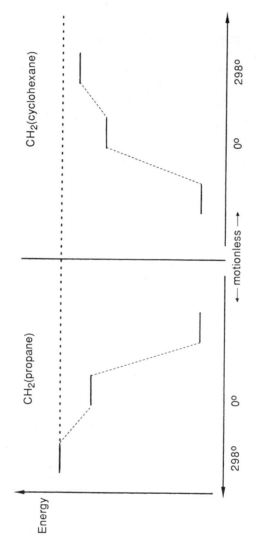

Figure 3-4. Schematic representation of the effects of zero-point energies and temperature on the energy of CH_2(propane) and CH_2(cyclohexane). The lowest energy levels correspond to the motionless molecules at 0 K; the first energy increase is caused by consideration of zero-point energies; the second energy increase reflects the temperature rise from 0 to 298 K.

These are reasonably described when polarization functions are included in the C basis (see, eg, 6-31G(d) results in Table 3-4). At this level of theory, SEs are already close to values derived from E(THEO), that is, values one would obtain if an infinitely large basis set were used and correlation, relativistic, and other corrections were incorporated.[67] However, an HF/6-31G(d) description of strained rings is still somewhat unbalanced, which becomes obvious when analyzing the properties of the H atoms in these molecules. A more reliable account is obtained when moving to a 6-31G(d,p) basis with polarization functions both in the C and the H basis (Table 3-4).[6]

B. The Puzzling Similarity of the Strain Energies of Cyclopropane and Cyclobutane

The energy and enthalpy data of Tables 3-3 and 3-4 reveal that for decreasing ring size, the SE values increase almost exponentially. If one assumes that this increase will be dominated by the increase in Baeyer strain caused by an increasing reduction in the CCC angle, then the SE should be parallel to SE(Baeyer) given in Equation 3-7 (for reasons of simplification, cyclopentane and cyclobutane are taken to be planar):

$$\text{SE(Baeyer)} = n \frac{k_\alpha}{2} \left[109.5° - \frac{180°(n-2)}{n} \right]^2 \qquad (3\text{-}7)$$

An analysis of this equation reveals that the Baeyer SE for (CH$_2$) increases with $1/n$ for $n = 5, 4, 3$. This implies a *monotonic* increase of the total SE when going from cyclohexane to cyclopropane. As a matter of fact, a monotonic function CSE = CSE(n) has been found for example in the case of cyclosilanes by Schleyer and co-workers.[11,47] This, however, is only partially true in the case of cycloalkanes. While CSE values increase for $n = 6$, 5, and 4, *the CSEs of cyclopropane and cyclobutane are almost the same* (see Tables 3-3 and 3-4, and Figure 3-5). This puzzling fact has been disguised for a long time by discussing CSEs per CH$_2$ group ("normalized CSEs"[32]) rather than total CSE values. If, however, both sets of CSEs are plotted as a function of the ring size n (Figures 3-5 and 3-6), it becomes immediately clear that the CSE of cyclopropane, whether normalized or not, should be considerably larger than the one observed.

In principle, there are three explanations possible for the striking anomaly in the CSEs of cyclopropane and cyclobutane. First, the CSE of cyclopropane could signal a stabilizing effect that reduces its Baeyer strain. This would mean that the CSE of cyclobutane is normal (see curves I in Figures 3-5 and 3-6). Second, the CSE of cyclobutane could be abnormally high while that of cyclopropane is normal (curves II in Figures 3-5 and 3-6). Finally, both CSEs may contain effects not arising from ring strain (curve III in Figure 3-5).

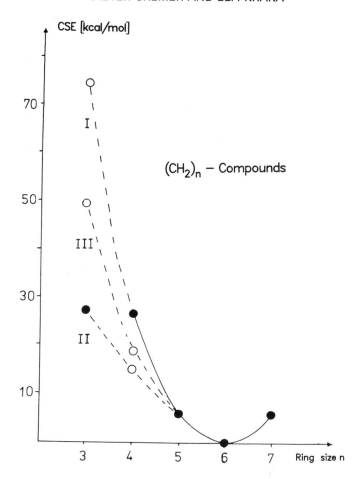

Figure 3-5. Dependence of the conventional strain energies CSEs (solid circles) of cycloalkanes on the ring size n. Curve I: extrapolated to CSE(3) utilizing CSE values for $n = 4, 5, 6$. The extrapolated value for $n = 3$ is denoted by the open circle. Curve II: based on the CSE values for $n = 3, 5, 6$. The interpolated CSE for $n = 4$ is denoted by the open circle. Curve III: based on the CSE values for $n = 5$ and 6 and the calculated SEs (Baeyer plus Pitzer SEs) for $n = 3$ and $n = 4$ given in Table 3-12.

To find out which of these three explanations is correct, one needs a better understanding of the bonding situation in cyclopropane and cyclobutane. As mentioned in Section 2, the concept of strain is inseparably connected to the concept of bonding. Therefore, one has to examine whether the bonding in cyclopropane and cyclobutane is correctly described in terms of a classical (two-center) electron pair bonding scheme on which the concept of strain is based. This implies an analysis of the electronic structure of the two ring compounds. We will do this by first discussing the wave function (and molecular orbitals) of cyclopropane and cyclobutane and then analyzing the total electron density distribution $\rho(\mathbf{r})$ of these molecules.

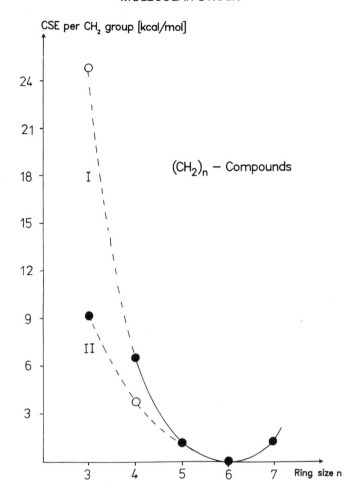

Figure 3-6. Dependence of the strain energy per CH_2 group on the ring size n. Curves I and II have the same meaning as in Figure 3-5.

6. MOLECULAR ORBITAL APPROACH TO STRAIN

In general, the bonding σ MOs of an alkane are symmetric about the interatomic connection line. This can be nicely demonstrated by using localized rather than delocalized MOs. Upon bending of a CC bond, the bond orbital is no longer symmetric with regard to the interatomic connection line. In a strained molecule the hybrid orbitals of an atom try to maintain the tetrahedral directions at the costs of bonding overlap. *Thus strain is closely related to hybrid orbital bending and the concomitant defect in bonding overlap.*

This can be seen when analyzing the localized MOs of cyclopropane obtained from self-consistent field (SCF) MOs by a Boys localization, in which localized orbitals are constructed from the delocalized SCF MOs under the constraint that the distances between the centroids of charge, determined for all occupied orbitals, become maximal. The CC bonding hybrid orbitals are directed outward, by 28° relative to the CC connection line; that is, they enclose an interorbital angle of 116°.[31] Using the Müller–Pritchard equation,[73] hybrid orbitals and interorbital angles can also be derived from ^1J(CH) NMR coupling constants. Lüttke and co-workers[74] have obtained a CCC interorbital angle for cyclopropane somewhat smaller than the ab initio value (102.6°, see Table 3-5), but still substantially larger than the geometrical angle α of cyclopropane. Most important, the hybridization degree n of sp^n hybrid orbitals is considerably larger than 3 (Table 3-5) while that of the CH hybrid orbitals of cyclopropane is essentially 2.

This result was first anticipated by Coulson and Moffitt,[75] who established the bent bond model of cyclopropane by elaborating ideas first proposed by Förster.[76] These authors used two types of sp^n hybrid orbital to describe CC and CH bonding in small cycloalkanes. They minimized the energy of a molecule with respect to the hybridization ratios using different constraints. For cyclopropane they obtained sp^4(CC) and sp^2(CH) hybrid orbitals enclosing angles of 104° (CCC) and 116° (HCH) (Table 3-5). Bonding and antibonding CC bent bond orbitals of cyclopropane are shown in Figure 3-7.

There are various other ways of ascertaining the hybridization ratio n. Randić and Maksić[53] have applied the criterion of maximum overlap connected with a proper weighting of CC and CH bond energies. They obtained sp^5, sp^2 hybrid orbitals for cyclopropane (Table 3-5).

Taking all these clues from the bent bond orbital picture, three important results emerge:

1. Strain is reflected by the nature of the bond orbitals. Bending of the orbitals and a concomitant decrease of bonding overlap is indicative of strained bonds.[75]
2. The hybridization ratios of the bent bond orbitals reveal that the CC bonds in cyclopropane are severely strained while those of cyclobutane are already close to normal.
3. Most important, the CH bond orbitals of cyclopropane differ markedly from those in cyclobutane. They are made up from sp^2 hybrids, which suggests that the CH bonds of cyclopropane are definitely stronger than those in the higher cycloalkanes (CH$_2$)$_n$ with n = 4, 5, 6.

Result 3 suggests that the exceptional similarity in the CSEs of three- and four-membered ring may result from extra-stabilizing effects in the case of cyclopropane. However, before investigating the energetic consequences of these effects, we will consider the Walsh model[77] of cyclopropane and cyclobutane. Since Walsh MOs have been discussed extensively in the literature,[78] we will confine ourselves to pointing out only some essential

TABLE 3-5. Hybridization and Interorbital Angles in Cycloalkanes

Parameter	Bond	C_3H_6				C_4H_8		C_5H_{10}	C_6H_{12}
		Coulson and Moffitt, 1949 Ref.: 75	Randić and Maksić, 1965 53	Newton, 1977 31	Honegger et al., 1982 80a	Wardeiner et al., 1982 74	Coulson and Moffitt, 1949 75	Newton, 1977 31	Coulson and Moffitt, 1949 75	Wardeiner et al., 1982 74
n in sp^n	CC	4.12	4.91	3.38	3.21	4.58	3.24	2.44	3.0	2.99
	CH	2.28	2.02	1.86	1.94	2.12	2.79	2.54 ax	3.0	3.01
								2.35 eq		
Interorbital angle	CCC	104	101.7	115	117.4	102.6	108	111.4	108	109.6
	HCH	116	119.6	117.1	115.9	118.2	111	113.3	109.5	109.4

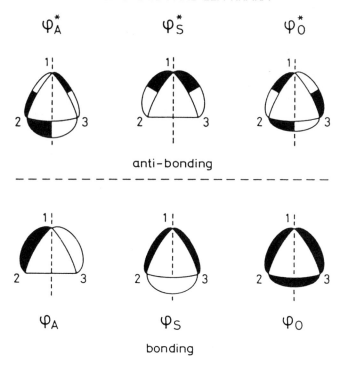

Figure 3-7. The Förster–Coulson–Moffitt bent bond MOs of cyclopropane. (Reproduced with permission from Reference 80b. Copyright © 1986 from *Nouveau Journal de Chimie*.)

features of the MOs of the three- and four-membered ring (compare with Figure 3-8).

There are two distinct sets of Walsh MOs, the r set, which consists of linear combinations of radially (toward the ring center) oriented sp^2 hybrid orbitals, and the t set, which consists of linear combinations of tangentially (with respect to the ring parameter) oriented p orbitals. The r orbitals always form a Hückel system while the t orbitals lead to a Möbius system for cyclopropane (n odd) and a Hückel system for cyclobutane (n even). As shown in Figure 3-8, the final orbitals are obtained by combining r and t orbitals of appropriate symmetry. There exists always a totally symmetric, doubly occupied, low-lying r MO (a_1' and a_{1g} in Figure 3-8) resulting from an in-phase overlap of all sp^2 orbitals inside the ring. The nature of this orbital changes dramatically with the size of the ring (Figure 3-9): For cyclopropane, it is a *surface orbital* covering the ring surface due to effective overlap of the sp^2 hybrid orbitals inside the ring. Increase of the ring size leads to an exponential decrease of orbital overlap. The surface orbital changes to a *ribbon orbital*, which facilitates electron delocalization along the carbon skeleton of cycloalkane similar to a π orbital in cyclopolyenes (Figure 3-9). For rings with $n > 4$, the ribbon MO is topologically equivalent to a π orbital while the t-set orbitals correspond to σ orbitals.

MOLECULAR STRAIN 93

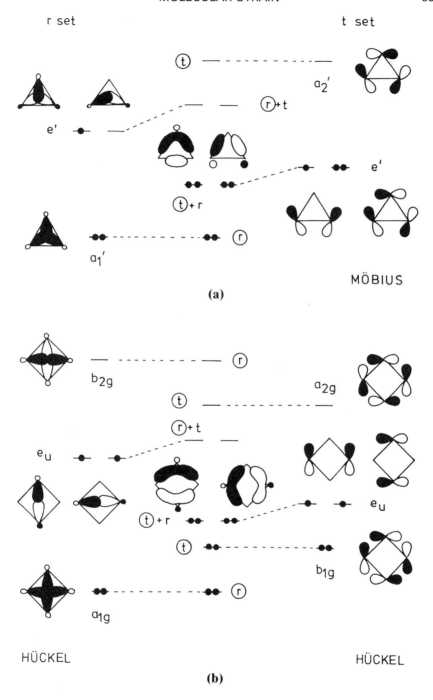

Figure 3-8. The Walsh MOs of cyclopropane (*a*) and cyclobutane (*b*). The predominant nature of the final MOs is indicated by a circled *r* or *t*, respectively. (Reproduced with permission from Reference 6. Copyright © 1986 from the American Chemical Society.)

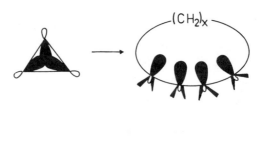

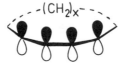

Figure 3-9. Equivalence of the r MOs in a large n-membered ring and the p_π MOs of a cyclopolyene. (Reproduced with permission from Reference 6. Copyright © 1986 from the American Chemical Society.)

For cyclopropane, this correspondence is reversed. Now, the r and t orbitals can be classified as σ and π orbitals as becomes immediately clear by comparison with the MOs of ethylene. For cyclobutane, a classification of r and t orbitals is no longer possible, since both enclose angles of 45° with the CC connection lines.[6]

From the Walsh MOs of cycloalkanes one can draw some important conclusions:

1. *Due to its topology, cyclopropane differs from all other cycloalkanes by possessing a surface orbital. Occupation of this MO leads to a two-electron, three-center bond* similar to that in H_3^+, B_2H_6, or other electron-deficient compounds. Hence, CC bonding in cyclopropane may be much stronger than one would expect in view of the poor overlap of the t orbitals (Figure 3-8).

2. Since the t orbitals of cyclopropane resemble the π MOs of ethylene, the former should possess properties typical of alkenes.[79]

3. Mixing of r and t orbitals for $n > 3$, apart from improving 1, 2-bonding interactions in the ring (see Figure 3-8), leads to antibonding interactions across the ring. They are strongest in cyclobutane, causing relatively large 1,3-CC nonbonded repulsion.

The Walsh MOs provide further clues for the electronic effects operating in cycloalkanes and, possibly, affecting their thermodynamic stability. It is evident from the discussion that nonbonded repulsion plays an important role in cyclobutane by enhancing total strain. In the case of cyclopropane, a two-electron, three-center bond may reduce ring strain effectively.[6]

The Walsh and the Coulson–Moffitt bent bond models provide complementary pictures of small rings employing two different sets of basis orbitals. As has been demonstrated by Heilbronner and co-workers,[80] they are equivalent provided complete orbital sets, including bonding and antibonding MOs, are considered. It seems that the Coulson–Moffitt model easily offers a possibility of estimating the energetic consequences of strain in cycloalkanes, while the Walsh model provides a basis of drawing a connection between strain and reactivity. Of course, neither of the two orbital models leads to a direct assessment of the SE or of the energetic effects increasing or decreasing it. Even if the existence of a two-electron, three-center bond is established for cyclopropane, its energetic consequences cannot be readily established through an orbital analysis. This has to do with the present lack of a rigorous solution to the problem of obtaining a unique definition of the chemical bond based on orbital theory. Since any conceptual approach to molecular strain is inevitably connected to a unique description of chemical bonding, one must spend some time establishing a physically reasonable definition of the chemical bond before assessing the energetic effects of CH bond strengthening or three-center bonding in cyclopropane. We will tackle this objective by investigating the electron density distribution in cycloalkanes before returning to the energetic consequences of strain.

7. ELECTRON DENSITY APPROACH TO STRAIN

Contrary to the nonobservable molecular orbitals, the total electron density distribution $\rho(\mathbf{r})$ of a molecule is an observable quantity that can be determined both experimentally and theoretically.[81] As shown by Hohenberg and Kohn,[82] the energy of a molecule in a nondegenerate ground state is a functional of $\rho(\mathbf{r})$. All physical and chemical properties of a molecule depend in some way on the electron density distribution. Therefore, it may be possible to derive useful information about molecular strain from $\rho(\mathbf{r})$.

Figure 3-10 depicts the calculated electron density distribution of cyclopropane in the form of a perspective drawing with regard to the plane of the three C nuclei. The distribution $\rho(\mathbf{r})$ is maximal at the positions of the nuclei and decreases exponentially in off-nucleus directions. It seems that the exponential decay of $\rho(\mathbf{r})$ obscures all details of the density distribution that relate to the peculiar bonding situation or to the effects of strain in cyclopropane. Hence, the main problem in analyzing $\rho(\mathbf{r})$ is to find the right tool to uncover bonding features like the bent bond character of a strained CC bond.

A. Description of Bent Bonds with Difference Densities

A popular way of eliminating the dominant exponential decay in the off-nucleus direction and unearthing the details of $\rho(\mathbf{r})$ in the bonding region is based on the difference density distribution[83]:

$$\Delta\rho(\mathbf{r}) = \rho[\text{molecule}] - \rho[\text{promolecule}] \qquad (3\text{-}8)$$

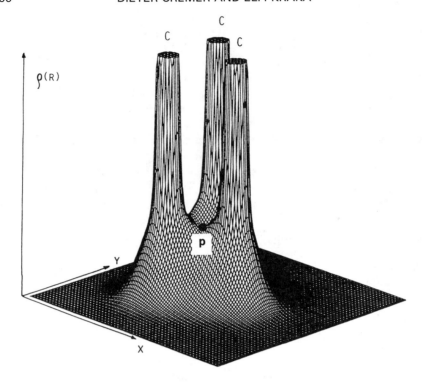

Figure 3-10. Perspective drawing of the calculated [HF/6-31G(d, p)] electron density distribution $\rho(\mathbf{r})$ in the ring plane of cyclopropane; **p** denotes the position of the bond critical point between two neighboring carbon atoms. (For better presentation, values above 14 e/Å^3 are cut off.)

The promolecular density is conventionally constructed by summing over spherically averaged atomic densities, with the atoms kept in the positions that they adopt in the molecule. A positive difference density in the internuclear region is generally considered to be indicative of bonding.

X-ray diffraction studies of various compounds containing strained three-membered rings have led to difference electron density maps which reveal $\Delta\rho(\mathbf{r})$ maxima displaced up to 0.3 Å from the internuclear axes[84] (see Figure 3-11). These maxima are generally interpreted as arising from the bent character of the strained bonds. Connecting their location with the nuclei, the resulting CCC angles are 104–108°,[84] which is reminiscent of the interorbital angles of the bent bond orbitals (see above: Table 3-5). However, it must be stressed that interorbital angles and angles derived from difference electron densities are not comparable quantities. Also, these angles are *not* related to the true interbond angles needed for a description of strained bonds.

The use of $\Delta\rho(\mathbf{r})$, although convenient from the point of view of the crystallographer, implies some serious problems insofar as it depends on a hypo-

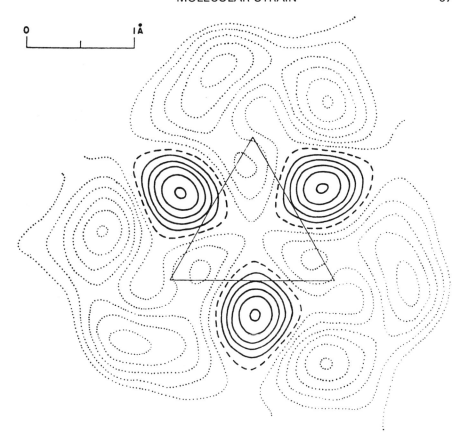

Figure 3-11. Contour line diagram of the difference density in the ring plane of cis,cis-,1,2,3-tricyanocyclopropane (X-ray diffractometric study). Solid lines are in regions with positive difference density, dotted lines are in regions with negative difference density. Dashed lines correspond to zero values. (Reproduced with permission from Reference 84a. Copyright © 1986 from *Acta Crystallographica*.)

thetical reference state, the promolecule. For example, negative rather than positive difference electron densities have been found for several bonds.[85] A new method for deriving suitable difference electron densities from "oriented" rather than spherically averaged atomic densities proposed by Schwarz and co-workers[86] may alleviate these problems. Nevertheless, it would be far better to analyze the observable quantity itself rather than a model-dependent quantity, derived from the observable quantity. A physically meaningful description of the chemical bond is more likely to be developed from an analysis of $\rho(\mathbf{r})$ rather than any arbitrarily defined $\Delta\rho(\mathbf{r})$. Therefore, we will briefly describe the topological analysis of $\rho(\mathbf{r})$, a method that is quantum mechanically justified,[57] and does not suffer from the difficulty of choosing a proper reference state.

B. The Topological Analysis of the Electron Density Distribution

The topological analysis of $\rho(\mathbf{r})$, developed by Bader and co-workers,[57] is based on the investigation of the critical (stationary) point $\mathbf{p}_s$ of $\rho(\mathbf{r})$. These are the sources and sinks of the gradient paths (trajectories) of the gradient vector field $\nabla\rho(\mathbf{r})$. The gradient vector always points into the direction of a maximum increase in $\rho(\mathbf{r})$.

An analysis of the gradient vector field $\nabla\rho(\mathbf{r})$ is more than just a way of describing the distribution $\rho(\mathbf{r})$ with appropriate mathematical tools. It directly leads to a *quantum mechanically* based definition of molecular subspaces, which in turn can be used to define an *atom in a molecule*, a *chemical bond*, and the *molecular structure*—that is the network of bonds connecting the atoms in a molecule. These definitions are of general chemical importance, but they become particularly important when trying to determine the SE of a molecule from in situ bond energies or in situ atomic energies (see Subsection B of Section 3).

To facilitate a brief discussion of the essence of the topological analysis, the gradient vector field $\nabla\rho(\mathbf{r})$ corresponding to the distribution $\rho(\mathbf{r})$ of Figure 3-10 is shown in Figure 3-12. Three types of trajectory of the vector field $\nabla\rho(\mathbf{r})$ can be distinguished. First, there are those that end at one of the three nuclei starting either at infinity or at the center of the ring (type I). Second, there are trajectories that start at a point **p** between the C nuclei and terminate at one of the nuclei in question (type II). The point **p** is shown in Figure 3-10 for the CC bond in front. The electron density assumes a minimal value at **p** in the internuclear direction but a maximal value at **p** in all directions perpendicular to that direction; that is, **p** is a saddle point in three dimensions. Exactly, three saddle points **p** can be found in the $\nabla\rho(\mathbf{r})$ field of the carbon ring, each being located in one of the three CC bond regions (Figure 3-12). There are just two type II trajectories per saddle point **p**. They connect the neighboring C nuclei and describe a *path of maximum electron density (MED path)*. Any lateral displacement from the MED path leads to a decrease in $\rho(\mathbf{r})$. Finally, there are trajectories that originate at infinity and terminate at the saddle point **p** (type III). In three dimensions, these trajectories form a surface S separating the two nuclei that are linked by the MED path. The flux of $\nabla\rho(\mathbf{r})$ vanishes for all surface points:

$$\nabla\rho(\mathbf{r}) \cdot \mathbf{n}(\mathbf{r}) = 0 \qquad \mathbf{r} \in S \qquad (3\text{-}9)$$

where **n** is the unit vector normal to the surface S. The surfaces S have been named *zero-flux surfaces*.[57]

As shown in Figure 3-12, the zero-flux surfaces partition the molecular space into subspaces, each containing one and only one atomic nucleus. This observation has been made for many molecules and, therefore, it is reasonable to consider the subspaces derived from zero-flux surfaces as atomic subspaces.[57] All type I trajectories terminating at a given nucleus define its

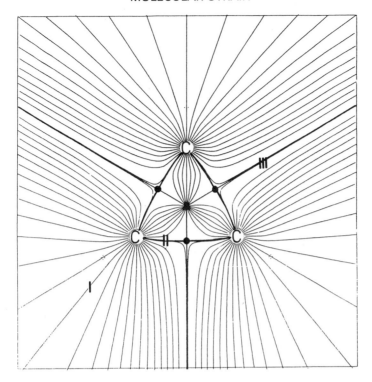

Figure 3-12. Gradient vector field of the [HF/6-31G(d, p)] electron density distribution $\rho(\mathbf{r})$ calculated for the plane of the cyclopropane ring. Bond critical points are denoted by dots. Type I trajectories start at infinity or the center of the ring and end at a carbon nucleus; type II trajectories (heavy lines) define the bond path linking two neighboring carbon atoms; type III trajectories form the three zero-flux surfaces between the C atoms (in the two-dimensional display, only their traces can be seen). They terminate at the bond critical points.

basin. The nucleus and its associated basin define the atom in the molecule.[57] Hence, the zero-flux surfaces correspond to *interatomic* surfaces.

C. The Definition of a Chemical Bond

Investigation of the ring opening of cyclopropane has revealed that the MED path vanishes upon bond rupture and reappears when the bond is formed again.[87] These observations have led Cremer and Kraka[88] to consider the existence of a saddle point **p** and hence a MED path linking the two nuclei of the adjoining atomic subspaces as a necessary condition for the existence of a chemical bond.

However, MED paths are also found for any ensemble of weakly or noninteracting atoms and molecules or for the dissociation products of a molecule.[88] To distinguish between covalent chemical bonds and closed-shell interactions as found in the case of van der Waals molecules, hydrogen bonding, or electrostatic (ionic) interactions, it has been suggested that the properties of the energy density[88]:

$$H(\mathbf{r}) = G(\mathbf{r}) + V(\mathbf{r}) \qquad (3\text{-}10)$$

are analyzed.[57] Here, $G(\mathbf{r})$ and $V(\mathbf{r})$ correspond to *a* local kinetic energy density and to *the* local potential energy density, respectively. Since $G(\mathbf{r})$ is always positive and $V(\mathbf{r})$ is always negative, the sign of the energy density $H(\mathbf{r})$ reveals whether accumulation of electron density at a point $\mathbf{r}$ is stabilizing [$H(\mathbf{r}) < 0$] or destabilizing [$H(\mathbf{r}) > 0$]. Analysis of a variety of different bonds[88,89] suggests that covalent bonding is characterized by a predominance of the local potential energy density $V(\mathbf{r})$ at $\mathbf{p}$, hence $H(\mathbf{p}) < 0$. In contrast, closed-shell interactions lead to $H(\mathbf{p}) \geq 0$. Therefore, Cremer and Kraka[88] suggested that a negative local energy density $H(\mathbf{r})$ at the minimum $\mathbf{p}$ of the MED path be considered as *sufficient* condition for covalent bonding (Table 3-6). In this case, the MED path provides an image of the covalent chemical bond and, therefore, may be called a "bond path." Accordingly, the saddle point $\mathbf{p}$ corresponds to a "bond critical point."

A clear definition of the (covalent) chemical bond is of paramount importance when describing strained bonds. In this connection, two further definitions are useful, namely those of the bond length and the bond angle. The former is equal to the bond path length r_b. It is not necessarily identical with

TABLE 3-6. Description of Atoms in Molecules and Chemical Bonds in Terms of the Properties of $\rho(\mathbf{r})$

Chemical term	Terms used in density analysis	Comment
Atom	Nucleus + basin	Basin defined by virial partitioning[57]
Interatomic surface	Zero-flux surface	Using zero-flux surfaces defined by Equation 3-9
Covalent bond	Bond path	Necessary condition: Existence of MED path linking the bonded atoms
		Sufficient condition: Negative energy density $H(\mathbf{p})$ at the minimum of the MED path
	Bond critical point $\mathbf{p}$	Saddle point of ρ, identical with the minimum of the MED path
Bond length	Bond path length r_b	r_b is larger than the geometrical distance r_e for bent bonds
Bond angle	Interpath angle β	β is (normally) larger than the geometrical angle α in strained rings
Bent bond character		Deviation d of bond path from the interatomic connection line
Bond order		Evaluated from ρ at $\mathbf{p}$ according to Equation 3-11
π Character	Bond ellipticity ε	Defined by the curvatures of ρ at $\mathbf{p}$ perpendicular to the bond (see Equation 3-12)

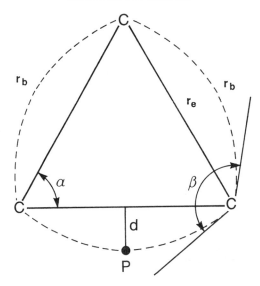

Figure 3-13. Geometrical and bond parameters of cyclopropane. Bond paths are indicated by dashed lines and α is the geometrical angle, β the interpath angle, r_e the geometrical distance, and r_b the bond path length; d denotes the perpendicular distance between the bond critical point **p** and the internuclear connection line.

the geometrical distance r_e between two bonded atoms. As a matter of fact r_b is larger than r_e in the case of bent bonds (Table 3-6; see also Reference 75). The bond angle is equal to the interpath angle β, which also can differ considerably from the geometrical angle α for strained molecules. Thus the curvature of a bond is described by both the ratio r_b/r_e and the interpath angle β. Another useful quantitative measure of bond bending is the perpendicular distance d between the bond critical point **p** and the internuclear connection line (see Figure 3-13).

Other useful bond features (e.g., the bond order n and the π character) can be extracted from the properties of $\rho(\mathbf{r})$ at the bond critical point.[87,88] For hydrocarbons, the bond order is determined with the relationship:

$$n(CC) = \exp\{a\,[\rho(\mathbf{p}) - b]\} \qquad (3\text{-}11)$$

where $b = \rho(\mathbf{p}, \text{ethane})$ and a is adjusted to lead to $n \approx 2$ (3) for ethylene (acetylene). The bond order thus defined depends on a local property of $\rho(\mathbf{r})$; hence, care must be taken when discussing the bond strength, which actually depends on both the total electron density in the internuclear region and the forces exerted on this density.

The π character of a bond can be related to the anisotropy of $\rho(\mathbf{r})$ at the bond critical point. It is measured with the aid of the three curvatures of $\rho(\mathbf{r})$ along the principal axes at **p** corresponding to the eigenvalues λ_i ($\lambda_1 \leq \lambda_2 \leq \lambda_3$) and the eigenvectors $\mathbf{v}_i$ of the matrix of second derivatives $\partial^2\rho/\partial x_i \partial x_j$ (i,

$j = 1, 2, 3$).[87] For the standard CC bond in ethane, the charge distribution at **p** is isotropic, yielding $\lambda_1 = \lambda_2 < 0$. In the case of ethylene, however, the density falls off less rapidly in the π direction ($\mathbf{v}_2$) and $\lambda_1 < \lambda_2 < 0$. The density distribution at the bond critical point of the CC double bond (or other double bonds) is *anisotropic*. The degree of anisotropy is measured by the bond *ellipticity* ε[87]:

$$\varepsilon = \frac{\lambda_1}{\lambda_2} - 1 \qquad (3\text{-}12)$$

Since values of ε larger than zero are found for bonds with (partial) π character, it is useful to define the π character of a bond by its ellipticity ε at **p**. The bond properties defined with the aid of $\rho(\mathbf{r})$ are summarized in Table 3-6.

D. Bonding in Cyclopropane and Cyclobutane

In Table 3-7, bond parameters of some strained rings are compared with the corresponding geometrical parameters. For three-membered rings the bond lengths r_b are up to 0.02 Å longer than the r_e values, while $r_b \approx r_e$ for cyclobutane. The shift parameters d, however, range from 0.04 to 0.08 Å with the smaller values for the four-membered ring. Even more pronounced are the differences between the angles α and β (Table 3-7), being roughly 19° for cyclopropane and 6–7° for cyclobutane.[4,6]

The importance of this result can be highlighted by inserting β and α values in the Hooke equation (3-3, above). Assuming a force constant of unity, a Baeyer strain energy of 58 kcal/mol is evaluated from β(CCC) of cyclopropane while an SE of 161 kcal/mol results from the geometrical angle.

TABLE 3-7. Description of Strained Rings in Terms of the Properties[a] of $\rho(\mathbf{r})$

Molecule	Bond	r_e	r_b[b]	d[c]	ε[d]	Angle	α	β[b]
Cyclopropane	CC	1.497	1.506	0.060	0.49	CCC	60	78.8
Aziridine	CC	1.470	1.486	0.080	0.39	CCN	59.5	77.3
	CN	1.449	1.455	0.043	0.50	CNC	60.9	76.4
Oxirane	CC	1.453	1.476	0.094	0.31	CCO	58.8	72.8
	CO	1.401	1.404	0.004	0.88	COC	62.4	75.8
Bicyclobutane	CC[e]	1.484	1.502	0.089	0.36	CCC[e]	58.8	73.6
	CC	1.513	1.522	0.056	0.49	CCC	60.6	77.4
Cyclobutane	CC	1.544	1.547	0.038	0.02	CCC	88.6	95.6

[a] Distances in angstrom units, angles in degrees. All values from HF/6-31G(d, p) calculations[4,6] and unpublished results of the authors.
[b] Compare with Figure 3-13 and Table 3-6.
[c] Bent bond character: see Table 3-6.
[d] Bond ellipticity (π character) defined in Equation 3-12.
[e] Central bond and angle opposite to central bond, respectively.

Clearly, *the latter value is meaningless because it is derived from a value that depends on the topology of the ring but bears no resemblance to the actual deformation of the CC bonds.*

Another interesting feature emerges from the density analysis. The calculated CC bond orders are literally identical with that of ethane ($n \approx 1$).[4] It must be remembered, of course, that the bond order is derived from a local property, the value of $\rho(\mathbf{r})$ at **p**. It is unlikely that any local quantity is a useful means of assessing a global property of a molecule. This also applies to the interorbital angle, but not to the orbital overlap and the interpath angles. Hence, the latter quantities will lead to a useful description of strain (see Table 3-8).

TABLE 3-8. Descriptors of Strain

Property	MO Approach	ρ Approach	Assessment of strain
Local	Interorbital angle	Bond order (density at **p**)	No
Global	Overlap	Interpath angle	Yes

For cyclopropane, CC bond ellipticities (0.49) comparable to those of ethylene (0.45) have been calculated (Table 3-7[4,87]). Not surprisingly, the soft curvature of $\rho(\mathbf{r})$ is *in* the plane of the C_3 ring, indicating that the density extends from the bond critical point toward the ring center. Hence, the π character of the cyclopropane bonds, which has been substantiated in many experimental investigations,[79] is confirmed and quantified by the analysis of the electron density distribution. In addition, an important conclusion can be drawn from the density parameters that describe the bent bonds in cyclopropane: *bending of a formal CC σ bond leads to an admixture of π character.*[90]

The π character of the CC bonds of cyclopropane is connected with the extension of the density into the center of the ring. At the ring center, the value of $\rho(\mathbf{r})$ is more than 80% that at the CC bond critical points.[4-6] Obviously, electron density is smeared out over the whole ring surface, a phenomenon that has been termed *surface delocalization.*[4-6] For cyclobutane, the density at the center of the ring is just 30% of that found at the CC bond critical points. Also, the bond ellipticity is vanishingly small (0.02), revealing that both π character and surface delocalization are of no relevance for the four-membered ring.[4-6]

It is appealing to draw a connection between the surface delocalization of electrons revealed by the analysis of $\rho(\mathbf{r})$ and the existence of a surface orbital within the Walsh MO description of the three-membered ring. Obviously, the properties of $\rho(\mathbf{r})$ add further support to the existence of a two-electron, three-center bond in cyclopropane. On the other hand, there are remarkable differences between the MO and the ρ description of the three-membered ring. Interorbital and interpath angles (Tables 3-5 and 3-7) differ greatly. The value of $\rho(\mathbf{p})$ in the CH bonds of cyclopropane is not very different from that found for the CH bonds of cyclobutane (although a slight increase of the former value is found). A weakening of the CC bonds in cyclopropane is not reflected by the bond order.

A priori, one should expect any information about bonding and strain gained from an analysis of the MOs to be "hidden" also in the electron density. The only question is, How can this information be displayed? We tackle this problem next, by analyzing the Laplacian of $\rho(\mathbf{r})$.

8. A STEP TOWARD A UNIFIED DESCRIPTION OF STRAIN: THE LAPLACIAN OF THE ELECTRON DENSITY

The Laplacian of any scalar field $f(\mathbf{r})$, $\nabla^2 f(\mathbf{r})$, is given by the second derivatives of $f(\mathbf{r})$ with regard to $\mathbf{r}$. It can be obtained as the sum of the eigenvalues of the Hessian matrix [matrix of the second derivatives of $f(\mathbf{r})$ with regard to the three components of the vector $\mathbf{r}$, namely x, y, and z]. The Laplacian is negative where the scalar field concentrates.[91] It adopts a minimum where $f(\mathbf{r})$ possesses a maximum.[92] The Laplacian of the electron density distribution $\nabla^2 \rho(\mathbf{r})$ has been used to detect locations in molecular space at which electronic charge is concentrated ($\nabla^2 \rho(\mathbf{r}) < 0$) or is depleted ($\nabla^2 \rho(\mathbf{r}) > 0$).[89,93] This can be done *without* defining an arbitrary reference density.

Bader has shown that the Laplacian $\nabla^2_\rho(\mathbf{r})$ plays a key role in the quantum mechanical equations governing the behavior of $\rho(\mathbf{r})$.[57] For example, $\nabla^2 \rho(\mathbf{r})$ provides the link between electron density and energy density via a local virial theorem:

$$\frac{\hbar^2}{4m} \nabla^2 \rho(\mathbf{r}) = 2G(\mathbf{r}) + V(\mathbf{r}) \qquad (3\text{-}13)$$

where the sum of the kinetic energy density $G(\mathbf{r})$ and the potential energy density $V(\mathbf{r})$ equals the energy density $H(\mathbf{r})$ (Equation 3-10).[88]

An increase of $|V(\mathbf{r})|$ leads to enhanced concentration of electronic charge at $\mathbf{r}$, an increase of $G(\mathbf{r})$ to its depletion. Integrated over an atomic subspace defined by the zero-flux surfaces (Equation 3-9), or integrated over total molecular space, the Laplacian of $\rho(\mathbf{r})$ vanishes; that is, the fluctuations in $\nabla^2 \rho(\mathbf{r})$ are such that local depletion or concentration of electronic charge cancel each other, both for an atom in a molecule and for the molecule itself.

For an isolated atom with spherically averaged electron density, negative charge is concentrated in spheres, which in turn are separated by spheres of charge depletion.[93] [One must keep in mind that charge concentration and electron density are two different quantities and, as such, this finding does not contradict the exponential decay of $\rho(\mathbf{r})$, described above.] It is appealing to associate the spheres of the Laplacian with quantum shells. Then, for a first-row element, the inner concentration sphere is assigned to the 1s shell, the outer sphere to the valence shell. This is shown in Figure 3-14a for an isolated C atom with spherically averaged electron density.

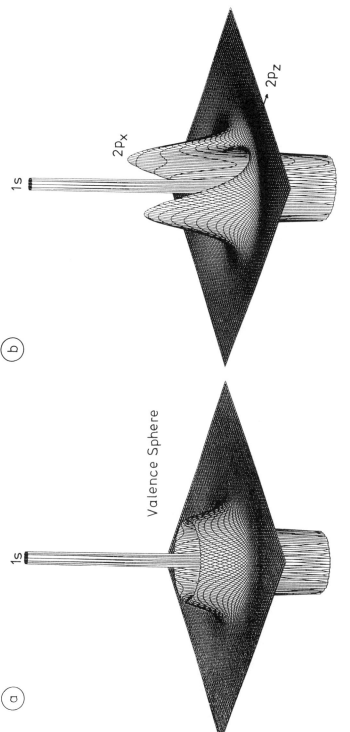

Figure 3-14. Perspective drawing of the calculated [UHF/6-31G(d)] Laplace concentration $-\nabla^2\rho(\mathbf{r})$ of the carbon atom. (*a*) C atom with spherically averaged electron density. (*b*) C atom in the ^3P ($1s^2\,2s^2\,2p_x\,2p_y$) ground state, depicted in the xz plane. The concentration peak at the position of the C nucleus is assigned to the 1s electrons, the outer concentration sphere to valence shell electrons. In (*b*), the lumps and holes in the valence sphere correspond to the occupied $2p_x$ and the empty $2p_z$ AO, respectively. (For a better presentation, values above 48 e/Å^5 and below −24 e/Å^5 are cut off.)

TABLE 3-9. Information Available from $\rho(\mathbf{r})$

Analysis of $\rho(\mathbf{r})$ via	Calculation of	Leads to	Chemical information about
Gradient vector field $\nabla\rho(\mathbf{r})$	First derivatives of $\rho(\mathbf{r})$	Critical points and electron density paths	Atoms in molecules, chemical bonds, molecular structure
(Scalar) Laplace field $\nabla^2\rho(\mathbf{r})$	Second derivatives of $\rho(\mathbf{r})$	Concentration lumps and holes	Reactive sites in a molecule

If the charge distribution of the isolated atom possesses isotropic and anisotropic components as in the case of the C atom in the ^{3}P ($1s^22s^22p^2$) ground state, then the Laplacian of $\rho(\mathbf{r})$ is no longer spherical. Local maxima (lumps) and minima (holes) develop in the valence shell, while the inner concentration shell remains unchanged. The lumps are in the direction of the occupied p orbitals, the holes in the direction of the empty p orbital. Hence, the lumps and holes of the Laplacian of $\rho(\mathbf{r})$ reflect the shape of the "frontier orbitals" of C(^{3}P).

The Laplacian bridges the gap between the orbital and density descriptions of electronic structure. Information extracted from the analysis of the orbitals is also obtained in $\rho(\mathbf{r})$ and can be revealed by the analysis of the Laplacian of ρ.

In the same way that sites of nucleophilic or electrophilic attack in a molecule can be predicted by analyzing the form of the frontier orbitals, an investigation of the Laplace concentration of $\rho(\mathbf{r})$ helps to identify the active sites of the molecule: sites with distinct electron concentration lumps are prone to an electrophilic attack, while deep concentration holes in the valence shell are prone to a nucleophilic attack.

The chemically relevant information extracted from $\rho(\mathbf{r})$ by analyzing either $\nabla\rho(\mathbf{r})$ or $\nabla^2\rho(\mathbf{r})$ is summarized in Table 3-9. It is important to keep in mind that only $\rho(\mathbf{r})$ is observable and that $\nabla\rho(\mathbf{r})$ and $\nabla^2\rho(\mathbf{r})$ must not be mixed up with $\rho(\mathbf{r})$ itself.

Figure 3-15 presents a perspective drawing of $\nabla^2\rho(\mathbf{r})$ for cyclopropane with respect to the plane of the carbon nuclei. The positions of the carbon atoms can be easily recognized by the 1s concentration peaks. The valence shells are distorted so that each carbon possesses four concentration lumps in the direction of the four carbon valences, two of which are hidden in Figure 3-15 because they are perpendicular to the reference plane. The distinct concentration lumps at the C atoms suggest that corner protonation is more likely than edge or face protonation of cyclopropane.[4]

At first sight, the concentration lumps in the CC bonding regions seem to disclose the location of the CC bond paths. A quantitative analysis, however, reveals that the concentration maxima in the valence region of C are

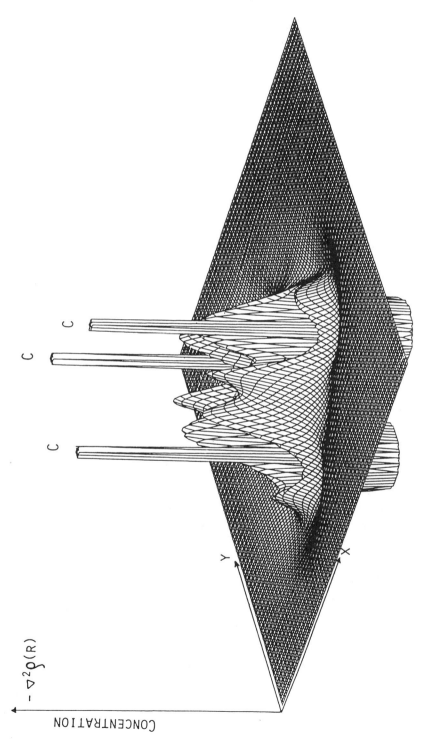

Figure 3-15. Perspective drawing of the calculated [HF/6-31G(d, p)] Laplace concentration $-\nabla^2 \rho(\mathbf{r})$ of cyclopropane, depicted in the ring plane. (For a better presentation, values above 48 e/Å^5 and below -24 e/Å^5 are cut off.)

significantly displaced away from the bond paths (compare with Figure 3-16). They enclose angles of 92° that are closer to the interorbital angles (102°) than the interpath angles (79°).[4] In the same way, as found for C(^{3}P), the lumps can be associated with the highest occupied molecular orbital (HOMO) of cyclopropane. Maximal concentration of negative charge is found where the e' MOs of cyclopropane (compare with Figure 3-8) possess their largest amplitude. Similarly, the holes at the C atoms (Figures 3-15 and 3-16) can be linked to the a_2' lowest unoccupied MO (LUMO) of cyclopropane (Figure 3-8).

Inspection of the contour line diagram of $\nabla^2\rho(\mathbf{r})$ shown in Figure 3-16a reveals that the σ electrons of the three-membered ring concentrate not only in the bonding region but also in the ring interior (contour lines with $\nabla^2\rho(\mathbf{r}) < 0$ are dashed in Figure 3-16). Hence the Laplacian of $\rho(\mathbf{r})$ also reflects the surface delocalization of electrons.[4] We attribute this to the occupation of the a_1' MO of cyclopropane, which is identified as a surface orbital (Section 6).[6]

Surface delocalization of σ electrons implies that the absolute value of the potential energy density is large inside the ring. Electrons "stay" longer in the ring center (relatively low kinetic energy), since they experience the stabilizing attraction of the three carbon nuclei. The electrostatic potential due to the nuclei is homomorphic with $\rho(\mathbf{r})$,[94,95] which indicates that nucleus–

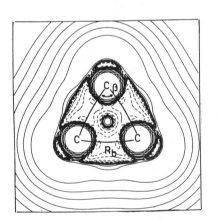
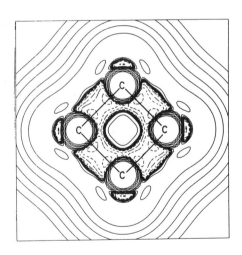

a   b

Figure 3-16. Contour line diagrams of the calculated [HF/6-31G(d, p)] Laplace concentrations $-\nabla^2\rho(\mathbf{r})$ of cyclopropane (a) and cyclobutane (b). Bond paths are indicated by heavy solid lines; R_b denotes the bond path length and β the interpath angle. Dashed and solid lines are in regions in which electronic charge is concentrated and depleted, respectively. Inner shell concentrations are not shown. (Reproduced with permission from Reference 6. Copyright © 1986 from the American Chemical Society.)

electron attraction is the dominant physical factor supporting σ-electron delocalization in cyclopropane.

There are some significant differences in Laplacian distribution for three-membered and four-membered rings. As can be seen from the contour line diagram of $\nabla^2\rho(\mathbf{r})$ for cyclobutane (Figure 3-16b), the CC concentration maxima are much closer to the CC bond paths, indicating that orbital overlap within the HOMO can follow the bond paths. The CC bonds are less strained. Contrary to cyclopropane, electronic charge is depleted from the interior of the four-membered ring (Figure 3-16b). Surface delocalization no longer plays any role in cyclobutane.[6]

Another feature of $\nabla^2\rho(\mathbf{r})$ provides information that is not easily obtained from $\rho(\mathbf{r})$ itself. A quantitative analysis of the carbon concentration lumps in the direction of the CH bonds reveals that these are larger for cyclopropane than for cyclobutane, which is consistent with the higher s character of the CH hybrid orbitals in the former case.

We conclude that a unified analysis of the electron density distribution and its associated Laplacian reveals all critical characteristics of strained cycloalkanes. The next step will be to attach energies to the various effects described and, in this way, to give a detailed accounting of the strain energy and the possible factors comprising it.

9. WAYS OF ASSESSING THE STRAIN ENERGY FROM QUANTUM CHEMICAL CALCULATIONS

The major advantage of the electron density analysis is that it leads to a precise and physically meaningful definition of *atoms in molecules, chemical bonds, molecular geometry, and molecular structure* (Section 7; see also Section 13). As a consequence, a calculation of CSEs or SEs becomes feasible starting either from in situ atomic energies, in situ bond energies, or deformation energies evaluated with Hooke's law and inserted into the Westheimer equation (compare Figures 3-1, 3-2, and 3-17). The last method, actually, should lead to a value close to the true SE.

If there are other effects acting in the molecule, their energetic consequences will become obvious in the difference CSE − SE. By applying a second or a third method for the calculation of CSE and SE, there is a good chance of unraveling the various effects contributing to the stability of the molecule. For example, there may be two electronic effects reducing the ring SE of cyclopropane, namely CH bond strengthening due to hybridization effects and σ-electron delocalization. The joint contribution of these effects to the molecular stability will show up in the difference CSE − SE. Using a method other than method 5 in Figure 3-17 leads to the energy change due to CH bond strengthening or σ-electron delocalization, allowing one to determine all energies contributing to the stability of cyclopropane. Obviously,

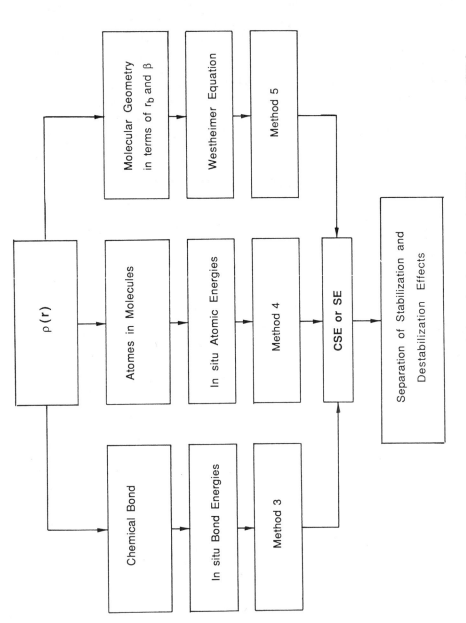

Figure 3-17. Schematic survey of possible ways to access conventional strain energies (CSE) or strain energies (SE) from the properties of the electron density distribution $\rho(\mathbf{r})$. (For explanation of methods, see Figures 3-1 and 3-2.)

method 3, which is based on in situ bond energies, is the best method for separating the energy contribution resulting from CH bond strengthening.

One could also think of calculating in situ bond energies from orbitals by relating their overlap to bond strength and bond energy.[96,97] Such a procedure, although operationally satisfactory for some bond types, suffers from the problem that atoms in molecules, chemical bonds, and molecular geometry cannot be defined in terms of orbitals in a rigorous way. Orbitals are nonobservables and change their form if subjected to an appropriate unitary transformation.

In the sections that follow, we will describe two of the three possible ways of determining CSE or SE of cycloalkanes from properties of the electron density distribution. The third method, based on the calculation of in situ atomic energies, has been discussed by Cremer and Gauss.[6] It shows that due to the higher s character of the CH hybrid orbitals in cyclopropane, there is a buildup of negative charge on the carbon atoms, which therefore possess lower atomic energies than in cyclobutane or in an alkane such as propane. The hydrogen atoms of cyclopropane, on the other hand, are destabilized relative to those in cyclobutane or in an alkane, since they have lost part of their negative charge to the C atoms. Their atomic energies entail CH_2 group energies of E(cyclopropane) $>$ E(cyclobutane) $>$ E(alkane) and SEs of 27.9 and 26.4 kcal/mol for cyclopropane and cyclobutane, respectively, in close agreement with experimentally based SE values.[6] Hence, the relative stabilities of small cycloalkanes appear to be a consequence of a destabilization of their H atoms. This is a novel description,[6] which, however, adds little to a conceptual understanding of ring strain and, therefore, is not discussed further.

10. CALCULATION OF THE STRAIN ENERGY FROM IN SITU BOND ENERGIES

As mentioned in Section 3, one way of calculating SEs or CSEs is based on the determination of in situ bond energies (see Figures 3-1 and 3-2). If these are known for cyclopropane, then the energetic consequences of CC bond bending can be directly determined with the aid of appropriate reference bond energies. Also, the importance of CH bond strengthening due to $C(sp^2)$—H bonding in cyclopropane should become obvious from in situ bond energies.

A. In Situ Bond Energies from Hybrid Orbitals

More than 50 years ago, Pauling[98] pointed out that larger overlap between two hybrid orbitals will give rise to a stronger bond. If the bond strength is in turn taken to be proportional to the bond energy, it will be possible to determine the latter from calculated orbital overlaps (S), provided a linear dependence on S can be assumed and an appropriate conversion factor can be found:

$$be = k\,S \qquad (3\text{-}14)$$

Kilpatrick and Spitzer[99] used this approach to evaluate the CC bond energy of cyclopropane. By assuming that $C(sp^2)$—H bonds possess the same bond strength as $C(sp^3)$—H bonds, i.e. that the stability of cyclopropane is solely determined by the strain in the CC bonds, these authors estimated the CC bond energy to be 70.1 kcal/mol, which is 8.7 kcal/mol smaller than the CC bond energy of ethane. This difference implies an SE of 26 kcal/mol, close to the experimental CSE of cyclopropane.

However, this approach did not consider CH bond strengthening. Randić and Maksić,[53] in a similar calculation, explicitly took the CH bond strengthening for cyclopropane into account, using as standard CC and CH bond energies 79.2 (from ethane) and 99.5 kcal/mol (from methane). With scaling factors $k(CC) = 121.37$ kcal/mol and $k(CH) = 142.67$ kcal/mol and calculated S values for cyclopropane, they obtained $be(CC) = 69.9$ and $be(CH) = 102.5$ kcal/mol. Hence, CC bond bending was estimated to cause an SE of 27.8 kcal/mol, while CH bond strengthening reduces this value by 6 × 3.1 kcal/mol, leading to a CSE of just 9.7 kcal/mol, one-third of the experimental CSE of cyclobutane. Obviously, this approach either underestimates CC bond weakening or overestimates CH bond strengthening.

In this connection we note that bond energies for strained molecules have also been obtained from quantum chemically computed valencies.[133] According to these calculations, the CC bonds of cyclopropane are weakened by just 3 × 6.4 = 19.2 kcal/mol; another 8.4 kcal/mol of strain is attributed to a weakening of the CH bonds. This, however, is contrary to all other descriptions of the CH bonds in cyclopropane and, therefore, casts doubts on the usefulness of the theoretical model employed.

B. Estimation of CH Bond Energies from Experimental Data

The CH bond dissociation enthalpy DH of cyclopropane (106.3 kcal/mol) is 11.2 kcal/mol higher than the one for the secondary CH bond of propane (95.1 kcal/mol).[100] It has been suggested that this difference be used to estimate the CH bond energy of cyclopropane.[10] However, DH values do not necessarily reflect the magnitude of bond energies (enthalpies), since they depend on both the stability of the reactant *and* the stability of the fission products. The large $DH(CH)$ value of cyclopropane simply reflects the significant increase in ring strain when the cyclopropyl radical is formed.

The CH stretching frequency of cyclopropane (3056 cm^{-1}) is almost identical with that of ethylene (3055 cm^{-1}) but about 100 cm^{-1} higher than that of alkanes (ethane: 2950 cm^{-1}, s-CH of propane 2920 cm^{-1}).[102,103] By relating the CH stretching frequency to the strength of the CH bond, a bond energy (102 kcal/mol) 3 kcal/mol higher than that of a normal alkane has been predicted for the CH bonds of cyclopropane.[104] This is in line with estimates derived from orbital overlap[53] and a comparison of CH bond lengths.[35,79] Roberts and Caserio[105] used the similarity of the CH bonds in cyclopropane and ethylene to predict an overall stabilization of 18 kcal/mol for the former molecule. Hence, the actual SE of the three-membered ring could be decreased by this amount to yield the CSE.

Although this estimate of the energetic consequences of hybridization effects appears to be reasonable, a caveat is appropriate. CH force constants and stretching frequencies reflect the curvature of the potential hypersurface at the minimum. The curvature, however, depends to some extent on the stability (lability) of the fission products: the higher their energy, the steeper the potential curve and the higher the force constants and stretching frequencies. This is reflected by the fact that experimental CH stretching frequencies correlate with dissociation enthalpies DH(CH), as demonstrated by McKean.[103,106] Therefore, bond energies cannot be unambiguously derived from spectroscopic data.

C. In Situ Bond Energies from Electron Density Analysis

The number of electrons in the bonding region provides a measure of the bond strength. This number can be assessed by integrating the total electron density distribution $\rho(\mathbf{r})$ over the zero-flux surface (see Equation 3-9) and relating the value thus obtained to the thermochemical bond energies of appropriate reference compounds.[56] Cremer and Gauss[6] have used this approach to evaluate in situ bond energies at 0 K from HF/6-31G(d,p) calculations using the atomization energies of CH_4 and C_2H_6 at 0 K (ZPE corrected) to derive suitable conversion factors. Their values are summarized in Table 3-10. Comparison with the appropriate bond energies of propane leads to ring strain energies of 34 and 29 kcal/mol for cyclopropane and cyclobutane, respectively. These energies, however, are reduced by 6.6 and 3.2 kcal/mol (Table 3-10) resulting from CH bond strengthening in both the three- *and* four-membered rings as was first predicted by Coulson and Moffitt.[75]

The approach by Cremer and Gauss[6] represents a successful attempt to derive the SE of small cycloalkanes with method 3, shown in Figure 3-1 above. It resolves part of the problem of assessing the SE in small molecules but not the whole. In Sections 6 and 7, it was pointed out that CC bonding in cyclopropane may be improved by delocalization of two electrons in the ring surface. Clearly, any stabilization resulting from σ-electron delocalization will be absorbed in the CC bond energies. Thus, the SE values given in Table 3-10 may be still too small. Other ways (Figure 3-17) must be used to examine SEs obtained from in situ bond energies.

TABLE 3-10. Bond Energies and Strain Energies (kcal/mol) from Hartree–Fock Calculations with a 6-31G(d,p) Basis Set

Molecule	Bond	Bond energy	SE(CC)[a]	STAB(CH)[b]	CSE[c]
Propane[d]	sec-C—H	105.5			
	C—C	81.9			
Cyclopropane[e]	C—H	106.6	34	6.6	27.5
	C—C	71.0			
Cyclobutane	C—H	105.9	29	3.2	26
	C—C	73.9			

[a] Strain energy derived from the difference in the CC bond energies and corrected for errors in the theoretical atomization energies (see Reference 6).
[b] Stabilization due to hybridization effects in CH bonding.
[c] Conventional strain energy.
[d] The CH_2 group of propane is used as a reference (see Section 3).
[e] To be compared with 69.9 and 102.5 kcal/mol found in Reference 53.
Source: Cremer and Gauss.[6]

11. QUANTUM CHEMICAL EVALUATION OF THE MOLECULAR STRAIN ENERGY USING THE WESTHEIMER APPROACH

As described in Section 7, the analysis of the electron density distribution $\rho(\mathbf{r})$ leads to a definition of the chemical bond and, thereby, a characterization of bonds in terms of π character, bent bond character, and so on. In addition, it leads to clear definitions of bond length and bond angle. In this way, it becomes possible not only to evaluate in situ bond energies for strained molecules but also to evaluate the Westheimer equation using quantum chemical methods (Equations 3-1 and 3-6; see also method 5 in Figure 3-1). For this purpose, one must define a set of suitable reference compounds that can be used to derive all constants needed for the various terms in the Westheimer equation.

Such a set is shown in Table 3-11.[6] It is based on ab initio calculations on molecules such as ethane, propane, cyclopropane, and cyclobutane complemented by known spectroscopic constants.[107] *The determination of a CCC bending constant k that does not lead to energy contributions actually arising from 1,3-CC nonbonded repulsion* is essential for the evaluation of the Baeyer SE. In molecular mechanics, this problem is solved by considering k as an adjustable parameter that is chosen to reproduce experimentally known molecular properties. Values between 0.45 and 0.8 mdyn-Å/rad² have been used[29,36,37,39] (i.e., values considerably smaller than the CCC bending force constant of propane: 1.071 mdyn-Å/rad²).[107]

Cremer and Gauss[6] solved the problem of determining an appropriate $k(CCC)$ by setting the Baeyer SE of cyclobutane in relation to its 1,3-CC

TABLE 3-11. Constants for an ab initio Evaluation of the Westheimer Equation for Cyclopropane and Cyclobutane[6]

Strain	Energy term	Constants	Reference value	Reference molecule	Comment
Stretching	ΔE_r	$k_r(CC) = 4.57$ mdyn/Å $k_r(CH) = 4.88$ mdyn/Å	$r(CC)^0 = 1.5268$ Å $r(CH)^0 = 1.0858$ Å	Ethane	k_r from experiment[107] r^0 from HF/6-31G(d,p)[6] *Note*: Bond path lengths r_b are used for bent bonds
Baeyer	ΔE_α	$k_\beta(CCC) = 0.583$ mdyn·Å/rad^2 $k_\alpha(CCH) = 0.656$ mdyn·Å/rad^2 $k_\alpha(HCH) = 0.550$ mdyn·Å/rad^2	$\alpha^\circ = 109.5^\circ$	Propane	k_β calculated for the absence of 1,3-CC repulsion[6]; k_α from experiment[107] *Note*: Bond path angles β are used for bent bonds
Pitzer	ΔE_τ	$V_3 = 3.0/3$ kcal/mol	$\tau_3 = 60^\circ$	Ethane	V_3 from HF/6-31G(p,d)[6]
Dunitz–Schomaker	ΔE_{nb}	Directly evaluated for cyclobutane	$l_{CC} = 3.7$ Å	Propane	From CNDO/2 calculations[108] *Note*: Results scaled in dependence of $k_\beta(CCC)$ to reproduce CSE and inversion barrier of cyclobutane[6]

nonbonded repulsion energy. For the latter they utilized appropriately scaled CNDO/2 energy differences calculated for cyclobutane with and without C,C-nonbonded repulsion.[108] Bending force constant and scaling factor were chosen to reproduce the CSE and the inversion barrier of cyclobutane. In this way, they obtained a $k(CCC)$ value (k^* in Reference 6) of 0.58 mdyn-Å/rad^2, applicable in the absence of 1,3-CC nonbonded repulsion and a Dunitz–Schomaker SE of 12 kcal/mol for cyclobutane.[6]

To describe the strain in cycloalkanes, Cremer and Gauss[6] chose the bond path length r_b and the interpath angle β rather than the geometrical distance r_e and the geometrical angle α. Clearly, the geometrical parameters are not of direct relevance when assessing the bending of the bond (i.e., the elastic spring in Baeyer's strain model).[1] As discussed in Section 7, the geometrical angle may be 60° in cyclopropane, but the CC bonds are bent by an angle β of only 79°.[4] Similarly, the actual bond path length is larger by a factor of 1.006 than the geometrical distance [HF/6-31G(d,p): r_e = 1.497 Å, r_b = 1.506 Å[4–6]]. This, of course, is of utmost importance when assessing the Baeyer SE of a small ring.

Baeyer and nonbonded SEs obtained by the procedure suggested in Reference 6 cannot be compared with energies used in molecular mechanics calculations. In molecular mechanics a physically meaningful definition of the chemical bond cannot be given and, therefore, information about bond bending, bending angles, and so on is not accessible in principle. By using geometrical distances and angles *without* defining the chemical bond, a consistent description of the factors contributing to molecular strain is impossible. *The various energy terms leading to the steric energy calculated in molecular mechanics cannot be related to the various strain energies of Equation 3-1 in a consistent and physically meaningful way.* They are only of operational value, namely to add up to a quantity that finally leads to the molecular enthalpy.

Table 3-12 lists the various strain energies calculated for cyclopropane and cyclobutane with the constants of Table 3-11.[6] The Baeyer SE of cyclopropane is 46 kcal/mol, including an estimated 5 kcal/mol from anharmonicity effects,[37] while the Baeyer strain of cyclobutane is just 13 kcal/mol. However, in the latter case, 12 kcal/mol is due to Dunitz–Schomaker strain, destabilizing the four-membered ring considerably. Pitzer strain adds in both cases just 4 kcal/mol.[6] The total SE of cyclopropane is 51 kcal/mol (ie, 21 kcal/mol larger than that of cyclobutane). We note that these values fit reasonably into the expected increase in the ring SE with decreasing ring size (Figure 3-5).

As noted in earlier Sections 3 and 9 (Figures 3-1 and 3-17), the calculated SEs must be corrected for stabilization or destabilization effects other than ring strain to get CSE values. From Figure 3-17 we see that this is possible only when determining the strain energy in a different way (e.g., from in situ bond energies). Calculated bond energies (see above: Table 3-10) show that both cycloalkanes are stabilized due to CH bond strengthening. Taking these energies (6.4 and 2.8 kcal/mol)[6] into account, CSEs of 44.4 and 27.1

TABLE 3-12. Ab initio Strain Energies and Stabilization Energies of Cyclopropane and Cyclobutane

	Strain energy (kcal/mol)	
Strain	Cyclopropane	Cyclobutane
Stretching	0.5	1.0
Baeyer[a]	46.3	13.0
Pitzer	4.0	3.9
Dunitz–Schomaker	0	12.0
Total	50.8	29.9
	− Stabilization energies (kcal/mol)	
CH strengthening	6.4	2.8
σ Delocalization	$x = 16.4$	0
CSE	$28.0 = 44.4 + x$	27.1

[a] Baeyer strain energy of cyclopropane calculated with Hooke's law (41.3 kcal/mol)[6] plus energy increase from anharmonicity effects calculated from a bending function with and without a cubic term[37] for $\beta = 79°$ (5 kcal/mol). [Note that the strain energy of propane (Reference 6, Table IX) has been set erroneously to 5.1 kcal/mol. This energy, however, is compensated by the increase in the CC bond energy relative to that of ethane, Table V, Reference 6.]

kcal/mol result. The latter value is in good agreement with both experimental and other theoretical CSE values for cyclobutane.

In the case of cyclopropane, however, there remains an energy difference of about 16 kcal/mol (x in Table 3-12), which can be attributed to another stabilizing effect, namely the delocalization of σ electrons in the ring surface[4-6] as indicated by both the Walsh MOs (Section 6) and the properties of $\rho(\mathbf{r})$.

12. PROS AND CONS OF σ AROMATICITY

According to Hückel theory, planar annulenes with (4q + 2) (q = 0, 1, 2, . . .) π electrons are aromatic.[109-111] The accepted empirical tests for aromatic character involve (among others) a determination of the molecular geometry and an NMR investigation of the compound in question. The shortening of formal CC single bonds leading to bond equalization[111] and the proton shifts arising from a diamagnetic ring current are considered to be indicative of aromatic character.[109] To assess the aromatic stabilization energy, a reference compound is defined that possesses the same number of π electrons in a localized form (suppression of π conjugation).[110]

Applying these tests to cyclopropane leads to the following picture.

1. The CC bonds of cycloalkanes such as cyclopropane or cyclobutane are all equivalent. As discussed in Section 6, all cycloalkanes possess "aromatic" subshells of electrons (compare with Figure 3-8) and, therefore, it seems to be trivial and of no particular advantage to term them σ aromatic. Comparing on the other hand, the CC bond lengths in cyclopropane and cyclobutane, it should be noted that both r_e and r_b (Table 3-7) are considerably shorter in the smaller ring. It seems that a particular force contracts the three-membered ring, thus enhancing the stabilizing electronic interactions.

2. Due to the $C(sp^2)$—H bond nature, one would expect the proton NMR signal for cyclopropane to appear downfield relative to the signals of the methylene protons in alkanes. In fact, it appears upfield by 1 ppm ($\delta = 0.22$ ppm).[112] Zilm and co-workers[113] have investigated the ^{13}C NMR spectrum of cyclopropane. They note that circulation of electrons in the ring plane leads to a most unusual upfield shift (~20 ppm) of the ^{13}C NMR signal of cyclopropane.

3. The analysis presented in Section 11 suggests that the cyclopropane ring is stabilized by at least 16 kcal/mol relative to a hypothetical three-membered ring in which surface delocalization of electrons is impeded.

Seeing points 1 through 3 in one context, it is appealing to consider cyclopropane as being σ aromatic.[2-6]

The description of cyclopropane as a system with six delocalized σ electrons dates back to the 1960s. For example, Brown and Krishna[3] calculated the excited electronic states of C_3H_6 by treating its σ electrons in the same way as the π electrons of benzene in a PPP description. These authors explicitly pointed out that there is a striking resemblance between the σ electrons of cyclopropane and the π electrons of benzene.[3]

Dewar[2] was probably the first to elaborate the idea of σ-electron delocalization. He stressed that the overlap, hence the resonance integrals between different hybrid orbitals of a given atom, will not vanish, even if the constituting atomic orbitals are orthogonal. The value of a resonance integral between sp^n hybrids in hydrocarbons is considerably larger than that of the π-resonance integral between adjacent 2p AOs in a polyene.[2,3] Taking this into account, Dewar concluded that the relative stabilities of alkanes can be rationalized only by considering σ-conjugative interactions. The latter should be even more important for small cycloalkanes. It is well known that in conjugated polyenes the $2p\pi$ AOs overlap with one another and the two-center MOs coalesce into a resonating system in which the π electrons can delocalize. Taking into account that the hybrid AOs of a given atom also overlap (an often neglected fact), then each two-orbital CH_2 unit in a parafin will play the same role as a two-orbital =CH—CH= unit in a conjugated polyene or cyclopolyene. Hence, cyclopropane can be considered to be isoconjugate with benzene (see Figure 3-18), hence σ aromatic, since both systems possess a six-electron ensemble that is delocalized along the ring framework.[2,3]

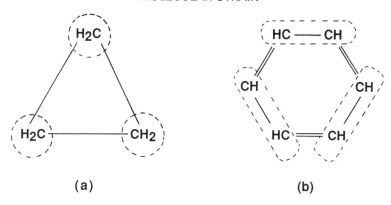

Figure 3-18. Analogy between cyclopropane (*a*) and benzene (*b*). Groups that interact in a σ-conjugated and in a π-conjugated system are circled.

Clearly, this description is based on a model, the model of hybrid orbitals. Within the Walsh model, the picture of a surface orbital with two delocalized electrons emerges (Section 6). Both descriptions are essentially equivalent within the limitations pointed out by Heilbronner.[80]

One might criticize the term "σ aromaticity," since aromaticity is connected strongly to the idea of π conjugation in cyclopolyenes. Comparison with an acyclic system consisting of the same number of conjugated π bonds is an inherent part of the definition of aromaticity. Such a comparison, of course, cannot be readily made in the case of cyclopropane. Certainly neither propene nor the trimethylene biradical constitutes an appropriate reference system. In addition, "aromatic" character usually implies a breakdown in a localized bonding picture—that is, the existence of more than one resonance structure with alternating single and double bonds for a given molecule, which is not true in case of cyclopropane. For these reasons, the present authors prefer the term *surface* or *σ-electron delocalization* and speak of a "*σ-delocalization energy*"[4–6] rather than an aromatic stabilization energy.

σ-Electron delocalization causes or, at least, influences a number of properties of cyclopropane, including the following:

1. The relatively low CSE
2. The relatively high electron density in the interior of the ring as reflected by both $\rho(\mathbf{r})$ and $\nabla^2\rho(\mathbf{r})$
3. The relatively short CC distances
4. The upfield shifts of its proton and ^{13}C NMR signals
5. The similarity of the activation energies needed to break a CC bond in cyclopropane (61 kcal/mol) and in cyclobutane (62.5 kcal/mol).[114]

Furthermore, σ-electron delocalization may play an important role in the interaction of cyclopropane with substituents.[5,115,117] This has been stressed by Cremer and Kraka.[5] Also, the incorporation of the three-membered ring in conjugated systems and the possibility of homoaromatic interactions has been discussed utilizing the idea of σ-electron delocalization.[4-6,8,7b]

Various authors have added support to this concept. For example, Coulson and Moffitt[75] were the first to note that there is a plateau of relatively high negative charge inside the C_3 ring. Over the years other authors have also pointed to the special role of the a_1' MO of cyclopropane and terms like "internal σ orbital"[59b] were coined. Recently, Schwarz and co-workers[8] have investigated MO composition and AO reorganization in cyclopropane and propane. They note that the total electron density is increased by 0.16 e/Å^3 in the center of the cyclopropane ring as compared to superimposed spherical free atoms. The topology of the ring supports favorable interference of three overlapping contragradient 2s or 2p$_r$ orbitals, which is constructive over the whole ring surface and increases the total electron density in the ring center. The response of the AOs to the strongly reduced kinetic energy density in the a_1' (surface) orbital is enhanced AO contraction, which restores the virial relation and lowers the total energy significantly.[8] This AO contraction is essential for the stabilization of the C_3 ring and, also, causes a CC bond length reduction. Contrary to their behavior in the central "super-σ bond," the AOs expand in the Walsh e' MOs, which is typical of π orbitals and supports the π character of the cyclopropane ring bonds.[8]

Ahlrichs and Ehrhardt[9] have calculated shared electron numbers for alkanes. While bonding is reflected in these compounds by two-center contributions and negligible contributions from three- and four-center terms, a CCC shared electron number of 0.3 is calculated for cyclopropane, which is indicative of three-center bonding.

Experimental observations indicative of special electronic effects active in the three-membered ring have been published by a number of authors.[7,136] For example, Verhoeven and co-workers[7] have found kinetic anomalies for the formation of small cycloalkanes that cannot be explained by the Ruzicka hypothesis.[130] In the latter approach, two competing effects are considered, namely increasing ring strain for decreasing ring size and an opposing entropy factor that favors ring closure for small rings. Contrary to these explanations, the closure of a four-membered ring is exceptionally slow when compared with that of a five- or three-membered ring. Verhoeven[7] has shown that this is due to irregularities in the activation enthalpies; that is, anomalies in the kinetics are due not only to a decrease of $\Delta S^{\ddagger}$ with decreasing n but have also an electronic reason. The latter can be elucidated when utilizing the Dewar–Zimmermann rules[131] for transition states of pericyclic reactions: a thermal pericyclic reaction is allowed (forbidden) for an aromatic (antiaromatic) transition state. Aromatic character requires the involvement of 4q + 2 electrons for a Hückel system and 4q electrons for a Möbius system.

If a cyclopropane ring is formed, a Hückel aromatic transition state with six electrons will be traversed. However, in case of the formation of a cyclobutane ring, a Hückel antiaromatic transition state with eight electrons results. This is schematically shown in Scheme II. As a consequence, the activation enthalpy for the closure of a cyclobutane ring is markedly higher than that for a cyclopropane (cyclopentane) ring. This has been found for carbanion cyclizations and other reactions.[7] (For a different opinion, see References 136).

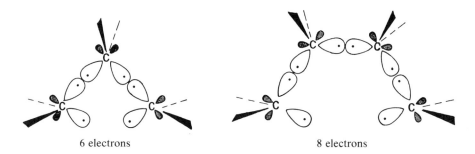

6 electrons   8 electrons

Scheme II. Cyclic transition rotates with Hückel aromatic (left) and Hückel antiaromatic electron ensembles (right).

In this connection, it is interesting to note that a description of the CC and CH MOs of cyclopropane in terms of Hückel and Möbius arrays has been given by Epiotis.[132] He describes bonding in cyclopropane as due to a "superaromatic" interaction between Hückel/Möbius-aromatic C_3 and $(CH_2)_3$ cycles.

The idea of σ-electron delocalization has not met with unanimous approval. For example, Schleyer[11,47] has questioned whether a delocalization of σ electrons in cyclopropane, if existing, entails any energetic consequences. Analyzing the various contributions to the CSE of cyclopropane and cyclobutane, he concludes that there is no need to invoke σ aromaticity to explain the relative stabilities of small cycloalkanes.[11] He suggests that the low CSE of cyclopropane is due to CH bond strengthening and assumes a stabilizing contribution of 10 kcal/mol from this source.

Certainly, if CH bond strengthening is not correctly described by the bond energies given in Table 3-10, the delocalization energy (x in Table 3-12) will change. This is shown in Figure 3-19. Assuming for example, values given by Schleyer[11] or Roberts and Caserio,[105] σ-delocalization energies of 13 and 5 kcal/mol, respectively, result from CSE and the strain energies listed in Table 3-12. A delocalization energy even smaller than zero would be obtained if the total effect of CH bond strengthening were larger than 23 kcal/mol. However, this is unlikely in view of the similarity of the sec-CH bonds in cyclopropane and propane.[6]

The σ-delocalization energy of cyclopropane also depends critically on the calculated Baeyer SE, which in turn depends on the value of the CCC bending force constant. If one uses the spectroscopic force constant of pro-

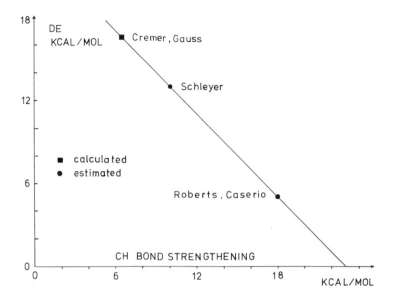

Figure 3-19. Dependence of the σ-delocalization energy DE on the CH bond strengthening (hybridization effect) in cyclopropane. Values of the latter have been taken from References 6, 11, and 23f, respectively. (Reproduced with permission from Reference 6. Copyright © 1986 from the American Chemical Society.)

pane, a delocalization energy of about 50 kcal/mol will be found[2,4] (see Figure 3-20). This value can be considered only as an upper bound to the σ-delocalization energy, since $k(CCC)_{propane}$ corresponds to CCC bending in the presence of strong H,H- and C,C-nonbonded repulsion.

The force constant determined in Reference 6 describes CCC bending in the absence of nonbonded repulsion and, therefore, is appropriate to evaluate the Baeyer SE of cyclopropane. Similar values of k are used in molecular mechanics. They will lead to comparable σ-delocalization energies as can be seen from Figure 3-20.

Schleyer[11] adjusts the Baeyer SE to 33 kcal/mol to reproduce the CSE of cyclopropane. According to Hooke's law, this implies a CCC bending force constant of 0.20 mdyn-Å/rad^2, which is probably far too low. By applying the same method to determine the Baeyer SE of cyclobutane, Schleyer obtains a value of 10 kcal/mol, which corresponds to $k(CCC) = 0.27$ mdyn-Å/rad^2. If the latter force constant is used for cyclopropane, a Baeyer SE of 43 kcal/mol will be obtained, entailing a σ-delocalization energy of 10 kcal/mol. Hence, the estimate of various contributions to the CSEs of cyclopropane and cyclobutane given by Schleyer[11] makes it difficult to draw *any* conclusions about the energetic consequences of σ-electron delocalization.

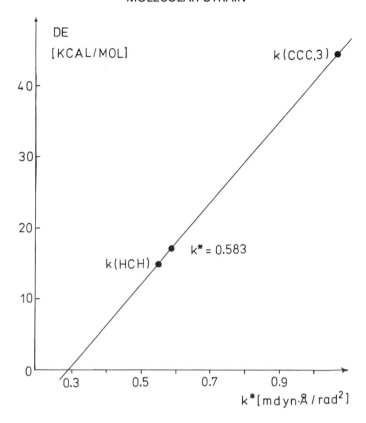

Figure 3-20. Dependence of the σ-delocalization energy DE on the value of the CCC bending force constant k^* in cyclopropane: k(CCC) and k(HCH) are from propane, k^* has been calculated for CCC bending in the absence of 1,3-CC-nonbonded repulsion. (Reproduced with permission from Reference 6. Copyright © 1986 from the American Chemical Society.)

13. LIMITATIONS OF THE CONCEPT OF STRAIN

The concept of strain is applied to molecules with covalent bonds. It looses its usefulness when discussing, for example, ionic bonding. The strength of the ionic bond is best determined by Coulomb's law. A description of deformed ionic bonds in terms of stress and strain appears superfluous. Of course, this applies also to those bonds that possess strong ionic character. The transition from covalent to ionic bonding is continuous, and the pure covalent (ionic) bond is more the exception than the rule. One might also ask whether a description of molecular stability in terms of strain could also be done from an electrostatic point of view.[118] This question, of course, can be rephrased in a more fundamental form: Will the concept of strain still be useful if one describes strained molecules strictly with the tools and aids of quantum chemistry, abolishing the model taken from classical mechanics?

A quantum chemical assessment of molecular strain implies the calculation of the stress tensor $\overleftrightarrow{\sigma}(\mathbf{r})$ that determines in a stationary state the force density $\mathbf{F}(\mathbf{r})$[57]:

$$\mathbf{F}(\mathbf{r}) = -\nabla \cdot \overleftrightarrow{\sigma}(\mathbf{r}) \qquad (3\text{-}15)$$

Analysis of the force density in strained bonds probably will lead to a quantum mechanical definition of what is exceptional in these bonds. However, it remains to be seen whether such a definition can be expressed in terms of a simple physical model. Much work remains to be done to clarify this point.

The concept of strain becomes questionable also when considering nonclassical bonding. It makes little sense to discuss the strain of a penta- or hexavalent carbon atom. But even in the case of tetravalent carbon, one may doubt the validity of applying the concept of strain in all situations. We will exemplify this for three-membered rings.

A. The Relationship Between Three-membered Rings and π Complexes

Following ideas first proposed by Dewar,[119] Walsh[120] suggested in 1947 a π-complex formula for three-membered rings:

$$\begin{array}{ccc} \mathrm{CH_2{=}CH_2} & \mathrm{CH_2\!-\!CH_2} & \mathrm{CH_2\ \ CH_2} \\ \downarrow & \leftrightarrow \ \ \diagdown\!\diagup \ \ \leftrightarrow & \diagdown\!\!\!\diagup\!\!\!\diagup \\ \mathrm{CH_2} & \mathrm{CH_2} & \mathrm{CH_2} \end{array}$$

Later he renounced this description because the molecular properties of cyclopropane were difficult to describe with the π-complex model.

Dewar[121] revived the π-complex description by showing that two kinds of interaction must be considered in three-membered rings. First, an electron donation from the ethylene π orbital (basal group) to a vacant orbital of the apical group X (CH_2, NH, O, etc.). Second, back-donation of electrons from a filled p orbital of X into an antibonding π^* MO of the ethylene double bond. The apical group X and the basal group C_2H_4 are doubly linked by two opposed dative bonds. Depending on which interaction is stronger, three situations are possible.

1. Donation and back-donation are of comparable magnitude. A stable three-membered ring is formed.
2. Donation prevails over back-donation. A three-membered ring results, which possesses partial π-complex character.
3. There is just donation to the apical group. A π complex is formed.

Dewar and Ford[122] have suggested geometry- or orbital-based parameters that describe the degree of π-complex character. They have shown that a change in the electronegativity difference between the apical and basal groups leads to a continuous transition from the classical ring structure (small electronegativity difference) to a π complex (large electronegativity difference).

Cremer and Kraka[4] have approached the problem of distinguishing between a three-membered ring and a π complex by analyzing the electron density distribution $\rho(\mathbf{r})$ (see Section 7). As schematically shown in Figure 3-21$a(b)$, donation from (to) the basal group via an a_1 MO leads to a buildup of electron charge along the C_2 axis of a system A_2X (A_2: CH_2CH_2, HCCH, CC, etc) with C_{2v} symmetry. The resulting bond path[123] (Section 7) connects the apical group X with the midpoint of A_2. The T-structure of a π complex is formed (Figures 3-21a and 3-21b).

As soon as back-donation, e.g., from an occupied b_2 MO of X occurs, electron density builds up along the lines A, X. Together with the accumulation of density along the C_2 axis, a plateau of relatively high negative charge is formed between A_2 and X. Depending on the electronic nature of A_2 and X, or more specifically on the energy gap between the b_2 MOs and, hence, the amount of electronic charge back-donated, bond paths are formed between A and X. Strong back-donation leads to outwardly curved (*convex*) AX bond paths (Figure 3-21c), while relatively weak back-donation yields inwardly curved (*concave*) AX bond paths (Figure 3-21d). Concave bond paths are indicative of partial π-complex character of A_2X. Hence, *an analysis of the electron density reveals in a quantitative way the extent to which a given molecule should be classified as either a three-membered ring or a π complex.*[4]

Figure 3-22 shows the calculated bond paths by means of heavy lines (bond critical points are indicated by dots), as well as the Laplace concentrations (dashed contour lines indicate charge concentration) of six molecules: cyclopropane, oxirane, protonated oxirane, F-bridged fluoroethyl cation, beryllocyclopropane, and beryllocyclopropyne.[4] As can be seen, the CX bond paths gradually change with increasing electronegativity of group X, that is, in the series:

$X = Be, CH_2, NH(\text{not shown}), O, NH_2^+(\text{not shown}), OH^+, F^+$

from convex to concave bending. In the extreme case, the two CX bond paths coincide largely ($X = F^+$) or completely as in BeC_2 (Figure 3-22f; see References 4 and 124 for a detailed discussion). These molecules are π complexes and, therefore, possess a T structure.

Figure 3-22 reveals the very dubious nature of the concept of strain when applied to three-membered rings other than cyclopropane. Certainly, no chemist would ever think of discussing the "strain of a π complex." But on the other hand, it appears to be completely legitimate to most chemists to

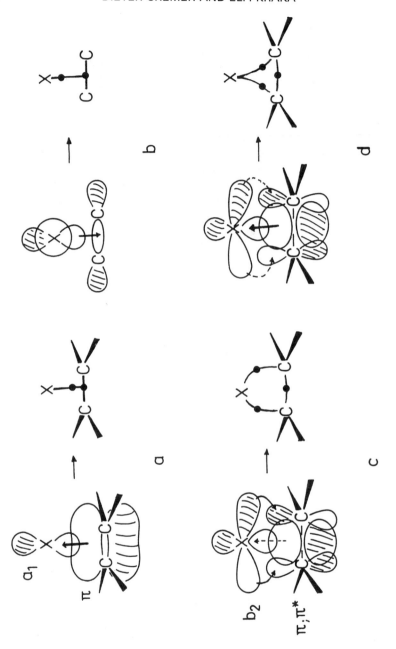

Figure 3-21. Donor–acceptor interactions between A_2 (C_2H_4 or C_2) and X, showing relevant orbitals (a_1 or b_2 symmetry) on the left and the corresponding molecular graphs on the right. (*a*) T structure of an ethylene π complex. (*b*) T structure of BeC_2. (*c*) Convex-shaped three-membered ring. (*d*) Concave-shaped three-membered ring. The direction of the charge transfer is indicated by arrows (reduced charge transfer by dashed arrows); dots denote bond critical points. (Reproduced with permission from Reference 4. Copyright © 1986 from the American Chemical Society.)

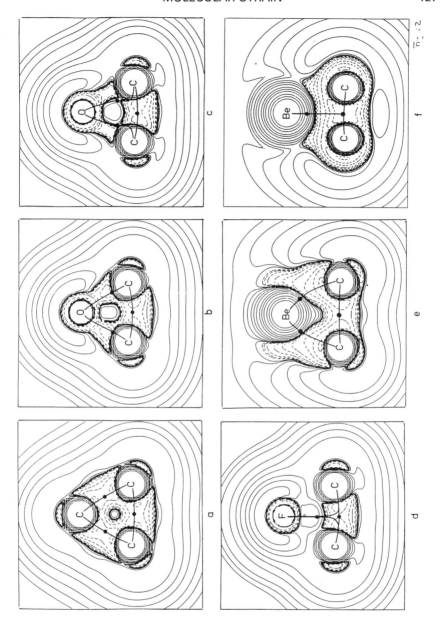

Figure 3-22. Molecular graphs and Laplace concentrations $-\nabla^2\rho(\mathbf{r})$ [HF/6-31G(d) calculations] of cyclopropane (*a*), oxirane (*b*), protonated oxirane (*c*), F-bridged fluoroethyl cation (*d*), beryllocyclopropane (*e*), and beryllocyclopropine (*f*). Bond paths are indicated by heavy lines and bond critical points by dots. Dashed and solid contour lines of $\nabla^2\rho(\mathbf{r})$ are in regions in which electronic charge is concentrated and depleted, respectively. Inner shell concentrations are not shown. (Reproduced with permission from Reference 4. Copyright © 1986 from the American Chemical Society.)

rationalize the low stability of, for example, protonated oxirane in terms of strain. This molecule, however, possesses concave bent bonds (Figure 3-22c). The OCC interpath angle of protonated oxirane is just 20°.[4] With an assumed[125] force constant of 0.35 mdyn-Å/rad^2, a Baeyer SE of 121.6 kcal/mol just for the two basal OCC angles results. This makes no sense at all, and it indicates that the concept of strain should be discarded for three-membered rings with partial π-complex character or any strained molecule possessing concave rather than convex bent bonds.

Clearly, MO theory cannot provide sufficient information about the bonding situation in such systems. Therefore, it is advisable to first analyze the electron density distribution of a potentially strained molecule and then to decide whether its stability can be rationalized in terms of strain. Of course, one can always evaluate a homodesmotic reaction energy. Any destabilization revealed e.g., in the case of protonated oxirane must not be mistaken as an indication of strain. It simply shows that the CO bonds are weakened due to reduced back-donation from a highly electronegative OH$^+$ group.

The application of the concept of strain will become even more problematic if one considers not the thermodynamic but the kinetic stability of potentially strained molecules. In discussions, one is often confronted with the following statement: "The formation of this or that molecule is impeded because of its strain." In Section 4 we stressed that from an energetic point of view, such a statement is true only in exceptional cases. Approaching the question of a relationship between reactivity and strain from structural theory suggests that there may be no such relationship at all. This is briefly discussed in the next section.

B. Conversion of a π Complex into a Three-membered Ring: The Structure Diagram

The network of all MED paths linking the nuclei (see Section 7 for the distinction between MED and bond path) is called a molecular graph.[126] Molecular graphs that possess the same number of MED paths are equivalent. *Each class of equivalent graphs corresponds to a certain molecular structure.*[127] For a given molecular state, there exist a finite number of structures (classes of equivalent graphs) and associated regions of nuclear configuration space. Accordingly, the equivalence relation of molecular graphs can be used to partition nuclear configuration space into structural regions. Within a structural region thus defined, small changes in geometry do not change the molecular structure. However, at the boundaries of the structural regions, a differential change in geometry leads to a discontinuous change in molecular structure that is best described by catastrophe theory.[128]

In the structure diagram of A$_2$X (Figure 3-23a), the coordinates u and v are chosen to describe two independent coordinates of X. For every (u, v) point, the third degree of freedom of A$_2$X is considered to be optimized. In this

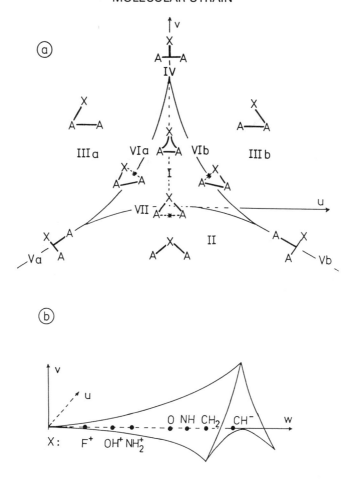

Figure 3-23. Structure diagram of system A_2X shown in (u, v, w) space. (a) control parameter w = constant. For each of the structural regions (I–VII) a molecular graph is given. Asterisks denote catastrophe points. (b) Three-dimensional display of the bifurcation set of the elliptic umbilic in (u, v, w) space for $w \geq 0$. The meaning of w as a chemical control parameter is illustrated by giving the approximate positions of three-membered rings of type C_2H_4X on the positive w axis. (Reproduced from Reference 4. Copyright © 1986 from the American Chemical Society.)

way, v has the meaning of the reaction coordinate of a C_{2v} symmetrical approach (decreasing v) of X toward A_2 (a_1 symmetrical displacement of X) and u describes b_2 symmetrical displacements of X from this reaction path.

If X approaches A_2 along the v axis, a T structure will be formed. In catastrophe theory the T form is called a conflict structure.[128] Any infinitesimal displacement of X from this v axis leads to one of the acyclic structures in region IIIa or IIIb. At a certain value of v, the conflict structure unfolds to a three-membered ring i.e., A_2X enters the hypocycloid-shaped structural region. Since this structural change involves a large redistribution of electron density, it should be close to the transition state in energy space. We note that at this point A_2X is far from possessing the classical ring structure with fully developed (convex) bent bonds. Hence, a discussion of the transition state structure in terms of strain (bond angle strain, stretching strain, etc) again is useless. The same is true for structural changes of A_2X when moving along region Va or Vb (Figure 3-23a).

On the other hand, the concept of strain may be useful when describing the structural changes of A_2X from region I to either II or III. This, however, implies that the three-membered-ring region I is large enough to encompass classic ring geometries. As indicated by Cremer and Kraka,[4] region I will become smaller as the electronegativity difference between A_2 and X increases (Figure 3-23b). In this way, the concept of strain becomes useless to describe the ring structure of A_2X; it is all the more useless for describing any structural changes of the ring.

Taking all aspects together, we conclude that the application of the concept of strain to other than normal hydrocarbons in the ground state requires considerable care and additional information about bonding in the molecule in question.

14. CONCLUSIONS

Investigation of the strain energies of small cycloalkanes has led to an improved understanding of the factors contributing to the relative stabilities of these molecules. *Due to its topology, cyclopropane differs from all other cycloalkanes.* Surface delocalization of σ electrons leads to enhanced stability of the three-membered ring, compensating in part the destabilizing effect of Baeyer strain. Cyclobutane is also exceptional, as it is the cycloalkane with the largest Dunitz–Schomaker strain (Table 3-12).

These observations entail a number of chemically important conclusions for molecules containing three- or four-membered rings (Figure 3-24).

1. Substituents with σ-electron acceptor capacity withdraw electrons from the surface orbital of cyclopropane, thus destabilizing the ring. For cyclobutane, however, a σ-electron acceptor withdraws charge from the 1,3-antibonding MO (Figure 3-8), thus leading to a stabilization of the molecule. As an example, the strain energy of hexafluorocyclopropane is found to be remarkably high relative to that of the parent cyclopropane. In contrast, the strain energy of octafluorocyclobutane is likewise low when compared to that of cyclobutane[134] (see also discussion in Reference 6).

2. When replacing the C atoms of cyclopropane by Si or Ge atoms, the threefold overlap inside the ring is reduced by (a) an increase of the ring dimensions and (b) a decreasing tendency of the atoms to form sp^2-hybrid orbitals. Also, H has to be considered as a relatively strong σ-electron

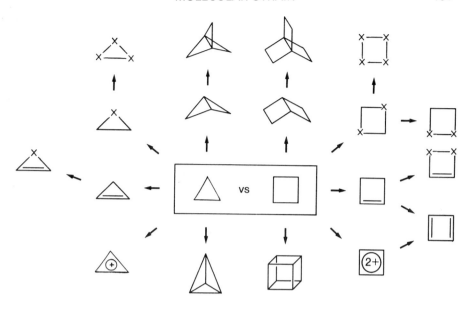

Figure 3-24. Schematic representation of compounds for which the electronic effects influencing structure and stability of cyclopropane and cyclobutane may be of importance.

acceptor if bound to Si or Ge. As a consequence, σ-electron delocalization is of little importance in $(SiH_2)_3$ or $(GeH_2)_3$. Thus, the CSE should be larger in trisilacyclopropane than in cyclopropane. For similar reasons, 1,3-repulsion should be lower in the Si, Ge, etc analogues of cyclobutane; hence the CSE (SE) should be smaller in these systems. Work by Schleyer and co-workers[129] confirms these predictions as illustrated in Figure 3-25. Calculated CSEs of cyclosilanes $(SiH_2)_n$ show the expected increase with decreasing n ($n \leq 6$) as originally expected for $(CH_2)_n$ (compare with Figures 3-5 and 3-6).

3. Although in some cases the SEs of bicyclic and tricyclic compounds containing cyclopropane and cyclobutane as subunits equal the sum of the SEs of its subunits, this cannot be expected to be true in general. Interactions between the subunits may either enlarge or reduce the SE arising from the subunits. For example, the SE of bicyclobutane is about 8 kcal/mol larger than twice the SE of cyclopropane.[49]

4. Electron delocalization may occur not only in one dimension (ribbon delocalization of π electrons) or in two dimensions (surface delocalization of σ electrons) but also in three dimensions (volume delocalization of σ electrons).[6] In a cage compound like tetrahedrane, there is a totally symmetrical a_1 MO that leads, when occupied, to a two-electron, four-center bond. This can improve CC bonding in tetrahedrane and, as a consequence, reduce destabilizing effects caused by strain. Two-electron, four-center bonding is also invoked to describe the structure of the neutral *closo*-boron compounds B_4R_4 (R = Cl, *t*-Bu, etc).[135]

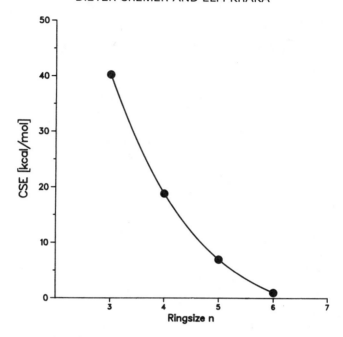

Figure 3-25. Homodesmotic SEs of cyclosilanes $(SiH_2)_n$ in dependence of the ring size n. (Unpublished [HF/3-21G//HF/3-21G] results of P. v. R. Schleyer[29a].)

5. Discussion of a compound like [1.1.1]propellane in terms of strain should be postponed until the nature of the central CC bond has been clarified.

Further investigations are needed to show the energetic consequences of strain and σ-electron delocalization in detail. Theory has advanced in understanding the strain in small rings. Nevertheless, there is still a long way to go to assess all facets and aspects of the chemical behavior of strained molecules. Certainly, textbooks claiming that the strain in small rings is nowadays well understood ought to have their sections on strain revised.

ACKNOWLEDGMENTS

Fruitful discussions with Professor P. v. R. Schleyer are acknowledged. The authors wishes to thank Dr. T. H. Dunning, Jr., Dr. L. B. Harding, and Dr. R. Duchovic for their helpful suggestions and comments during the preparation of this manuscript. Valuable assistance by J. Gauss is gratefully acknowledged. This work was supported by the Deutsche Forschungsgemeinschaft and the Fonds der Chemischen Industrie and was performed under the auspices of the Office of Basic Energy Sciences, Division of Chemical Sciences, U.S. Department of Energy, under contract W-31-109-Eng-38.

REFERENCES

1. (a) Baeyer, A. *Chem. Ber.* **1885**, *18*, 2269. (b) For a recent evaluation of von Baeyer's work see: Huisgen, R. *Angew. Chem. Int. Ed. Engl.* **1986**, *25*, 297.
2. (a) Dewar, M. J. S. *J. Am. Chem. Soc.* **1984**, *106*, 669. For earlier work see: (b) Dewar, M. J. S. *Bull. Soc. Chim. Belg.* **1979**, *88*, 957. (c) Dewar, M. J. S.; McKee, M. L. *Pure Appl. Chem.* **1980**, *52*, 1431.
3. Brown, R. D.; Krishna, V. G. *J. Chem. Phys.* **1966**, *45*, 1482.
4. Cremer, D.; Kraka, E. *J. Am. Chem. Soc.* **1985**, *107*, 3800.
5. Cremer, D.; Kraka, E. *J. Am. Chem. Soc.* **1985**, *107*, 3811.
6. Cremer, D.; Gauss, J. *J. Am. Chem. Soc.* **1986**, *108*, 7467.
7. (a) Van der Kerk, S. H.; Verhoeven, J. W.; Stirling, C. J. M. *J. Chem. Soc. Perkin Trans. 2* **1985**, 1355. (b) Verhoeven, J. W.; Pasman, P. *Tetrahedron* **1981**, *37*, 943. (c) Verhoeven, J. W. *Rec. Trav. Chim. Pays-Bas* **1980**, *99*, 369. (d) Verhoeven, J. W. *Rec. Trav. Chim. Pays-Bas* **1980**, *99*, 143.
8. Pan, D. K.; Gao, J.-N.; Liu, H.-L.; Huang, M.-B.; Schwarz, W. H. E. *Int. J. Quantum Chem.* **1986**, *29*, 1147.
9. Ahlrichs, R.; Ehrhardt, C. *Chem. Z.* **1985**, *19*, 120, and personal communication.
10. Delocalization of σ electrons leading to σ aromaticity was previously used to rationalize the conformational behavior of dimethylcarbene. Cremer, D.; Binkley, J. S.; Pople, J. A.; Hehre, W. J. *J. Am. Chem. Soc.* **1974**, *96*, 6900.
11. Schleyer, P. v. R. In "Substituent Effects in Radical Chemistry," Viehe, H. G.; Janusek, R.; and Merenyi, R., Eds.; NATO Advanced Study Institutes Series C. Reidel,: Dordrecht, 1986, p. 61, and personal communication.
12. Love, A. E. H., "A Treatise on the Mathematical Theory of Elasticity." Dover: New York, 1944.
13. We note that in German, von Baeyer's mother tongue, the word *Spannung* (English: strain) is often used for stress rather than for the deformations. This, however, does not comply with the original formulation of Hooke's law: *Ut tensio sic vis*. Thus, von Baeyer speaks in his seventh sentence about the *stress* (and not the strain) that is proportional to the relative deviation (strain) from the tetrahedral angle. See Reference 1a.
14. (a) Pitzer, K. S. *J. Chem. Phys.* **1937**, *5*, 469; 473. (b) Pitzer, K. S. *Science* **1945**, *101*, 672.
15. (a) Brown, H. C. *Rec. Chem. Prog.* **1953**, *14*, 83. (b) Brown, H. C. *J. Chem. Soc.* **1956**, 1248.
16. Kehrmann, F. *J. Prakt. Chem.* **1889**, *40*, 257.
17. Stoll, M.; Stoll-Comte, G. *Helv. Chim. Acta* **1930**, *13*, 1885. In German the term *Pressung* is used, which indicates that stress and strain are mixed up.
18. Dunitz, J. D.; Schomaker V. *J. Chem. Phys.* **1952**, *20*, 1703.
19. Westheimer, F. H. In "Steric Effects in Organic Chemistry"; Newman, M. S., Ed.; Wiley: New York, 1956, p. 523.
20. Olah, G. A.; Liang, G.; Schleyer, P. v. R.; Engler, E. M.; Dewar, M. J. S.; Bingham, R. C. *J. Am. Chem. Soc.* **1973**, *95*, 6829.
21. (a) Köbrisch, G. *Angew. Chem. Int. Ed. Engl.* **1973**, *12*, 464. (b) Buchanan, G. L. *Chem. Soc. Rev.* **1974**, *3*, 41, see also Reference 22 and references cited therein.
22. Liebman, J. F.; Greenberg, A. *Chem. Rev.* **1976**, *76*, 311.
23. (a) Beyer, W. "Lehrbuch der Organischen Chemie." S. Hirzel Verlag: Stuttgart, 1984. (b) Hendrickson, J. B.; Cram, D. J.; Hammond, G. S. "Organic Chemistry." McGraw-Hill: New York, 1970. (c) Carey, F. A.; Sundberg, R. J. "Advanced Organic Chemistry," Part A. Plenum Press: New York, 1977. (d) Maskill, H. "The Physical Basis of Organic Chemistry." Oxford University Press: New York, 1985. (e) Morrison, R. T.; Boyd, R. N. "Organic Chemistry," 3rd ed. Allyn & Bacon: Boston, 1974. (f) Roberts, J. D.; Caserio, M. C. "Basis Principles of Organic Chemistry." Benjamin: New York, 1965. (g) Solomons, T. W. G. "Organic Chemistry." Wiley: New York, 1984. (h) Streitwieser, A., Jr.; Heathcock, C. H. "Introduction to Organic Chemistry." Macmillan: New York, 1976.

24. (a) Van Nostrand's Scientific Encyclopedia, 5th ed. Van Nostrand Reinhold: New York, 1976. (b) Kingzett's Chemical Encyclopedia; Hey, D. H., Ed.; London, 1966.
25. Le Noble, W. J. "Highlights of Organic Chemistry." Dekker: New York, 1974.
26. Newman, M. S. "Steric Effects in Organic Chemistry." Wiley: New York, 1956.
27. (a) Eliel, E. L.; Allinger, N. L.; Angyal, S. J.; Morrison, G. A. "Conformational Analysis." Wiley: New York, 1965. (b) Eliel, E. L. "Stereochemistry of Carbon Compounds." McGraw-Hill: New York, 1962.
28. (a) Dale, J. "Stereochemistry and Conformational Analysis." Universitetsforlaget: Oslo, 1975. (b) Mislow, K. "Introduction to Stereochemistry." Benjamin: New York, 1965.
29. (a) Buckert, U.; Allinger, N. L.; "Molecular Mechanics," ACS Monograph 177. American Chemical Society: Washington, DC, 1980. (b) Boyd, D. B.; Lipkowitz, K. B. *J. Chem. Educ.* **1982**, *59*, 269. (c) Cox, P. J. *J. Chem. Educ.* **1982**, *59*, 275.
30. Gol'dfarb, Ya. I.; Belen'kii, L. I. *Russ. Chem. Rev.* **1960**, *29*, 214.
31. Newton, M. D. In "Applications of Electronic Structure Theory," Vol. 4, Schaefer, H. F., III, Ed.; Plenum Press: New York, 1977. See also: Newton, M. D.; Switkes, E.; Lipscomb, W. N. *J. Chem. Phys.* **1970**, *53*, 2645.
32. Greenberg, A.; Liebman, J. F. "Strained Organic Molecules." Academic Press: New York, 1978. See also Reference 22.
33. Wiberg, K. B. *Angew. Chem. Int. Ed. Engl.* **1986**, *25*, 312.
34. (a) Rüchardt, C.; Beckhaus, H.-D. *Angew. Chem. Int. Ed. Engl.* **1985**, *24*, 529. (b) Rüchardt, C.; Beckhaus, H.-D. *Top. Curr. Chem.* **1985**, *130*, 1.
35. (a) Goldish, E. *J. Chem. Educ.* **1959**, *36*, 408. (b) Almenningen, A.; Bastiansen, O.; Skancke, P. N. *Acta Chem. Scand.* **1961**, *15*, 711.
36. Chang, S.-J.; McNally, D.; Shary-Tehrany, S.; Hickey, M. J.; Boyd, R. H. *J. Am. Chem. Soc.* **1970**, *92*, 3109.
37. Allinger, N. L.; Tribble, M. T.; Miller, M. A.; Wertz, D. H. *J. Am. Chem. Soc.* **1971**, *93*, 1637. See also discussion in Reference 29.
38. Schleyer, P. v. R.; Williams, J. E.; Blanchard, K. R. *J. Am. Chem. Soc.* **1970**, *92*, 2377.
39. Engler, E. M.; Andose, J. D.; Schleyer, P. v. R. *J. Am. Chem. Soc.* **1973**, *95*, 8005.
40. Ermer, O. *Struct. Bond.* **1976**, *27*, 161.
41. (a) See, eg, Pauling, L. "The Nature of the Chemical Bond," 3rd ed. Cornell University Press: Ithaca, NY, 1960. (b) "Lange's Handbook of Chemistry", Dean, J. A., Ed.; McGraw-Hill: New York, 1979.
42. Franklin, J. L. *Ind. Engl. Chem.* **1949**, *41*, 1070.
43. Benson, S. W.; Buss, J. H. *J. Chem. Phys.* **1958**, *29*, 546.
44. (a) Benson, S. W.; Cruickshank, F. R.; Golden, D. M.; Hangen, G. R.; O'Neal, H. E.; Rodgers, A. S.; Shaw, R.; Wash, R. *Chem Rev.* **1969**, *69*, 279. (b) Benson, S. W. "Thermochemical Kinetics." Wiley: London, 1976.
45. (a) Van Vechten, D.; Liebman, J. F. *Isr. J. Chem.* **1981**, *21*, 105. (b) Liebman, J. F.; Van Vechten, D. In "Molecular Structure and Energetics," Vol 2; Liebman, J. F.; and Greenberg, A., Eds.; VCH Publishers: Deerfield Beach, FL 1987, p. 315.
46. Wiberg, K. B.; Ellison, G. B.; Wendoloski, J. J. *J. Am. Chem. Soc.* **1976**, *98*, 1212.
47. Schleyer, P. v. R. Proceedings of the Eleventh Austin Symposium on Molecular Structure, Austin, 1986, and personal communication.
48. (a) George, P.; Trachtman, M.; Bock, C. W.; Brett, A. M. *Tetrahedron* **1976**, *32*, 317. (b) George, P.; Trachtman, M.; Bock, C. W.; Brett, A. M. *Theor. Chim. Acta* **1975**, *38*, 121.
49. Cox, J. D.; Pilcher, G. "Thermochemistry of Organic and Organometallic Compounds." Academic Press: New York, 1970. These authors introduced the operational term "conventional strain energy" to stress that the SE is derived either from energies or (more often) from enthalpies at T K.
50. Sanderson, R. T. In "Chemical Bonds and Bond Energy," Loebl, E. M., Ed.; Academic Press: New York, 1976.
51. Leroy, G. *Adv. Quantum Chem.* **1985**, *17*, 1.
52. Hay, J. M. *J. Chem. Soc. B* **1970**, 45.

53. Randić, M.; Maksić, Z. *Theor. Chim. Acta* **1965**, *3*, 59.
54. Ehrhardt, C.; Ahlrichs, R. *Theor. Chim. Acta* **1985**, *68*, 231.
55. Zhixing, C. *Theor. Chim. Acta* **1985**, *68*, 365.
56. Bader, R. F. W.; Tang, T.; Tal, Y.; Biegler-König, F. W. *J. Am. Chem. Soc.* **1982**, *104*, 946.
57. (a) Bader, R. F. W.; Nguyen-Dang, T. T. *Adv. Quantum Chem.* **1981**, 63, and references cited therein. (b) Bader, R. F. W.; Nguyen-Dang, T. T.; Tal, Y. *Rep. Prog. Phys.* **1981**, *44*, 893, and references cited therein. (c) Bader, R. F. W. *Acc. Chem. Res.* **1985**, *9*, 18. Bader has shown that the virial theorem applies to properly defined atomic subspaces. Therefore, the partitioning of the molecular space into atomic subspaces is called the virial partitioning method.
58. (a) Foote, C. S. *J. Am. Chem. Soc.* **1964**, *86*, 1853. (b) Schleyer, P. v. R. *J. Am. Chem. Soc.* **1964**, *86*, 1856.
59. (a) Gleiter, R. *Top. Curr. Chem.* **1979**, *86*, 197. (b) Basch, H.; Robin, M. B.; Kübler, N. A. *J. Chem. Phys.* **1969**, *51*, 52.
60. (a) Duddeck, H. In "Topics in Stereochemistry," Vol. 16; Eliel, E. L.; Wilen, S. H.; and Allinger, N. L., Eds.; 1986, 219. Wiley-Interscience, New York. (b) Günther, H. "NMR-Spektroskopie". Thieme: Stuttgart, 1973.
61. Wiberg, K. B.; Krass, S. R. *J. Am. Chem. Soc.* **1985**, *107*, 988.
62. (a) Wiberg, K. B.; Burgmaier, G. J. *Tetrahedron Lett.* **1969**, 317; *J. Am. Chem. Soc.* **1972**, *94*, 7396. (b) Gerson, F.; Müllen, K.; Vogel, E. *J. Am. Chem. Soc.* **1972**, *94*, 2924. (c) Lemal, D. M.; Dunlap, L. H., Jr. *J. Am. Chem. Soc.* **1972**, *94*, 6562. (d) Greenberg, A.; Liebman, J. F.; Van Vechten, D. *Tetrahedron* **1980**, *36*, 1161.
63. We note that Dale[28a] has defined strain entropies for cycloalkanes.
64. Klumpp, G. W. "Reactivity in Organic Chemistry." Wiley: New York, 1982.
65. (a) Stirling, C. J. M. *Pure Appl. Chem.* **1984**, *12*, 1781. (b) Stirling, C. J. M. *Tetrahedron*, **1985**, *41*, 1613. (c) Stirling, C. J. M. *Isr. J. Chem.* **1981**, *21*, 111. (d) Ferguson, L. N. *J. Chem. Educ.* **1970**, *47*, 46.
66. Czarnik, A. In "Mechanistic Principles of Enzyme Chemistry," Liebman, J. F.; and Greenberg, A., Eds. VCH Publishers: New York, NY, 1988, pp. 75–117.
67. Cremer, D. *J. Comp. Chem.* **1982**, *3*, 165.
68. (a) Snyder, L. C.; Basch, J. *J. Am. Chem. Soc.* **1969**, *91*, 2189. (b) Radom, L.; Hehre, W. J.; Pople, J. A. *J. Am. Chem. Soc.* **1971**, *93*, 289.
69. Actually, HSE and DSE will deviate slightly at the HF level of theory due to somewhat different basis set errors of propane, ethane and cyclohexane with a given finite basis set.[70]
70. Cremer, D.; Wallasch, M. unpublished results.
71. STO-3G: Hehre, W. J.; Stewart, R. F.; Pople, J. A. *J. Chem. Phys.* **1969**, *51*, 2657. 4-31G: Ditchfield, R.; Hehre, W. J.; Pople, J. A. *J. Chem. Phys.* **1971**, *54*, 724. 6-31G(d,p): Hariharan, P. C.; Pople, J. A. *Theor. Chim. Acta* **1973**, *28*, 213.
72. Lathan, W. A.; Radom, L.; Hariharan, P. C.; Hehre, W. J.; Pople, J. A. *Top. Curr. Chem.* **1973**, *40*, 15.
73. Müller, N.; Pritchard, D. E. *J. Chem. Phys.* **1959**, *31*, 1471.
74. (a) Wardeiner, J.; Lüttke, W.; Bergholz, R.; Machinek, R. *Angew. Chem. Int. Ed. Engl.* **1982**, *21*, 872. (b) For a discussion of hybrid orbitals see Bingel, W. A.; Lüttke, W. *Angew. Chem. Int. Ed. Engl.* **1981**, *20*, 899.
75. Coulson, C. A.; Moffitt, W. E. *Phil. Mag.* **1949**, *40*, 1.
76. Förster, T. *Z. Phys. Chem. B* **1939**, *43*, 58.
77. Walsh, A. D. *Trans. Faraday Soc.* **1949**, *45*, 179.
78. (a) Walsh, A. D. *Trans. Faraday Soc.* **1949**, *45*, 179. (b) Hoffmann, R. *Tetrahedron Lett.* **1970**, 2907. (c) Hoffmann, R.; Davidson, R. B. *J. Am. Chem. Soc.* **1971**, *93*, 5699. (d) Günther, H. *Tetrahedron Lett.* **1970**, 5173. (e) Jorgensen, W. L.; Salem, L. "The Organic Chemist's Book of Orbitals." Academic Press: New York, 1973.
79. (a) Ferguson, L. N. "Highlights of Alicyclic Chemistry," Part 1. Franklin: Palisades, NJ, 1973; Chapter 3. (b) Charton, M. In "The Chemistry of Alkenes," Vol. 2; Zabicky, J.,

Ed.; Wiley-Interscience: New York, 1970, p. 511. (c) Wendisch, D. In "Methoden der Organischen Chemie", Vol. 4; (Houben-Weyl) E. Müller; Thieme: Stuttgart, 1971; p. 3. (d) Lukina, M. Y. *Russ. Chem. Rev.* **1962**, *31*, 419.
80. (a) Honegger, E.; Heilbronner, E.; Schmelzer, A.; Jian-Qi, W. *Isr. J. Chem.* **1982**, *22*, 3. (b) Honegger, E.; Heilbronner, E.; Schmelzer, A. *Nouv. J. Chim.* **1982**, *6*, 519.
81. (a) Becker, P. "Electron and Magnetization Densities in Molecules and Crystals," NATO Advanced Study Institutes Series B: Physics, Vol. 48. Plenum Press: New York, 1980. (b) Coppens, P.; Hall, M. B. "Electron Distributions and the Chemical Bond." Plenum Press: New York, 1981.
82. Hohenberg, P.; Kohn, W. *Phys. Rev. B* **1964**, *136*, 864.
83. Coppens, P.; Stevens, E. D. *Adv. Quantum Chem.* **1977**, *10*, 1.
84. (a) Hartman, A.; Hirshfeld, F. L. *Acta Crystallogr.* **1966**, *20*, 80. (b) Fritchie, C. J., Jr. *Acta Crystallogr.* **1966**, *20*, 27. (c) Ito, T.; Sakurai, T. *Acta Crystallogr.* **1973**, *B29*, 1594. (d) Matthews, D.A.; Stucky, G. D. *J. Am. Chem. Soc.* **1971**, *93*, 5954.
85. (a) Dunitz, J. D.; Seiler, P. *J. Am. Chem. Soc.* **1983**, *105*, 7056. (b) Savariault, J. M.; Lehman, M. S. *J. Am. Chem. Soc.* **1980**, *102*, 1298.
86. (a) Schwarz, W. H. E.; Valtazanos, P.; Ruedenberg, K. *Theor. Chim. Acta* **1985**, *68*, 471. (b) Schwarz, W. H. E.; Mensching, L.; Valtazanos, P.; von Niessen, W. *Int. J. Quantum Chem.* **1986**, *29*, 909.
87. (a) Bader, R. F. W.; Slee, T. S.; Cremer, D.; Kraka, E. *J. Am. Chem. Soc.* **1983**, *105*, 5061. (b) Cremer, D.; Kraka, E.; Slee, T. S.; Bader, R. F. W.; Lau, C. D. H.; Nguyen-Dang, T. T.; Mac Dougall, P. J. *J. Am. Chem. Soc.* **1983**, *105*, 5069.
88. Cremer, D.; Kraka, E. *Croat. Chem. Acta* **1984**, *57*, 1265.
89. Cremer, D.; Kraka, E. *Angew. Chem. Int. Ed. Engl.* **1984**, *23*, 627.
90. See, e.g., Charton, M. In "The Chemistry of Alkenes" Vol. 2; Zabicky, J., Ed.; Wiley-Interscience: New York, 1970. Chapter 2.
91. This becomes obvious when considering the second derivatives of the scalar function f in the one-dimensional case:

$$\lim_{\Delta x \to 0} \{f(x) - \tfrac{1}{2}[f(x - \Delta x) + f(x + \Delta x)]\} = -\tfrac{1}{2} \lim_{\Delta x \to 0} \{[f(x + \Delta x)$$

$$-f(x) - [f(x) - f(x - \Delta x)]\} = -\frac{1}{2} \frac{d^2 f}{(dx)^2} (dx)^2$$

If the second derivative, hence, the curvature of f, is negative at x, then f at x is larger than the average of f at all neighboring points. See Morse, P. M.; Feshbach, H. "Methods of Theoretical Physics," Vol. 1; McGraw Hill: New York, 1953, p. 6.
92. For example, this is used in derivative spectroscopy to detect the maxima of overlapping bands. See, e.g., Drago, R. S. "Physical Methods in Chemistry." Saunders: London, 1977, p. 323.
93. (a) Bader, R. F. W.; Essén, H. *J. Chem. Phys.* **1984**, *80*, 1943. (b) Bader, R. F. W.; Mac Dougall, P. J.; Lau, C. D. H. *J. Am. Chem. Soc.* **1984**, *106*, 1594.
94. Tal, Y.; Bader, R. F. W.; Erkku, *J. Phys. Rev. A* **1980**, *21*, 1.
95. Politzer, P.; Zilles, B. A. *Croat. Chem. Acta* **1984**, *57*, 1055.
96. Randić, M.; Borćić, S. *J. Chem. Soc. A* **1967**, 586.
97. (a) Maksić, Z. B.; Randić, M. *J. Am. Chem. Soc.* **1973**, *95*, 6522. (b) Maksić, Z. B.; Kovačević, K.; Eckert-Maksić, M. *Tetrahedron Lett.* **1975**, 101.
98. Pauling, L. *J. Am. Chem. Soc.* **1931**, *53*, 1367.
99. Kilpatrick, J. E.; Spitzer, R. *J. Chem. Phys.* **1946**, *14*, 46.
100. (a) McMillen, D. F.; Golden, D. M. *Ann. Rev. Phys. Chem.* **1982**, *33*, 493. (b) Baghal-Vayjooee, M. H.; Benson, S. W. *J. Am. Chem. Soc.* **1979**, *101*, 2840. (c) Tsang, W. *J. Am. Chem. Soc.* **1985**, *107*, 2872.
101. Ingold, K. U.; Maillard, B.; Walton, J. C. *J. Chem. Soc. Perkin Trans. 2* **1981**, 970.

102. Shimanouchi, T. "Tables of Molecular Vibrational Frequencies." National Bureau of Standards: Washington, DC, 1972.
103. McKean, D. C. *Chem. Soc. Rev.* **1978**, *7*, 399.
104. Ferguson, K. C.; Whittle, E. *Trans. Faraday Soc.* **1971**, *67*, 2618.
105. See, e.g., Reference 23f, p. 113.
106. Zhixing[55] correlated MNDO in situ bond energies with McKean's CH stretching frequencies and obtained a poor correlation (correlation coefficient: 0.93, root mean square deviation: 1.4 kcal/mol).
107. (a) Snyder, R. G.; Schachtschneider, J. M. *Spectrochim. Acta* **1965**, *21*, 169. (b) Snyder, R. G.; Zerbi, G. *Spectrochim. Acta A* **1967**, *23*, 39.
108. Bauld, N. L.; Cesak, J.; Holloway, R. L. *J. Am. Chem. Soc.* **1977**, *99*, 8140.
109. See, e.g., (a) Garrat, P. J., "Aromaticity." McGraw-Hill: London, 1971. (b) Stevenson, G. R. In "Molecular Structure and Energetics," Vol. 3; Liebman, J. F.; and Greenberg, A., Eds.; VCH Publishers: Deerfield Beach, FL, 1986, p. 57.
110. (a) Bergman, E. D.; Pullman, B., Eds. "Aromaticity, Pseudo-Aromaticity, Antiaromaticity," Proceedings of an International Symposium, Jerusalem, 1971. (b) Dewar, M. J. S. "The Molecular Orbital Theory of Organic Chemistry." McGraw-Hill: New York, 1969. (c) Lloyd, D. "Non-Benzenoid Conjugated Carbocyclic Compounds." Elsevier: Amsterdam, 1984.
111. For a reconsideration of bond equalization of aromatic compounds, see (a) Shaik, S. S.; Hiberty, P. C. *J. Am. Chem. Soc.* **1985**, *107*, 3089. (b) Shaik, S. S.; Hiberty, P. C.; Lefour, J. M.; Ohanessian, G. *J. Am. Chem. Soc.* **1987**, *109*, 363. These authors point out that bond equalization may be enforced by the σ electrons.
112. Emsley, J. W.; Feeney, J.; Sutcliffe, L. H. "High Resolution Nuclear Magnetic Resonance Spectroscopy." Pergamon Press: Oxford, 1966, p. 690.
113. Zilm, K. W.; Beeler, A. J.; Grant, D. M.; Michl, J.; Chou, T.; Allred, E. L. *J. Am. Chem. Soc.* **1981**, *103*, 2119.
114. (a) Berson, J. A.; Pedersen, L. D.; Carpenter, B. K. *J. Am. Chem. Soc.* **1976**, *98*, 122. (b) See Reference 44b.
115. Dill, J. D.; Greenberg, A.; Liebman, J. F. *J. Am. Chem. Soc.* **1979**, *101*, 6814.
116. Clark, T.; Spitznagel, G. W.; Klose, R.; Schleyer, P. v. R. *J. Am. Chem. Soc.* **1984**, *106*, 4412.
117. Greenberg, A.; Stevenson, T. A. In "Molecular Structure and Energetics," Vol. 3, Liebman, J. F.; and Greenberg, A., Eds.; VCH Publishers: Deerfield Beach, FL, 1986, p. 193.
118. In this connection, it is interesting to note that the tetrahedral geometry of methane seems to be a result of the nuclear repulsion of the H atoms; see Reference 46.
119. (a) Dewar, M. J. S. *Nature* **1945**, *156*, 748. (b) Dewar, M. J. S. *J. Chem. Soc.* **1946**, *406*, 777.
120. Walsh, A. D. *Nature* **1947**, *159*, 165; 712.
121. (a) Dewar, M. J. S. *Bull. Soc. Chim. Fr.* **1951**, C71. (b) Dewar, M. J. S.; Marchand, A. P. *Annu. Rev. Phys. Chem.* **1965**, *16*, 321.
122. Dewar, M. J. S.; Ford, G. P. *J. Am. Chem. Soc.* **1979**, *101*, 183.
123. We facilitate the discussion by assuming that A_2X forms a stable molecule with covalent bonds. Accordingly, all maximum electron density (MED) paths correspond to bond paths (see Table 3-6 and Reference 88).
124. Koch, W.; Frenking, G.; Gauss, J.; Cremer, D.; Sawaryn, A.; Schleyer, P. v. R. *J. Am. Chem. Soc.* **1986**, *108*, 5732.
125. Bending force constants of cyclopropane, oxirane, and protonated oxirane have been calculated at the HF/6-31G(d) level of theory (Gauss, J.; Cremer, D. unpublished results) and then been calibrated for bending in the absence of 1,3-repulsion by utilizing $k_\beta(CCC) = 0.583$ mdyn-Å/rad^2 (Table 3-11 and Reference 6).
126. Bader, R. F. W.; Tal, Y.; Anderson, S. G.; Nguyen-Dang, T. T. *Isr. J. Chem.* **1980**, *19*, 8.
127. Note that each structure comprises an infinite number of different molecular geometries, one (or several) of them may correspond to (an) equilibrium geometry (geometries).
128. (a) Thom, R. "Structural Stability and Morphogenesis." Benjamin/Cummings: Reading,

MA, 1975. (b) Poston, T.; Stewart, I. "Catastrophe Theory and Its Applications." Pitman: Boston, 1981.
129. (a) Schleyer, P. v. R.; Sax, A. F.; Kalcher, J.; Janoschek, R. *Angew. Chem.* **1987,** *99,* 374, and unpublished work. (b) Sax, A. F. *Chem. Phys. Lett.* **1986,** *127,* 163; *129,* 66.
130. DeTar, D. F.; Luthra, N. *J. Am. Chem. Soc.* **1980,** *102,* 4505, and references cited therein.
131. See, e.g., (a) Dewar, M. J. S. *Angew. Chem.* **1971,** *83,* 859. (b) Zimmermann, H. E. *Acc. Chem. Res.* **1971,** *4,* 272.
132. Epiotis, N. D. "Unified Valence Bond Theory of Electronic Structure, Applications," Lecture Notes in Chemistry, Vol. 34. Springer Verlag: Berlin, 1983, p. 383.
133. Siddarth, P.; Gopinahtan, M. S. *J. Mol. Struct.* (THEOCHEM) **1986,** *148,* 101.
134. Liebman, J. F.; Dolbier, W. R., Jr.; Greenberg, A. *J. Phys. Chem.* **1986,** *90,* 394.
135. Morrison, J. A. In "Advances in Boron and the Boranes." Liebman, J. F.; Greenberg, A.; Williams, R. E., Eds.; VCH Publishers: New York, 1988, pp. 151–189.
136. (a) Casadei, M. A.; Galli, C.; Mandolini, L. *J. Am. Chem. Soc.* **1984,** *106,* 1051. (b) Cerchelli, G.; Galli, C.; Lillocci, C.; Luchetti, L. *J. Chem. Soc. Perkin Trans. 2* **1985,** 725. (c) Martino, A.; Galli, C.; Gargano, P.; Mandolini, L. *J. Chem. Soc. Perkin Trans. 2* **1985,** 1345. (d) Casadai, M. A.; Martino, A.; Galli, C.; Mandolini, L. *Gazz. Chim. Ital.* **1986,** *116,* 659.

CHAPTER 4

Twisted Bridgehead Bicyclic Lactams

Arthur Greenberg

Chemistry Division, New Jersey Institute of Technology, Newark, New Jersey

CONTENTS

1. Introduction... 139
2. Structure of the Amide or Lactam Linkage 142
3. Nonplanar Amides and Lactams.......................... 149
4. Syntheses of Bridgehead Lactams and Related Molecules.... 151
5. Spectroscopic Studies 156
6. Calorimetric and Crystallographic Data 168
7. Structure, Strain, and Reactivity: Conclusions 172
8. New Antibiotics?....................................... 172
9. Model Substrates for Proteases 175
References 175

1. INTRODUCTION

Among the bicyclic (l, m, n) bridgehead lactams (**BBL**), those containing a zero bridge ($m = 0$) have been extensively investigated because they include many unstrained molecules as well as penicillins, cephalosporins, and a variety of other bioactive compounds. These compounds appear to derive most of their activity from the strain in the β-lactam ring, although deviations from the "ideal" planar lactam geometry also play a role.[1] Bicyclo[l, m, n] lactams where $l \neq 0$, $m \neq 0$, $n \neq 0$, have received much less attention.

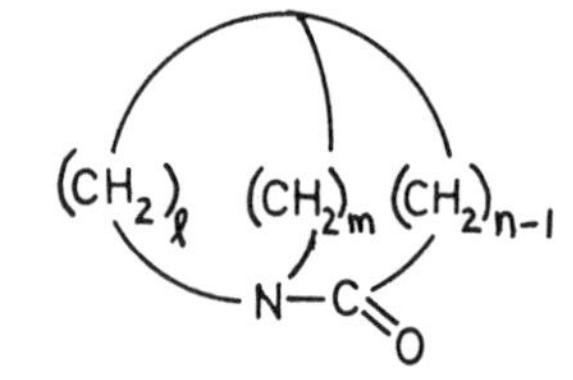

$$(CH_2)_l \ (CH_2)_m \ (CH_2)_{n-1}$$
$$N-C=O$$

BBL

This is surprising to us because they are synthetically quite accessible and their study can provide insights into a number of areas (see Figure 4-1). Nonplanar lactam linkages may be compared with twisted olefin linkages and used to study basic reactions of olefins, amides, and lactams. Studies of enthalpies of reaction and formation as well as structural studies could provide data useful in assessing feasibility of ring-opening polymerization as well as for parameterizing calculational studies based on both quantum mechanical and molecular mechanics methods. Such studies may be useful in understanding the structure and function of peptides, proteins, and enzymes. It is also possible that some of these compounds could be useful pharmaceutical agents.

In 1938 Lukes[2] recognized that Bredt's rule predicts that small bicyclic bridgehead lactams should be very destabilized, and attempts at synthesis appeared to "freeze" for 20 years. A 1957 report[3a] of the synthesis of 1-azabicyclo[2,2,2]octan-2-one (2-quinuclidone) (**1**) is now accepted with reservations, since characterization of the claimed lactam was incomplete; it should be noted, however, that this group has a considerable track record in this research area.[3b–d] However, the same reaction soon was employed successfully to synthesize derivatives **2–4**, which were isolated.[4–6] The parent compound, **1**, was characterized by infrared spectroscopy but could not be isolated in a pure state.[6] The tetramethyl derivative **5** was reported in 1970.[3b–d]

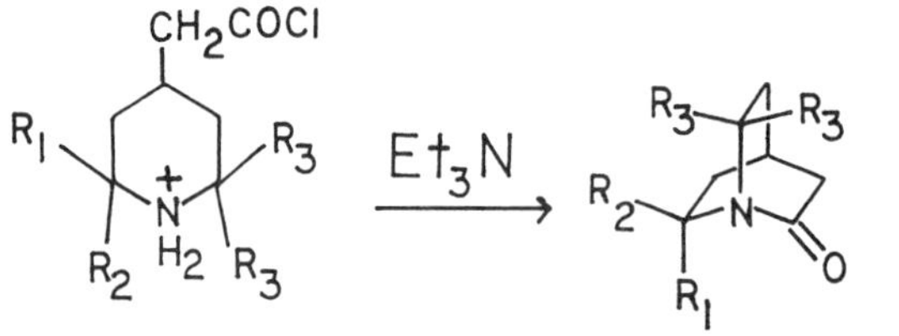

1 $R_1 = R_2 = R_3 = H$
2 $R_1 = R_2 = H$; $R_3 = CH_3$
3 $R_1 = H$; $R_2 = R_3 = CH_3$
4 $R_1 = R_3 = CH_3$; $R_2 = H$
5 $R_1 = R_2 = R_3 = CH_3$

It is interesting that these first bridgehead lactam linkages remain the most highly strained known. Their aminoketonelike properties and high sensitivity to moisture have made them great examples in organic chemistry texts of the results of loss of resonance. However, we wonder whether these extreme properties initially discouraged the large body of heterocycle chemists

Figure 4-1. Areas now or potentially related to twisted bridgehead lactam research.

from further exploring the group in a systematic manner; indeed, the literature citations for this class are few. A very fine and fairly recent review considered the field through about 1981–1982[7] with particular emphasis on polymerization. (Heats of polymerization can provide fairly decent estimates of the strain in the monomers if one makes some simplifying assumptions and, indeed, Hall and El-Shekeil note that the ease of polymerization is a sensitive diagnostic tool for strain in lactams.[7]) However, a recent surge in activity on bridgehead lactams as well as our own interests in a wide variety of properties of the linkage provide impetus for the present chapter. An earlier review article,[8] not so readily available, concentrated on the spectral properties, especially circular dichroism, of nonplanar amide linkages.

2. STRUCTURE OF THE AMIDE OR LACTAM LINKAGE

The statement published in 1960 that ". . . the structural information that has been obtained for [the amide group] is more extensive and more reliable than other group of comparable complexity"[9] undoubtedly still holds true today, in large part due to the work of Pauling, Corey, and collaborators, and subsequent work by many groups including those of Dunitz, Scheraga, and Ramachandran. A set of standard dimensions for *trans*-peptides[10] and a set of corresponding dimensions for *cis*-peptides[11] are depicted in Figures 4-2*a* and 4-2*b*.[12a,b] Figure 4-2*c* shows the average crystallographic structure of 16 noncyclic tertiary amides (sp^3 C attached to N).[12c] Figure 4-2*d* lists geometric parameters for the amide linkages in *cis*-lactams (unsubstituted on N); one sees a short carbonyl for propiolactam due to hybridization.[12d] The resonance stabilization in amides or lactams has been estimated at about 21 kcal/mol,[9] and this is comparable to the values of the rotational barriers[13,14] of simple acyclic amides such as *N*,*N*-dimethylformamide. The classic explanation is the presence of two principal resonance contributors (**A** and **B**) in the ground state planar structure with the loss of **B** in the rotational transition state. The transition state should, presumably, somewhat resemble 2-quinuclidone (**1**).

A   **B**

Although conventional views of resonance depict the ground state amide structure as ideally planar, it seems that this might not be true for even the simplest example, formamide. Thus, a high-level ab initio calculation (6-311-

A

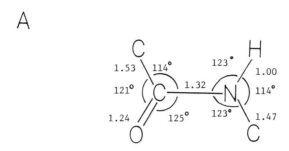

B

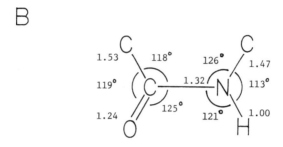

C

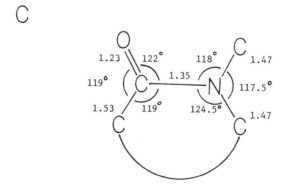

Figure 4-2. (a) Standard dimensions for *trans*-peptides.[10] (b) Standard dimensions for *cis*-peptides.[11] (c) Average crystallographic dimensions for 16 noncyclic tertiary amides.[12c] (d) Structures of various *cis*-lactams.[12d] Another comprehensive source on X-ray studies of amides, lactams, and peptides is: Vankatesan, K.; Ramakumar, S. "Structural Studies of Molecular Biological Interest", Dodson, G.; Glusker, J. P.; and Sayre, D., Eds.; Oxford University Press: New York, 1981, pp. 137–153.

144 ARTHUR GREENBERG

D

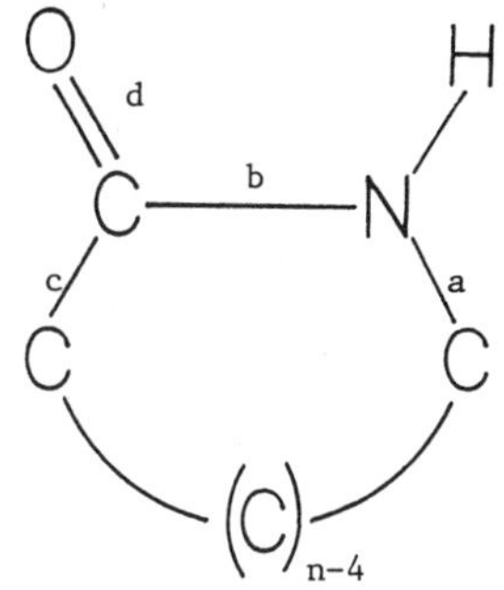

Compound	n	a	b	c	d	ab	bc	bd	cd
Enantholactam	8	1.462	1.334	1.510	1.246	127.8	119.0	120.0	121.0
Caprolactam	7	1.470	1.327	1.501	1.242	125.5	118.5	120.9	120.6
Valerolactam	6	1.462	1.333	1.512	1.243	126.4	117.9	122.0	120.1
"Various"	5	1.455	1.335	1.514	1.232	114.9	108.4	125.8	125.7
Propiolactam	4	1.467	1.333	1.522	1.226	96.2	91.7	132.4	135.9

Figure 4-2. Cont.

N**C**O**H**)[15] is in very good agreement with one microwave study[16] (Figure 4-3) that indicated nonplanarity mostly due to bending of the amino group. However, a more recent structural study of formamide[16b] indicated coplanarity at nitrogen. A very recent microwave study of 1-pyrrolidine carboxaldehyde indicated a highly pyramidalized nitrogen.[16c] The calculational study cited[15] very nicely explores the ability of different levels of ab initio calculations to provide amide structures. Thus, the 4-21G basis set, so reliable for bond length prediction (suitably scaled), predicts coplanarity, while small basis sets augmented by polarization functions on nitrogen alone greatly exaggerate pyramidality. Only when fairly sizable, flexible basis sets are augmented by polarization functions on all atoms of the linkage are reliable geometries obtained.[15]

The need to abandon the assumption of absolute planarity in peptide linkages was stressed by Ramachandran[17] among others. Subsequently, out-of-plane bending at the nitrogen and carbonyl carbons was carefully analyzed and these modes of distortion were quantitated.[18] Shortly thereafter, Winkler and Dunitz[12a,b] noted the redundancy in these distortion modes and defined the parameters, χ_C and χ_N (measures of the pyramidality at the carbonyl carbon and nitrogen, respectively) and τ, the amide twist angle (actually the torsional angle between the bisectors of χ_C and χ_N). These independent parameters uniquely describe the shape of any nonplanar amide group. Figure 4-4 illustrates the use of these parameters defined according to Winkler and Dunitz for a hypothetical amide constructed using the calculational parameters obtained for formamide by Boggs and Niu[15] (which clearly agree with the microwave data of formamide obtained by Costain and Dowl-

TWISTED BRIDGEHEAD BICYCLIC LACTAMS

Experimental value (theoretical value)

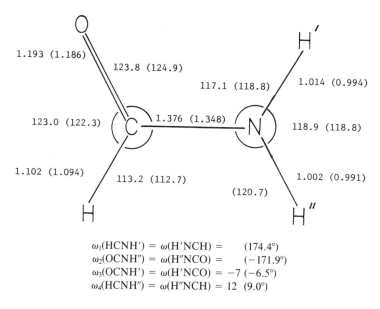

$\omega_1(HCNH') = \omega(H'NCH) = $ (174.4°)
$\omega_2(OCNH'') = \omega(H''NCO) = $ (−171.9°)
$\omega_3(OCNH') = \omega(H'NCO) = -7$ (−6.5°)
$\omega_4(HCNH'') = \omega(H''NCH) = 12$ (9.0°)

Figure 4-3. Results of an experimental microwave study of formamide[16a] and comparison with high-level ab initio calculations in parentheses.[15]

ing[16]). We have chosen to substitute a *trans*-amide for which τ should be near 180°; for a *cis*-amide, τ should be near 0°. It is clear that the distortion from planarity at N is significant, much larger than the pyramidality at C', and is the largest contributor to nonplanarity in this molecule. The twist angle ($\tau = +181.25°$) is not to be confused with conventional torsional angles. Winkler and Dunitz further showed that the use of a classical twofold torsional barrier (Equation 4-1) where K_ω is taken as the amide rotational barrier, estimated at 20 kcal/mol, and ω is the pure torsional angle about the C—N bond, incorrectly estimated the enthalpy difference between *cis*- and *trans*-caprylolactam. This is further evidence for the existence of other distortional modes. Dunitz and Winker[12b] provided an expression for the distortion of amides in terms of χ_N and τ' (neglecting χ_C, since it is small: Equation 4-2). The parameter $\tau' = \omega_1 + \omega_2 = 2\tau$ and is the spectroscopic equivalent used to analyze out-of-plane vibration of amides. This parameter, on the same scale as χ_C and χ_N, does not differentiate cis from trans amides. It is clear that τ' is analogous to the more classical 2ω (compare Equations 4-1 and 4-2).

$$V(\omega) = \tfrac{1}{2}[K_\omega(1 - \cos 2\omega)] \quad (4\text{-}1)$$

$$V(\tau', \chi_N) = \frac{V0}{2}(1 - \cos \tau') + p_N\chi_N^2 + q_N(1 - \cos \tau')[\exp(-a\chi_N^2) - 1] \quad (4\text{-}2)$$

The resulting τ', χ_N energy surface is shown in Figure 4-5.

Another interesting distortion from ideal geometry noted in lactams was the bond angle trend schematically depicted in Scheme I.[19] In small lactams such as β-lactams, ⟨cd − bd⟩ is greater than zero, whereas for larger lactams ⟨cd − bd⟩ is negative (see Scheme I). A more pronounced effect was observed in the lactones, and the general trend appears to be related to distortion, which relieves ring strain and tends to mimic the structure of an incipient acylium ion.

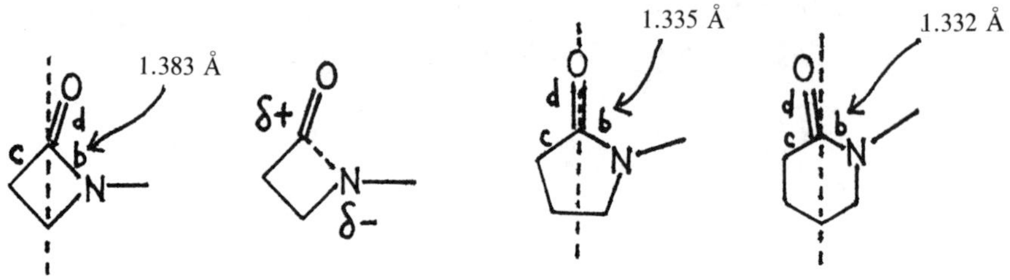

Scheme I

Another interesting distortion mode for amides has been suggested by Mock[20a] to account not only for facile cleavage of the amide linkage by proteolytic enzymes, but also for the stereospecificity of the enzyme attack. Thus, in the serine protease trypsin, syn distortion of a trans peptide induced by binding is thought to account for enzymatic attack on one face of the peptide (Scheme II). On the other hand, anti deformation may occur in the cleavage of a peptide linkage catalyzed by carboxypeptidase (Scheme III).

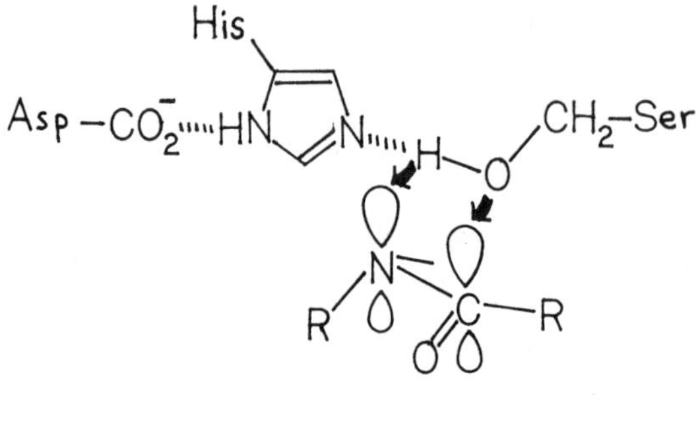

Scheme II

Detailed study[20b] of the out-of-plane distortions of about 300 amide bonds in the crystal structures of linear oligopeptides and aliphatic amides has provided interesting conclusions about coupled distortions. Pyramidality at the carbonyl group tends to increase in conformations where a C_α-substituted (allylic) bond can overlap significantly with the p-orbital on carbon and this pyramidalization tends to be antiperiplanar to the allylic bond. The

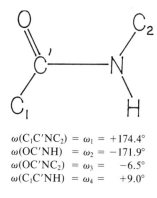

$\omega(C_1C'NC_2) = \omega_1 = +174.4°$
$\omega(OC'NH) = \omega_2 = -171.9°$
$\omega(OC'NC_2) = \omega_3 = -6.5°$
$\omega(C_1C'NH) = \omega_4 = +9.0°$

(a)

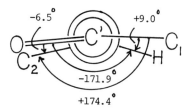

$\tau = \tfrac{1}{2}(\omega_1 + \omega_2)$

where $|\omega_1 - \omega_2| < \pi$ [must treat $-171.9°$ as $188.1°$]

$\chi_C = \omega_1 - \omega_3 + \pi = -\omega_2 + \omega_4 + \pi = +0.9°$ (mod 2π)
$\chi_N = \omega_2 - \omega_3 + \pi = -\omega_1 + \omega_4 + \pi = +14.6°$ (mod 2π)
$\tau = 181.25°$ and $\tau' = 2\tau = 2.50°$
$\omega_1 = \tau + \tfrac{1}{2}(\chi_C - \chi_N)$

(b)

Figure 4-4. Winkler–Dunitz parameters for defining nonplanar amide linkages.[12a,b] This hypothetical *trans*-amide, formed using the theoretical parameters calculated for formamide (Figure 4-3), is shown for illustrative purposes. The parameter τ' is a spectroscopic parameter related to $\tau(\tau' = 2\tau)$. One should recall that typical distortion equations are of the type:

$$V(\omega) = \tfrac{1}{2} k_\omega (1 - \cos 2\omega)$$

where ω = torsional angle

overall relationship between pyramidalization and rotation about the C_α—CO bond is reminiscent of the six-fold barrier for sp²-sp³ rotation. No such relation is found for corresponding pyramidalization at the amide nitrogen. Furthermore, highly substituted amide linkages tend toward syn distortion (Scheme II) because they are more polarized than less substituted link-

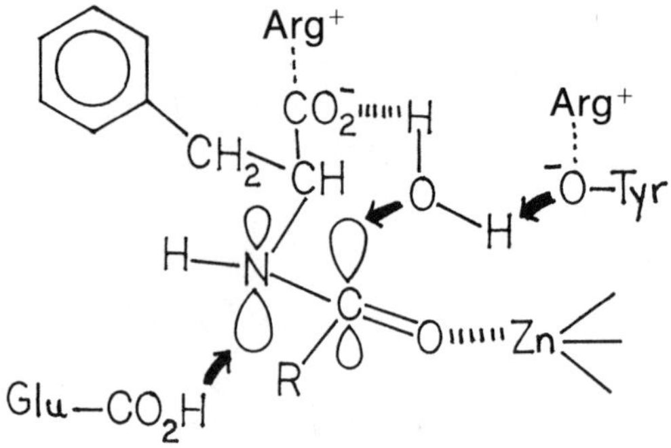

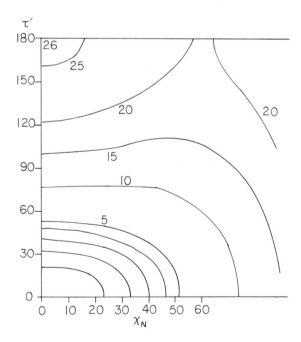

Scheme III

ages. Syn-distorted amide linkages tend to have shorter C—N bonds than corresponding anti linkages.

A very different picture of the amide linkage has recently been advanced by Wiberg and Laidig.[21] According to their calculational study, classical resonance need not be invoked to rationalize the preferred planarity of the amide linkage and the accompanying stabilization energy. They propose that the key factor may be the significant charge separation in the C—N bond in

Figure 4-5. Approximate energy for amide distortion as a function of χ_N and τ'; χ_C is assumed to be relatively small. (From Reference 12b.)

the planar structure due, in part, to a trigonal sp²-hybridized nitrogen, which is more electronegative than the pyramidal sp³ nitrogen in the rotated conformer. The extensive charge separation in the planar conformer provides a shorter, stronger, more polar covalent bond relative to the rotated conformers. This view is consistent with the prediction of near-equivalent carbonyl bond lengths in these conformers. Indeed, contrary to resonance predictions, nitrogen is calculated to be more *negative* in the planar conformer, while the charge on oxygen hardly changes. This iconoclastic view raises some interesting questions. For example, why does the amide nitrogen fail to adopt trigonal planar geometry in the rotational transition state?

3. NONPLANAR AMIDES AND LACTAMS

The amide or lactam group need not be confined to small bridgehead bicyclic systems in order to be twisted. Suitably congested acyclic molecules may show severe twisting. Thus, diamide 6 has markedly twisted and nonplanar linkages roughly corresponding to the Winkler–Dunitz parameters $|\chi_C| = 7.0°$, $|\chi_N| = 10.2°$, and $\tau = 11.6°$ of a model compound in the crystalline state.[21,22a]

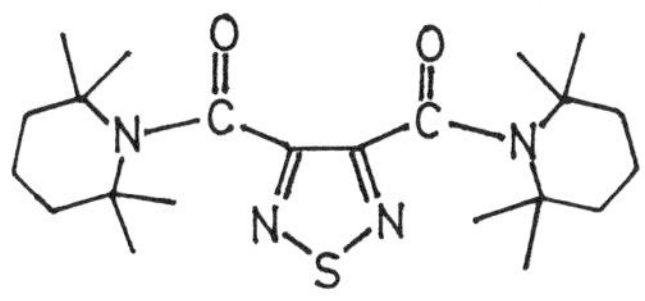

6

Small monocyclic lactams are exclusively cis, while the large rings revert to the trans structure characteristic of simple amides. Medium-ring lactams can adopt the trans structure, but this is accompanied by twisting of the lactam linkage. Caprylolactam is found as the trans isomer in the crystalline state, but the cis is less than 1 kcal/mol more stable in solution.[12a] In crystal-

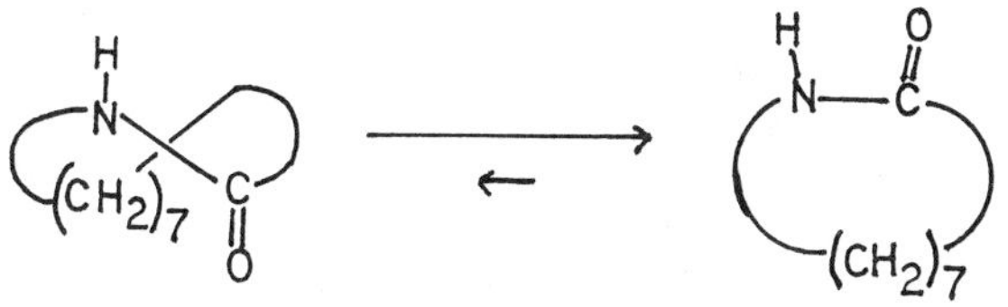

line *trans*-caprylolactam the value of $(C_2C'NC_8)$, the CO—N torsional angle, is 148.4°. The out-of-plane bending at C' ($\chi_C = -5.5°$) is again much smaller than that at N ($\chi_N = 23.1°$). The value of the twist angle is $\tau = 162.9°$.[12b] Winkler and Dunitz developed a force field to account reasonably for the enthalpy difference between *cis*- and *trans*-caprylolactam. Cyclodi-β-alanyl(1,5-diazacyclooctane-2,6-dione) is a *cis*-lactam, but to relieve Pitzer strain and minimize angle strain, it adopts the twist–boat conformation, **7**, which has considerable torsion in the amide linkage.[22b]

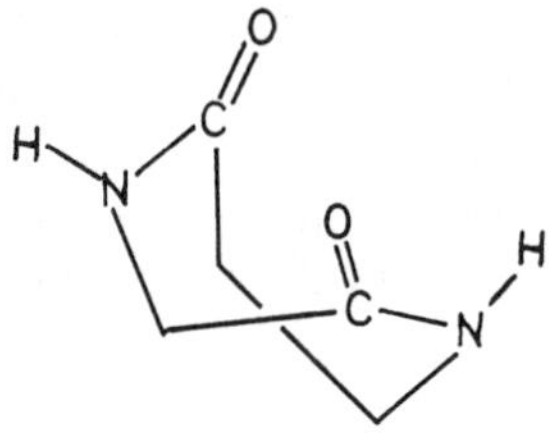

7

The inherent torsion in bicyclic lactams **8** and **9** twists their lactam linkages. Lactam **8** has pyramidal nitrogen ($\chi_N = 26.4°$) and considerable torsion ($\tau = -14.5°$).[8] A very interesting observation is that the circular dichroism spectra of these compounds (the twisted lactam linkage is an inherently dissymmetric chromophore) provide an indication of the dominant deformation mode. Thus, when the distortion is primarily torsional around CO—N, the $n - \pi^*$ and $\pi - \pi^*$ spectra show the same sign changes, while if N pyramidalization is the dominant deformation mode, the two bands show opposite signs. Compounds **8** and **9** as well as other lactams examined showed opposite signs for their respective $n - \pi^*$ and $\pi - \pi^*$ bands, showing the dominance of the N-pyramidalization mode over torsion in these compounds.[8] The carbonyl frequency of **8** (1710 cm^{-1}, CCl$_4$) is significantly higher than that of **10** (1698 cm^{-1}, CCl$_4$), again showing the distorted nature of the lactam ring. It is also noteworthy that while the $\nu_{C=O}$ is raised in the twisted lactam, the *cis*-amide II band corresponding to CO—N stretch is found at lower frequency (1405 cm^{-1} in **8**; 1425 cm^{-1} in **10**) due to reduced C—N bonding, and the N—H bond is also reduced in frequency (3422 cm^{-1} in **8**, 3440 cm^{-1} in **10**) due to hybridization changes.[23]

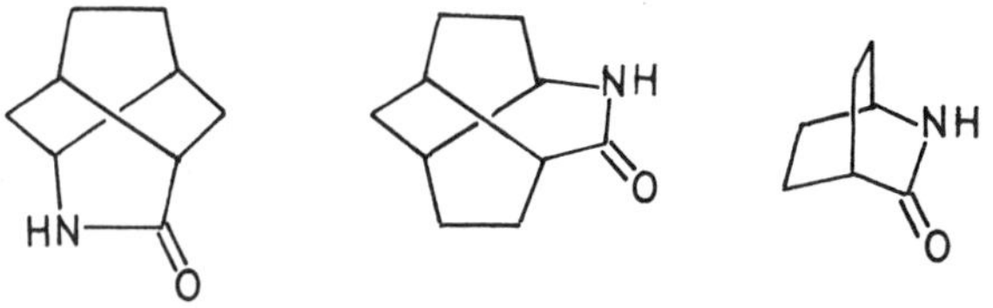

8   **9**   **10**

A related distortion technique involves fusing amide linkages into a spiro linkage as in 5,8-diaza-4,9-tricyclo [6.3.0.0^{1,8}]undecanedione (**11**).[24,25] The

11

$\nu_{C=O}$ in this compound, 1772 cm^{-1}, is significantly higher than that in N-methylpyrrolidone (1697 cm^{-1}). An X-ray structure indicated $\chi_C = -0.3°$, $\chi_N = -42.0°$, and $\tau = 21.3°$.[25]

4. SYNTHESES OF BRIDGEHEAD LACTAMS AND RELATED MOLECULES

Shortly after the Soviet report of 2-quinuclidone,[3a] complete characterization and isolation of four derivatives (**12–15**) of the 1-azabicyclo[3.2.1]octan-2-one series were documented.[26] The first of these, oxohemanthidine, was derived from the natural alkaloid hemanthidine via MnO$_2$ oxidation. The remaining compounds were derived from **12**. The unusual IR and UV properties were consistent with the bridgehead lactam structure. The parent compound, still unknown, is a more stable isomer of 2-quinuclidone and presumably has structure **16**, effectively a *trans*-caprolactam. As noted earlier, the 2-quinuclidones **1–5** began to appear in the literature in 1959.[3b–d,4–6]

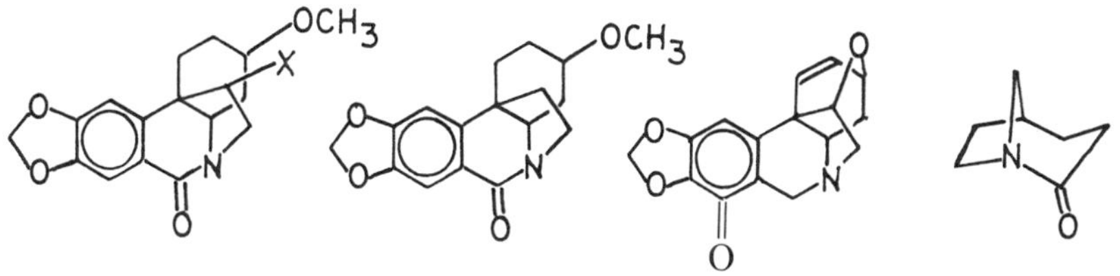

12, X = OH
13, X = OCOCH$_3$ **14**   **15**   **16**

The chemistry of 6,6,7,7-tetramethyl-2-quinuclidone (**5**) is instructive.[3b] As noted below, protic nucleophilic agents break the lactam linkage, while nucleophilic agents in aprotic media break the N—C(CH$_3$)$_2$ bond. On the other hand, quaternization reactions that leave the bicyclic ring structure intact occur with HCl and CH$_3$I.[3b] The importance of the parent ion in the mass spectrum of **5a** is explained in terms of the ring-opened radical cation.[3d]

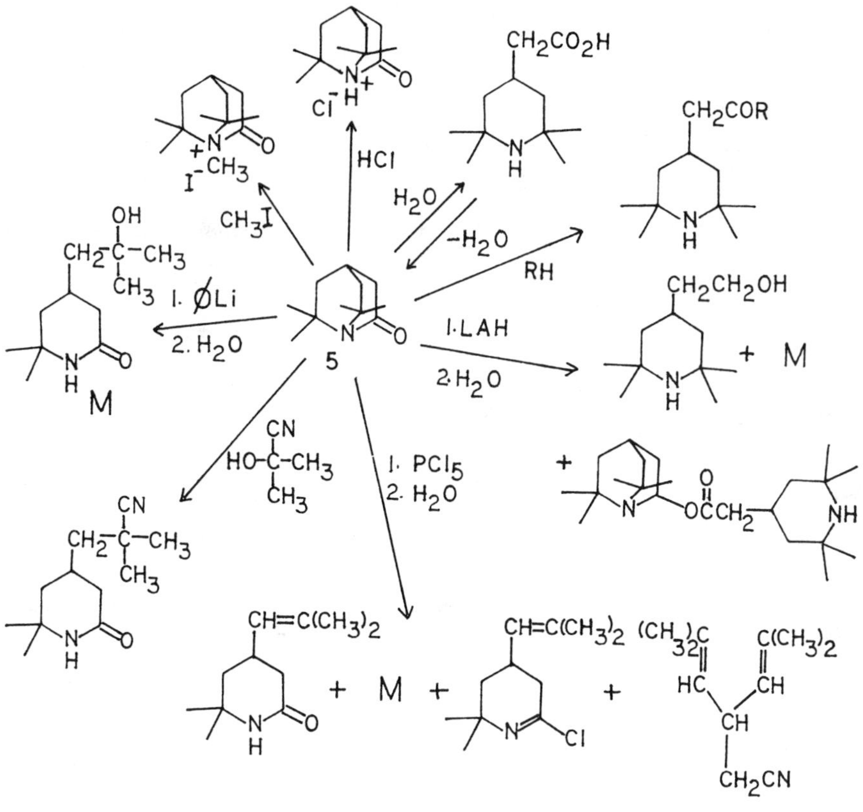

In 1969, new members of the family were reported. Use of the mixed anhydride synthesis provided compounds **17**, **18**, and **19**.[27]

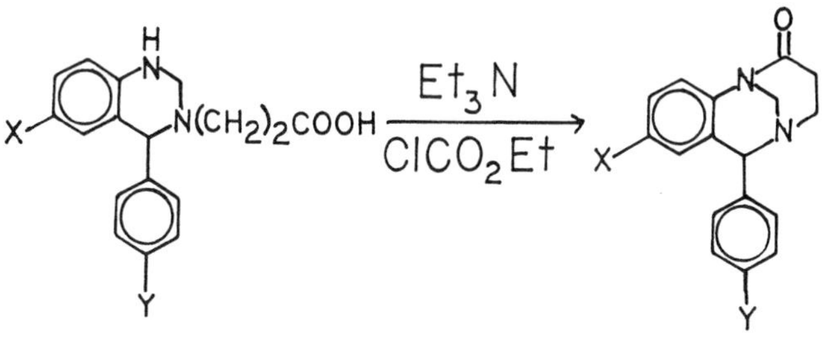

17 X = Cl, Y = H
18 X = H, Y = H
19 X = H, Y = Cl

The next entry into this system was the synthesis of the bridgehead bicyclic urea 3-isopropyl-1,3-diazabicyclo[3.3.1]nonan-2-one (**20**). This compound was quite stable to water, acid, and base.[28] The unsubstituted urea **21** was reported in 1980.[29]

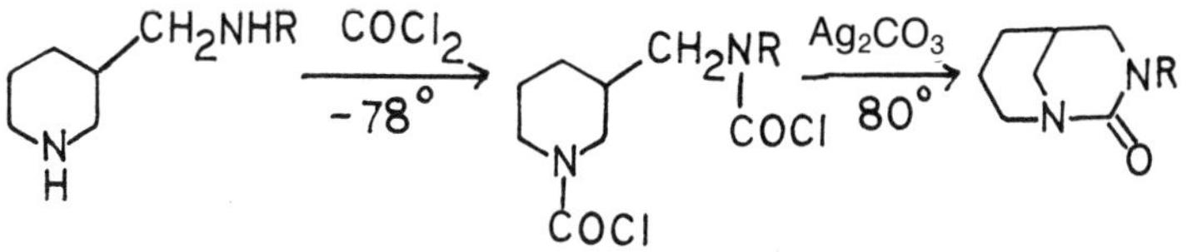

It is interesting that the synthesis of the related lactam, 1-azabicyclo[3.3.1]nonan-2-one (**22**), was not published until 1980.[30] While attempts at synthesis from the precursor amino acid through the acyl chloride–triethylamine route, successful for 2-quinuclidones, did not yield **22**, no attempt at the mixed anhydride route, feasible for **17–19**, was reported.[30] The synthesis of this compound was accomplished by heating the amino acid between 180 and 285°C at 0.05 torr and catching the lactam in a cold receiving flask.[30] The yield was about 7%. The conditions are fairly exact, since the lactam polymerizes readily. Subsequently, it was reported that di-*n*-butyltin oxide mediated ring closure of the amino acid afforded 77% yield,[31a] and we found that synthesis of **22** from the amino acid via the mixed anhydride route provides a comparably high yield of this compound.[31b] The dibutyltin (IV) oxide route was also used in the synthesis of optically-pure **22**.[31c] In contrast to the parent, Buchanan reported synthesis of 5-phenyl-1-azabicyclo[3.3.1]nonan-2-one (**22a**) from the amino acid via the acid chloride–triethylamine route (10% overall yield).[32a,b] Again, there was no evidence that the mixed anhydride route was tried. Spectroscopic evidence[32b] and subsequently X-ray crystallography[32c] established the conformation shown for **22a**, which has *trans*-lactam geometry for the eight-membered ring. Apparently, the phenyl group allows the acid chloride substituent to adopt the axial geometry that facilitates reaction.[32a–c]

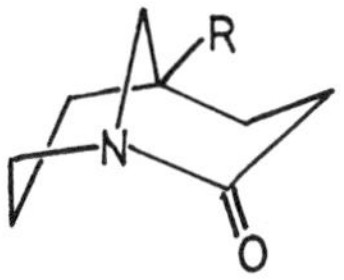

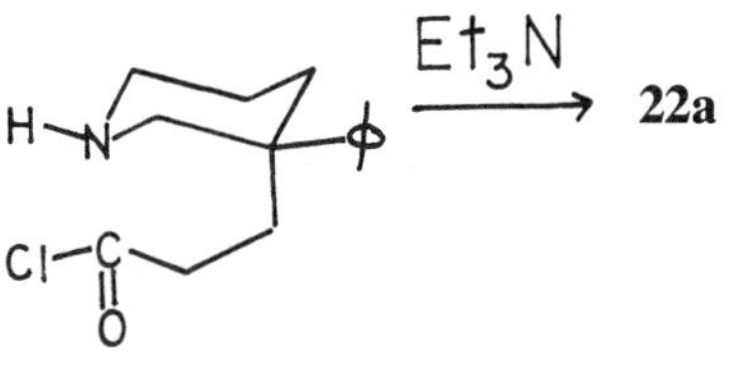

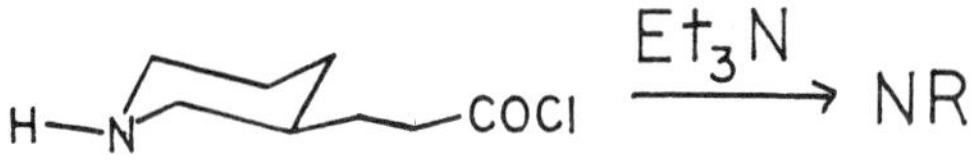

Recently, a new catalytic cyclization has been successfully employed to obtain benzo derivatives of **22** such as **23a–c**.[33] A related series of diones has also been reported.[34]

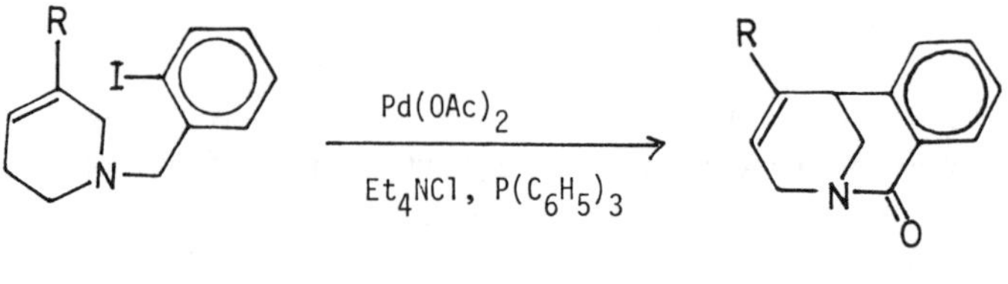

[Scheme: reactant with R, I, N → Pd(OAc)$_2$, Et$_4$NCl, P(C$_6$H$_5$)$_3$ → **23**]

23
a R = H
b R = C$_6$H$_5$
c R = CH$_3$

The same mixed anhydride route successful for **17-19** also yielded **24**, the first bridgehead γ-lactam.[27] An interesting and significant observation is

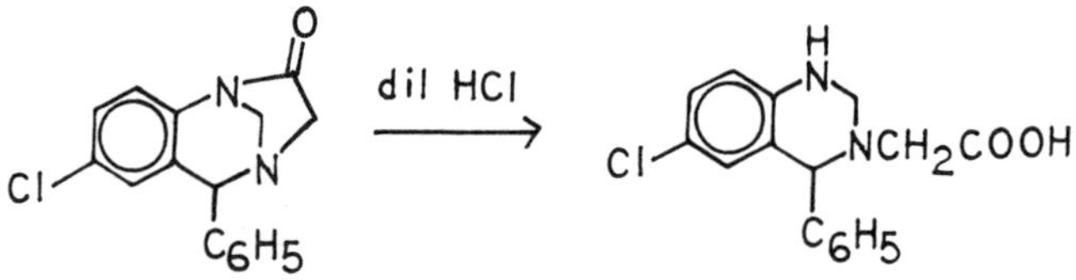

[Scheme: **24** (Cl-substituted bicyclic with C$_6$H$_5$) —dil HCl→ ring-opened product with NCH$_2$COOH]

24

that upon dissolution of **24**, rapid cleavage to the starting amino acid occurs. In contrast, **17** releases formaldehyde and is less reactive toward hydrolysis.

[Scheme: **17** —dil HCl→ benzodiazocine-type product]

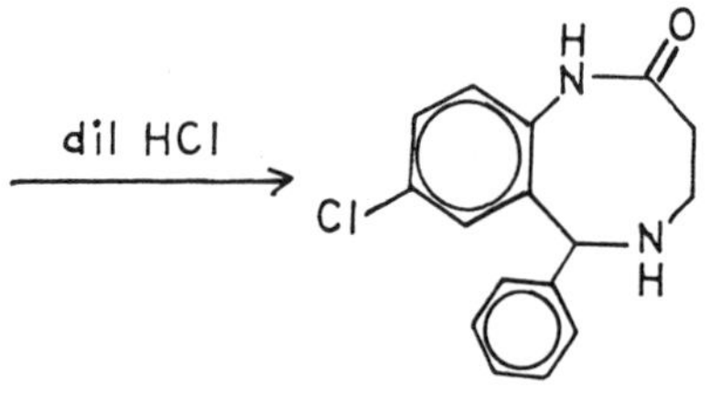

This is a clear indication of the strain difference between the two systems. The urethane, 1-aza-3-oxabicyclo[3.3.1]nonan-2-one (**25**), a relative of **22**, was obtained via the acid chloride–triethylamine closure shown below, and the more reactive smaller homologue **26** was obtained analogously.[35] The related bicyclic 2,4-oxazolidinediones **27** and **28** have also been reported.[36a] These authors failed to obtain the smaller 1-aza-6-oxabicyclo[3.2.1]octan-7,8-dione system.

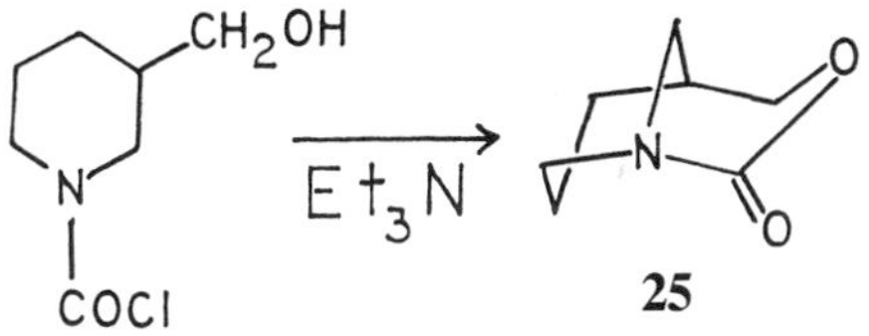

A related series of molecules, the 1,4-bridged pyrazolines (Scheme IV), has been reported to arise via ring opening of spiropyrazolium ylides.[36b] The infrared properties of this group of molecules reflect the decrease in lactam linkage torsion as the bridges are lengthened.

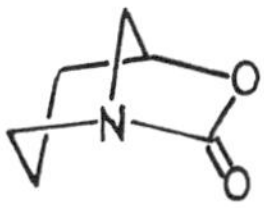

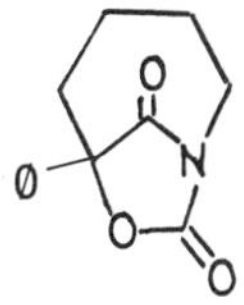

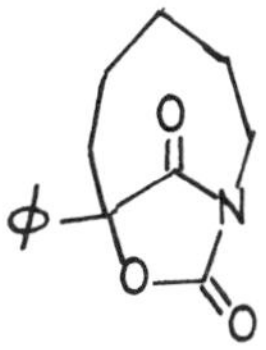

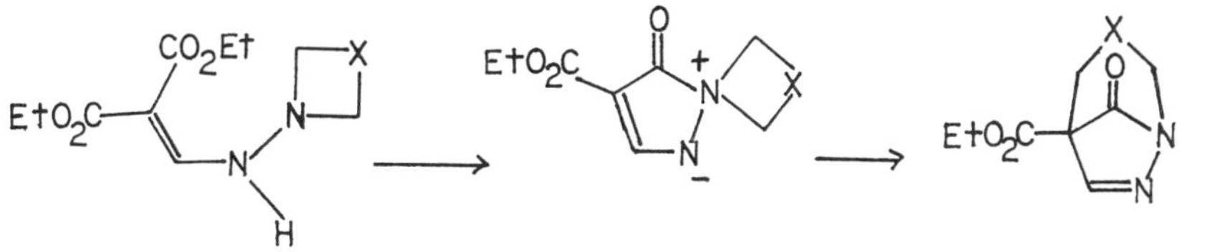

a $X = (CH_2)_2$
b $X = CH_2OCH_2$
c $X = (CH_2)_3$
d $X = (CH_2)_4$

Scheme IV

An X-ray structure of **27** indicated normal N—CO bond length between the 1- and 8-positions and a normal C=O bond length for the 8-carbonyl. The N—CO bond between the 1- and 9-positions is 0.038 Å longer and the 9-carbonyl C=O bond length is shorter by 0.014 Å, indicating reduced overlap.[36a]

A new lactam system, **29**, was obtained via dicyclohexylcarbodiimide (DCC) closure of the precursor amino acid at room temperature.[37a,b] The

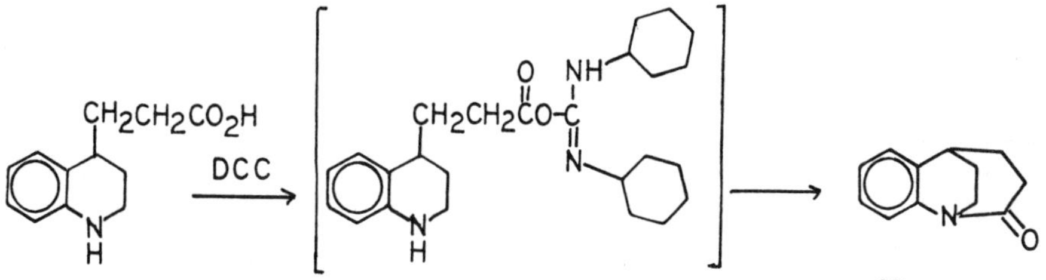

same route was successful in the synthesis of benzo-2-quinuclidones **30**–**33**.[37a] The unsubstituted compound **30** was previously made by the mixed anhydride techniques,[38] which Somayaji and Brown[37a] found less convenient than the DCC technique.

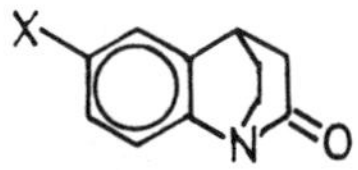

30	X = H
31	X = CH$_3$
32	X = OCH$_3$
33	X = Cl

In 1986[39] the first bridgehead β-lactam (**34**) was isolated and evidence provided for the detection of **35**, which rapidly polymerized.

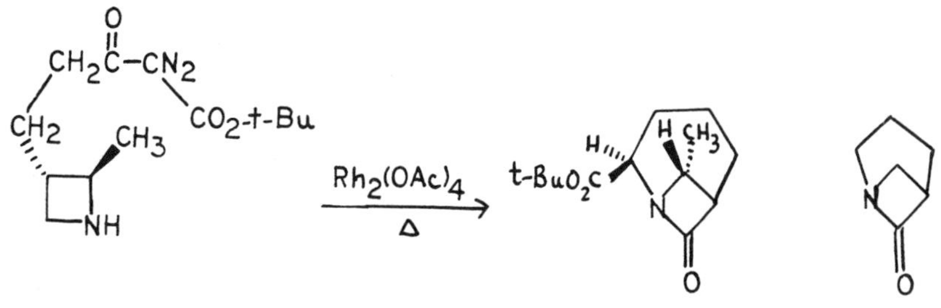

5. SPECTROSCOPIC STUDIES

A. Ultraviolet (UV) and Photoelectron Spectroscopy (PES)

The highest three occupied and the lowest vacant orbitals of formamide are sketched in Figure 4-6. These are termed π_N, n_O, π_{CO}, and π^*_{CO}, in accordance with the usage of Treschanke and Rademacher.[40–43] The lowest energy UV absorption of simple amides and lactams is the W band (λ_{max} = 210 –

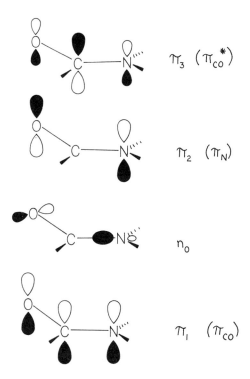

Figure 4-6. Highest four molecular orbitals for the amide linkage. (Adapted from Jorgensen W.; Salem, L. "The Organic Chemist's Book of Orbitals". Academic Press: New York, 1972.)

220 nm, $\varepsilon \simeq 100$) attributed to $n_O \rightarrow \pi^*_{CO}$, which is forbidden and therefore very weak. Frequently, it is obscured by the band $\pi_{CO} \rightarrow \pi^*_{CO}$ transition, ($\lambda_{max} = 170–188$ nm, $\varepsilon \simeq 8000$).[14] However, twisting shifts the W band bathochromically, absolutely and relative to the V band. Thus, in **8** a shoulder is seen at about 230 nm (cyclohexane),[8] while the values are still higher in **3** (246 nm), **4** (236 nm), and **5** (247 nm). Conjugation with a benzene ring in **30** raises the W band to 271 nm.[38] The UV spectra of **12** through **15** have three maxima, of which those in the 322–326 nm range may be the $n - \pi^*$ transition of this extended chromophore.[26] The band for the lowest energy orbital in typical amides or lactams is sharp because π_N is nonbonding. For tertiary lactams, vertical ionization potentials $[IP(\pi_N)_v$ and $IP(n_O)]$ decrease with increasing ring size (see Table 4-1), reaching limiting values of 8.70 and 9.10 eV, thus $\Delta IP_{1-2} = 0.40$ eV. The only bridgehead lactam of the type described here to be measured by PES is 1-azabicyclo[3.3.1]nonan-2-one.[41] The value for π_N was virtually equal to the normal limiting value, while that for n_O was found to be 0.36 eV higher, yielding a high ΔIP_{1-2} of 0.74 eV. The value of the $IP(\pi_{CO})$[41] also appeared to be somewhat high.

A more interesting quantitative comparison was made through correlation with infrared carbonyl frequencies. Using a relationship relating the ob-

TABLE 4-1. Experimental Ionization Potentials Corresponding to Highest Lying Molecular Orbitals for Selected Tertiary Lactams

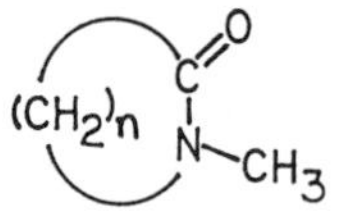	$IP(\pi_N)$	$IP_v(n_O)$	$IP_{ad}(n_O)$	$\Delta IP_{v(1-2)}$	IP_3
$n = 2$	9.29	9.80	9.40	0.51	12.50
$n = 3$	9.17	9.68	9.28	0.51	11.65
$n = 4$	8.92	9.36	9.00	0.44	11.66
$n = 5$	8.73	9.13	8.77	0.40	11.12
$n = 6$	8.76	9.16	8.84	0.40	11.05
$n = 7$	8.78	9.18	8.86	0.40	10.76
$n = 8$	8.72	9.12	8.80	0.40	10.63
$n = 9$	8.70	9.10	8.74	0.40	10.71
$n = 10$	8.74	9.10	8.78	0.36	10.41
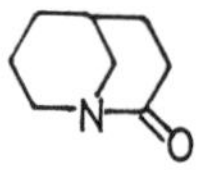	8.72	9.46	—	0.74	11.04
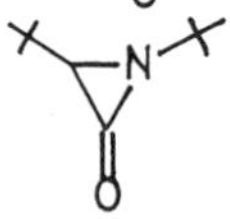	8.21	9.76	—	1.55	10.43

Source: Reference 41.

served carbonyl frequency ν_{CO} to the angle α of the carbonyl group $(CO(C)_2)$, Cook[44] calculated values for ν_{CO}^{120}, a hypothetical frequency for a given ketone if corrected to a carbonyl group angle of 120°:

$$\nu_{CO}^{120} = \frac{96(\nu_{CO} - 1439)}{216 - \alpha} + 1439 \text{ (cm}^{-1}) \qquad (4\text{-}3)$$

Treschanke and Rademacher then found a relationship between the calculated ν_{CO}^* (based on experimental ν_{CO} and α) from Equation 4-3 and the experimental $IP(n_O)$. This is represented by Equation 4-4 for 3° lactams (a similar equation was obtained for 2° lactams). The data for 3° lactams are plotted in Figure 4-7. It is quite clear that 1-azabicyclo[3.3.1]nonan-2-one lies well above the line indicating ν_{CO} higher than expected and, thus, reduced amide resonance. A similar observation is made for the α-lactam **36**,

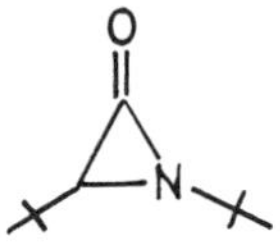

36

thus implying considerably reduced resonance. Modified neglect of differential overlap (MNDO) results indicated a nonplanar lactam linkage for the

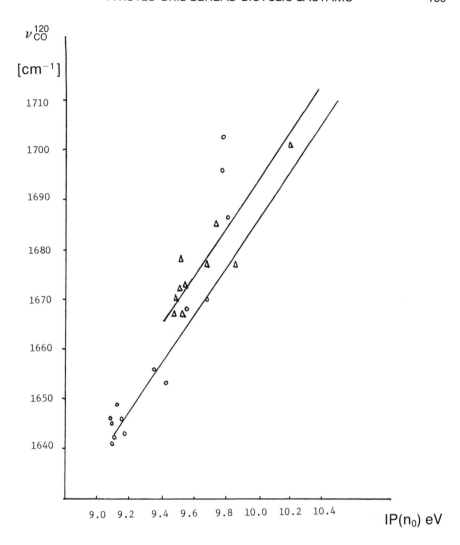

Figure 4-7. Treschanke–Rademacher relationship[41] between a corrected carbonyl frequency and ionization potential of nonbonding electrons in secondary (triangles) and tertiary (circles) lactams.

parent α-lactam.[40] One must note that the particularly low $IP(\pi_N)$ and the correspondingly high ΔIP_{1-2} in Table 4-1 are mostly a consequence of the nature of the t-butyl substituent on N, which strongly stabilizes radical cations. The first two bands of **22** and **36** are broader than the corresponding bands in other lactams. Deviation from the line for N-methyl β-lactam implies reduced resonance also, but β-lactam itself does not show this deviation. PES data could not indicate whether caprylolactam is cis, trans or a mixture in the gas phase.[41] Related studies of cyclic ureas also show how PES can successfully be employed to observe distorted linkages.[43] Trends in the PES spectra of substituted benzamides yield useful data on conjugation.[45]

$$\nu_{CO}^{120} = 50.00 \, IP(n_O) + 1187.2 \, (cm^{-1}) \qquad r^2 = 0.930 \qquad (4\text{-}4)$$

X-Ray photoelectron spectroscopy (ESCA) has not, to our knowledge, been used to probe nonplanar lactams. The sensitivity to differences in environment is clearly shown by the observation of three well-separated N 1s peaks in the compound below. However, considerable variation in the nature of R did not produce much change in $\Delta(N^+\text{—}NH)$.[46] A study relating

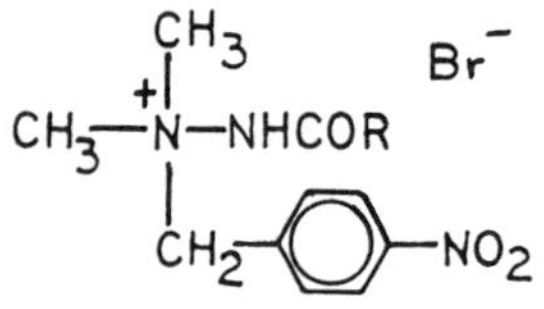

$$\begin{array}{c} CH_3 \\ | \\ CH_3\text{—}\overset{+}{N}\text{—}NHCOR \qquad Br^- \\ | \\ CH_2\text{—}\langle\bigcirc\rangle\text{—}NO_2 \end{array}$$

amide resonance interaction in N-ammonioimidates to ESCA data has also appeared.[47] An ESCA study[48] of simple amides and lactams found N_{1s} values ranging from 405.3 to 406.6 eV and O_{1s} values ranging from 536.4 to 538.0 eV. The amides and lactams did not fit into other correlations between $N_{1s}E_B$ and free energies of protonation because these compounds are O-protonated in the gas phase; however, they did show a relationship between $O_{1s}E_B$ and $\Delta G°$protonation.[48] ESCA studies of bridgehead lactams could be quite interesting in two regards. First, one would anticipate a crossover point at which lactams that are distorted enough become N-protonated and obey earlier correlations. Second, N_{1s} and O_{1s} shifts may be sensitive enough to act as distortion probes, and in this connection, ESCA may be useful for probing the novel charge predictions of Wiberg and Laidig.[21]

B. NMR Spectroscopy

Although proton NMR spectroscopy (^{1}H NMR) has been employed as a tool in assigning the boat–chair conformation of 1-azabicyclo[3.3.1]nonan-2-one, the potential use of ^{13}C, ^{15}N, and ^{17}O NMR approaches is still hardly tapped. Buchanan[32a] noted that the chemical shift of the carbonyl carbon in the 5-phenyl derivative **23** (184 ppm) is significantly higher than that of N-methylpiperidone (169 ppm). The expected shift for a cyclohexanone[49,50] is 208.8 ppm, and so it is clear that the lactam carbonyl in **23** has additional ketone character (Scheme V). The chemical shift in **29**[37] is 3.6 ppm higher than in **23**, but cycloheptanone has a shift 2.9 ppm higher than cyclohexanone. The chemical shifts in the benzoquinuclidones[37] **30** and **33** are higher still, yet more than 20 ppm lower than the ketone. Perhaps a value near 195 ppm is a

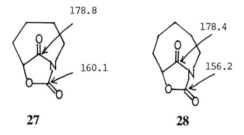

Scheme V

limiting value for an aminoketone. In addition, data have been obtained for the 2,4-oxazolidinediones **27** and **28** by ^{13}C NMR spectroscopy.

C. Infrared Spectroscopy

Infrared spectroscopy has been one of the most frequently employed means for discerning strain in bridgehead lactams, and it has been employed to discern "homolactam" resonance in compounds such as **37**.[51] Thus, 2-quinuclidones exhibit a ν_{CO} of 1750–1755 cm^{-1} (nonpolar solvent),[4–6] higher than the corresponding ketone by about 20 cm^{-1}, whereas N-methylpiperidone (1641 cm^{-1}, CCl$_4$) has a carbonyl frequency roughly 75 cm^{-1} lower than cyclohexanone (1716 cm^{-1}). Although it was said that ν_{CO} for 1-azabicyclo[3.3.1]nonanone (**22**) (1680 cm^{-1},[33] 1688 cm^{-1},[41] 1695 cm^{-1} for 5-phenyl derivative, **22**[32a])[b] is normal for tertiary amides, we will reexamine IR spectral trends for **22** and other known bridgehead lactams. Table 4-2 provides ν_{CO} IR data, in nonpolar solvents, for bridgehead lactams, monocyclic lac-

TABLE 4-2. Infrared Frequencies (in nonpolar solvents) for Carbonyl Groups of Selected Lactams and Their Corresponding Ketones, as Well as Difference in Frequencies

Lactam	ν_{CO} (cm^{-1})		
	Lactam	Ketone	$\Delta\nu_{CO}$
N-Methylvalerolactam	1641[a]	1716[b]	−75
N-Methylbutyrolactam	1697[a]	1748[b]	−51
N-Alkyl-β-lactam	1745–1760[c]	1791[b]	−31 to −46
1,3-Di-t-butyl-α-lactam	1835[d]	1822[e]	+13
1-Azabicyclo[3.3.1]nonan-2-one (**22**)	1688[f]	1720[g]	−32
1-Azabicyclo[3.2.2]octan-2-one (**29**)	1705[h]	1708[i]	−3
1-Azabicyclo[3.2.1]octan-2-one (**12**)	1695[j]	1717[b]	−22
1-Azabicyclo[2.2.2]octan-2-one (eg, **3**)	1750[k]	1731[b]	+19
1-Azabicyclo[3.2.1]octan-7-one (**24**)	1753[l]	1751[m]	+2
1-Azabicyclo[4.1.1]octan-7-one (**34**)	1795[n]	1783[o]	+12
1-Azabicyclo[3.1.1]heptan-6-one (**35**)	1795[n]	1783[p]	+12
(NR/O structure)	1830[q]	1782[r]	+48
N-methyl-2-pyridone	1681[s]	1666[t]	+15

[a] Smolikova, J.; Koblicova, Z.; Bláha, K. *Collect. Czech. Chem. Commun.*, **1973**, *38*, 532.
[b] Foote, C. S. *J. Am. Chem. Soc.* **1964**, *86*, 1853.
[c] Nakanishi, K.; Solomon, P. H. "Infrared Absorption Spectroscopy", 2nd ed, Holden-Day: San Francisco, 1977.
[d] Greene, F. D.; Stowell, J. C.; Bergmark, W. R. *J. Org. Chem.* **1969**, *34*, 2254.
[e] Greene, F. D.; Weinshenker, N. M. *J. Am. Chem. Soc.* **1968**, *90*, 506.
[f] This is the straight arithmetic average of the values reported in References 30 and 41, respectively, for the parent compound **22** (1680 and 1688 cm^{-1}) and for the 5-phenyl derivative **22a** (1695 cm^{-1}).[32a-c]
[g] Ferris, J. P.; Miller, N. C. *J. Am. Chem. Soc.* **1963**, *85*, 1325.
[h] See Reference 37a.
[i] The model ketone is $\Delta^{6,7}$-bicyclo[3.2.2]nonen-2-one [Berson, J. A.; Jones, M., Jr. *J. Am. Chem. Soc.* **1964**, *86*, 5019].
[j] See Reference 26.
[k] See Reference 6; note also that benzoquinuclidone (**30**) has ν_{CO} = 1755 cm^{-1} (Reference 38), and this difference mirrors that for the benzoketone (1735 cm^{-1}).
[l] The average value of ν_{CO} for **17–19** is 1715 cm^{-1} (CH$_2$Cl$_2$), while that for **24** is 1780 cm^{-1} (CH$_2$Cl$_2$). Adding this 65 cm^{-1} difference to that for the 1-azabicyclo[3.3.1]nonan-2-one system (1688 cm^{-1}) yields a value of 1753 cm^{-1} ± 8 + 5 = 1753 cm^{-1}.
[m] Used bicyclo[2.2.1]heptan-2-one (2-norbornanone) as estimate.
[n] See Reference 39.
[o] Erman, W. F.; Kretshmar, H. C. *J. Am. Chem. Soc.* **1967**, *89*, 3842.
[p] Assumed bicyclo[3.1.1]heptan-6-one has same ν_{CO} as bicyclo[4.1.1]octan-7-one (see note *o*).

q For *N*-phenylbenzo-2(1*H*)azetone, ν_{CO} is 1830 cm^{-1} (in benzene) [Burgess, E. M.; Milne, G. *Tetrahedron Lett.* **1966**, 93]. For *N*-phenyl-2,3-naphthoazetone ν_{CO} is 1805 cm^{-1} (KBr) or 1815 cm^{-1} (CH$_2$Cl$_2$) [Ege, G.; Beisiegel, E. *Angew. Chem. Int. Ed. Engl.* **1968**, *7*, 303], and we estimate a value of 1835 cm^{-1} (CCl$_4$ or benzene), since for *N*-methylbutyrolactam ν_{CO} is 1697 cm^{-1} (CCl$_4$) and 1676 cm^{-1} (CHCl$_3$). An earlier value for the simple monocyclic by Henery-Logan [Henery-Logan, K. R.; Rodricks, T. V. *J. Am. Chem. Soc.* **1963**, *85*, 3524] appears to be anomalously low.

r For 3,3-dimethylcyclobut-2-en-1-one, ν_{CO} is 1762 cm^{-1} (CDCl$_3$) [Kelly, T. R.; McNutt, R. W. *Tetrahedron Lett.* **1975**, 285]. This was corrected by 20 cm^{-1} to transfer to CCl$_4$; see note *q*.

s For *N*-methyl-2-pyridone ν_{CO} is 1666 cm^{-1} (CHCl$_3$) [Fuji, T.; Ohba, M.; Hiraga, T. *Heterocycles* **1981**, *16*, 1197]. This is corrected by 20 cm^{-1} for transfer to CCl$_4$ (note *q*).

t Estimated ν_{CO} (CCl$_4$) for 2,4-cyclohexadienone obtained by subtracting difference between aldehyde and α, β, γ, δ-unsaturated aldehyde (Gorden, A. J.; Ford, R. A. "The Chemist's Companion". Wiley: New York, 1972, p. 197) from cyclohexanone.

tams of ring size 3–6, and model ketones. It seems clear that in this comparison, loss of resonance in the $\overline{3}.3.1$ system (**22**) appears to be somewhat greater than in β-lactams (for the definition of $\overline{3}.3.1$, $\overline{3}.2.1$, etc, see Table 4-3). Increased strain is encountered in the $\overline{3}.2.1$ system (**12**), with further increases in the $\overline{3}.2.2$ system and the $3.\overline{2}.1$ system. The loss of resonance stabilization in the α-lactam ring appears to rival that in the bridgehead β-lactams **34** and **35**. The prediction that the carbonyl bond length in rotated formamide will be the same as in the planar conformer,[21] appears to be at odds with the reduced ν_{CO} for nonplanar lactams. While reduced charge separation between carbon and oxygen in the nonplanar conformer[21] may help, conventional resonance logic still appears to be most compelling.

To better quantify the relationship between loss of resonance stabilization and ν_{CO}, we have made an explicit analogy between the lactam resonance contributors on the one hand and methylenecyclo (bicyclo) alkanes and their isomeric methylcyclo (bicyclo) alkenes on the other. In the limit, virtual lack of a double bond in 1-bicyclo[2.2.2]octene would correspond to the virtual absence of resonance in 2-quinuclidone (Scheme VI).

Molecular mechanics (MM) calculations on bridgehead olefins have been published by three different groups.[52-54] Although Maier and Schleyer[53] em-

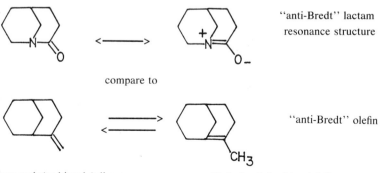

Scheme VI

ploy MM1 to calculate ΔH_f of bridgehead olefins including those having *trans*-cyclohexene rings, Warner and Peacock[54] demur from studies of these small rings, noting the considerable uncertainty in the larger systems. It has been suggested that these calculations exemplify "extrapolation beyond the limits of the parameterization set giving useful results."[55] Interestingly, a recent experimental value[56] for the strain in *trans*-cyclohexene is within the upper part of the calculated range for this system. The values of strain energy obtained by Ermer[52] are not accompanied by $\Delta H_f^\circ(g)$ values or the strainless group increments, so we do not employ them here. The values for ΔH_r (methylenecycloalkane → methylcycloalkene, Scheme VI) are listed in Table 4-3. A corresponding list of values for $\Delta\Delta H_f$ and $\Delta\nu_{CO}$ is found in Table 4-4 and graphed in Figure 4-8. The points denoted by crosses in Figure 4-8

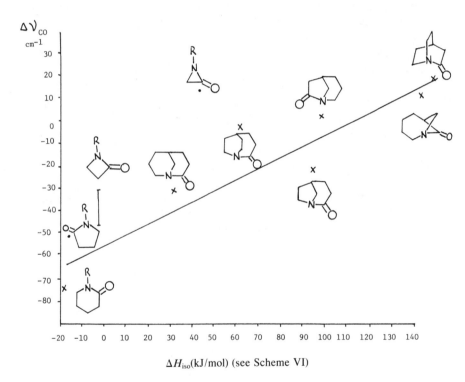

For $n = 7$ (Denoted by X): $\Delta\nu_{CO} = 0.501\ \Delta H_{isom} - 54.9 \qquad r = 0.93$

For $n = 6$ (Delete $\overline{3.2.2}$): $\Delta\nu_{CO} = 0.523\ \Delta H_{isom} - 60.3 \qquad r = 0.97$

Figure 4-8. Relationship between the carbonyl frequency difference between a lactam and the corresponding ketone and ΔH_{isom} (see Scheme VI), which provides an analogy to amide or lactam resonance stabilization.

TABLE 4-3. Enthalpies of Formation of Exo-olefin and Corresponding 1-Methyl Olefins (corresponding to analogous lactams) and ΔH_r for Isomerization (Scheme VI)

Abbreviations have the following meaning: $\bar{6}$ = δ-lactam; $3.\bar{2}.1$ = bridgehead lactam bond in 2-bridge

	ΔH_f° (kJ/mol)[a]		
	Exo-olefin	1-Methyl-endo-olefin	$\Delta\Delta H_f^\circ$ (kJ/mol) = ΔH_r
$\bar{6}$	−25.2	−43.3	−18.1
$\bar{5}$	12.0	−3.8	−15.8
$\bar{4}$	121.5	118.7[b]	−2.8
$\bar{3}$	200.5	243.6	+43.1
$3.\bar{3}.1$	−29.3	6.9[b,c]	+36.2 to +38.3
		9.0[b,d]	
$3.\bar{2}.2$	−8.1[e,f]	50.1[b,c]	+58.2 to +66.0
		57.9[b,d]	
$3.\bar{2}.1$	−3.1[e,f]	91.8[b,c]	+94.9 to +95.7
		92.6[b,d]	
$\bar{2}.2.2$	−9.2	140.7[b,c]	+149.9
$3.\bar{2}.1$	−12.9[e,g]	86.0[b,c]	+98.9
$4.\bar{1}.1$		219.8[b,c]	
$3.\bar{1}.1$	115.4[h]	259.9[b,c]	+144.5
β-lactam (azetidinone)	240[i]	420[j]	+180
2-pyridone	146.4[k]	50.4	−96

[a] Unless other noted, value taken from Pedley, J. B.; Naylor, R. D.; Kirby, S. P. "Thermochemical Data of Organic Compounds", 2nd ed. Chapman & Hall: London, 1986.
[b] To calculate $\Delta H_f^\circ(g)$ of 1-methylcycloalkene, took average of the difference $\Delta H_f^\circ(g)$ (1-methylcyclohexene) − $\Delta H_f^\circ(g)$ (cyclohexene) = −38.3 kJ/mol; $\Delta H_f^\circ(g)$ (1-methylcyclopentene) − $\Delta H_f^\circ(g)$ (cyclopentene) = −37.7 kJ/mol, and added this −38.0 kJ/mol to ΔH_f° of the parent olefin.
[c] $\Delta H_f^\circ(g)$ (bridgehead olefin) from Maier, W. F.; Schleyer, P. v. R. *J. Am. Chem. Soc.* **1981**, *103*, 1891 based on MM1.
[d] Warner, P. M.; Peacock, S. *J. Comput. Chem.* **1982**, *3*, 417, based on MM2.
[e] Obtained $\Delta H_f^\circ(g)$ (bicycloalkane) from compilation of MM1 data in Engler, E. M.; Andose, J. D.; Schleyer, P. v. R. *J. Am. Chem. Soc.* **1973**, *95*, 8005.
[f] Obtained $\Delta H_f^\circ(g)$ of bicyclo[3.3.1]nonane from Pedley et al. (note *a*) and added 98.2 kJ/mol, reflecting ΔH_f° (methylenecyclohexane-cyclohexane).
[g] Obtained $\Delta H_f^\circ(g)$ of bicyclo[3.2.1]octane from Engler et al. (note *f*) and added 88.4 kJ/mol reflecting ΔH_f° (methylenecyclopentane-cyclopentane).
[h] Obtained $\Delta H_f^\circ(g)$ of bicyclo[3.1.1]heptane from Engler et al. (note *f*) and added 93.5 kJ/mol reflecting ΔH_f° (methylenecyclobutane-cyclobutane).
[i] ΔH_f° (1-methyl-3-methylenecyclobutene) = 202 kJ/mol [Benson, S. W.; Cruickshank, F. R.; Golden, D. M.; Haugen, G. R.; O'Neal, H. E.; Rodgers, A. S.; Shaw, R.; Walsh, R. *Chem. Rev.* **1969**, *69*, 297]. To this value, +38 kJ/mol was added (see note *b*).
[j] ΔH_f° (cyclobutadiene) from high-level ab initio calculation = 458 kJ/mol [Melius, C. F.; Binkley, J. S. Sandia National Laboratory, unpublished results] added −38 kJ/mol (see note *b*) to obtain methyl derivative.
[k] ΔH_f° (5-methylene-1,3-cyclohexadiene) = 35 ±3 kcal/mol (146.4 kJ/mol) [Bartmess, J. E. *J. Am. Chem. Soc.* **1982**, *104*, 335].

TABLE 4-4. Correspondence Between $\Delta\Delta H_f$ for Exo-olefin and Corresponding 1-Methyl Olefin Analogous to a Lactam (Scheme VI: see Table 4-3) and the Difference in Carbonyl Frequencies Between That Lactam and the Corresponding Ketone (see Table 4-2)

	$\Delta\Delta H_f$ (kJ/mol)	$\Delta\nu_{CO}$ (cm^{-1})
$\bar{6}$	−18.1	−75
$\bar{5}$	−15.8	−51
$\bar{4}$	−2.8	−31 to −46
$\bar{3}$	+43.1	+13
$\bar{3}.3.1$	+36.2 to +38.3	−32
$\bar{3}.2.2$	+58.2 to +66.0	−3
$\bar{3}.2.1$	+94.9 to +95.7	−22
$2.\bar{2}.2$	+149.9	+19
$3.\bar{2}.1$	+98.9	+2
$3.\bar{1}.1$	+144.5	+12
α-lactam (azetidinone)	+180	+48
2-pyridone	−96	+15

are for bridgehead bicyclic lactams and for the standard N-methyl δ-lactam. It is clear that the more positive the value of ΔH_{isom}, the higher is $\Delta\nu_{CO}$ (lactam–ketone). If all seven points are employed, Equation 4-5 results. If the [$\bar{3}$.2.2] data point is removed, a significantly better correlation is obtained (Equation 4-6). It is quite clear that $\Delta\nu_{CO}$ is much higher for the α-lactam system than one would predict on the basis of this simple relationship. Interestingly, the $\Delta\nu_{CO}$ value for the 2(1H)-azetone is highest, corresponding to the highest ΔH_{isom}, thus reflecting the antiaromaticity of the cyclobutadiene system. On the other hand, 2-pyridone and N-alkyl derivatives would be expected to have $\Delta\nu_{CO} \simeq -100$ on the basis of Equation 4-5, unless values around −75 (N-methylvalerolactam, Table 4-2) are limiting values beyond which extra stabilization of the zwitterionic resonance structure can no longer contribute. Actually, the experimental $\Delta\nu_{CO}$ for N-methyl-2-pyridone is +15 cm^{-1}, some 115 cm^{-1} from the predicted value.

$$\Delta\nu_{CO} = 0.501 \times \Delta H_{isom} - 54.9 \qquad n = 7, r = 0.927 \qquad (4\text{-}5)$$

$$\Delta\nu_{CO} = 0.523 \times \Delta H_{isom} - 60.3 \qquad n = 6, r = 0.970 \qquad (4\text{-}6)$$

It is worth dwelling for a moment on the resonance stabilization in 2-pyridone. Thus, if one employs Equation 4-7 to obtain the resonance energy in N,N-dimethylacetamide, the value is 21 kcal/mol.[57]

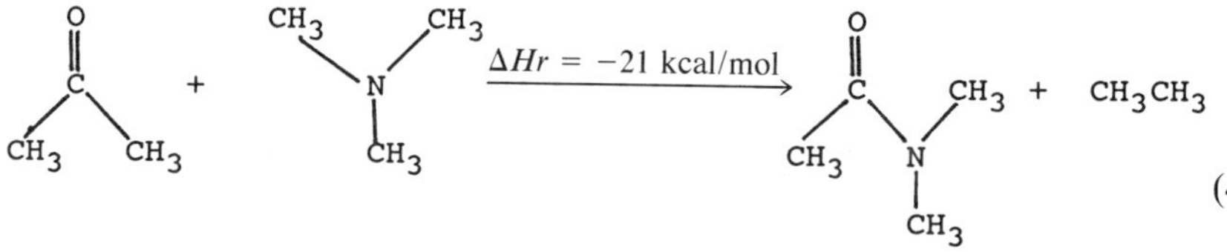

$$\text{(CH}_3\text{)}_2\text{C=O} + \text{(CH}_3\text{)}_2\text{NCH}_3 \xrightarrow{\Delta Hr = -21 \text{ kcal/mol}} \text{(CH}_3\text{)}_2\text{NC(O)CH}_3 + \text{CH}_3\text{CH}_3 \quad (4\text{-}7)$$

Equation 4-8 provides an analogous comparison for 2-pyridone. Here, since it is known that 2-pyridone and 2-hydroxypyridine are almost equienergetic,[58] the published $\Delta H_f^\circ(g)$ is employed[59] without concern over which isomer predominates. A value for $\Delta H_f^\circ(g)$ of 2,4-cyclohexadienone of -17 ± 3 kcal/mol has been recently reported.[60] While at first glance comparison of Equations 4-7 and 4-8 seems to indicate the same amide or lactam resonance energy, this neglects the resonance energy of 2,4-cyclohexadienone which, using the experimental value[60] and making comparison with cyclohexanone and two cyclohexenes, yields a resonance energy of 19.6 kcal/mol. Although this seems high, the comparison $(20.3 - 21 + 19.6 \simeq 19$ kcal/mol) would appear to provide a value for the aromatic stabilization above lactam stabilization in 2-pyridone. This is 17 kcal/mol below the commonly accepted and analogous value for benzene. It is also worthwhile making a comparison between resonance stabilization in 2-pyridone and benzene using Equations 4-9 and 4-10.[61] Similarly, the aromatic ring resonance energy in 2-pyridone is about 18 kcal/mol less than in benzene. Also, comparison of Equations 4-11 and 4-10[61] indicates an extra 12 kcal/mol

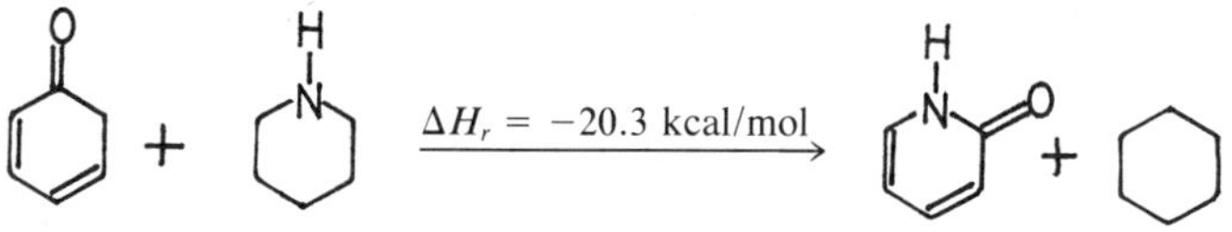

$$\text{cyclohexadienone} + \text{piperidine} \xrightarrow{\Delta H_r = -20.3 \text{ kcal/mol}} \text{2-pyridone} + \text{cyclohexane} \quad (4\text{-}8)$$

$$\text{2,4-hexadiene} + \text{(CH}_3\text{)}_2\text{NH (as drawn)} \xrightarrow{\Delta H_r = -28.7 \text{ kcal/mol}} \text{benzene} + 2\text{C}_2\text{H}_6 \quad (4\text{-}9)$$

$$\text{2,4-hexadiene} + \text{CH}_3\text{N(H)C(O)CH}_3 \xrightarrow{\Delta H_r = -11.2 \text{ kcal/mol}} \text{2-pyridone (N-H)} + 2\text{C}_2\text{H}_6 \quad (4\text{-}10)$$

$$\text{hexadiene} + \text{(CH}_3\text{)}_2\text{NC(O)CH}_3 \xrightarrow{\Delta H_r = +0.3 \text{ kcal/mol}} \text{N-methyl-2-pyridone} + 2\text{C}_2\text{H}_6 \quad (4\text{-}11)$$

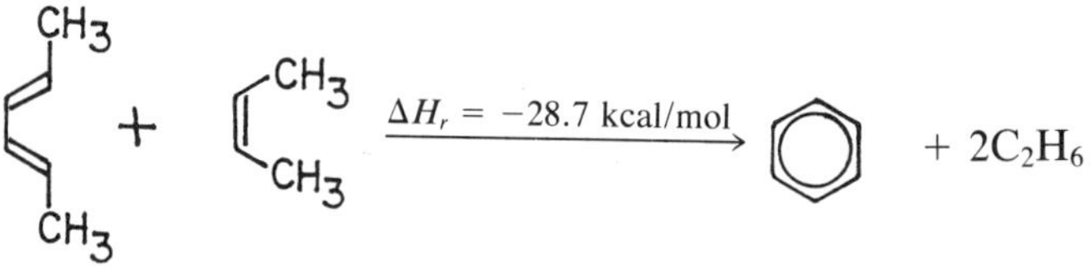

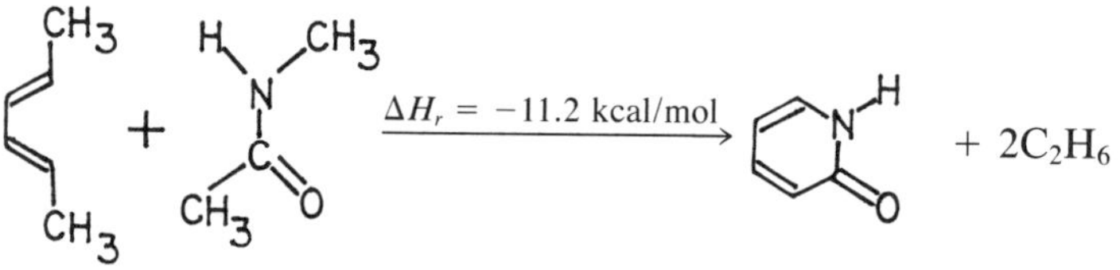

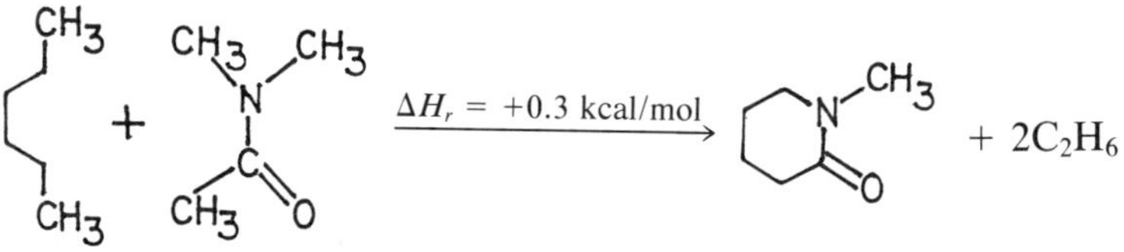

stabilization in 2-pyridone. It is clear that the extra 12 to 19 kcal/mol of resonance stabilization in 2-pyridone, compared to ordinary amides, is about half the aromatic stabilization in benzene.

It is worth noting briefly that the finding that methyl acetate's resonance energy (24 kcal/mol) is greater than that for N,N-dimethylacetamide need not contradict conventional logic.[57] First, if one includes the need to overcome an N-inversion barrier of roughly 6 kcal/mol, then the intrinsic resonance stabilization is greater in the amide in accord with the lesser electronegativity of nitrogen.[62] Furthermore, the low rotational barriers in esters are explained by significant resonance in the transition state.[57]

6. CALORIMETRIC AND CRYSTALLOGRAPHIC DATA

There are no calorimetric data on the bridgehead lactams. An MNDO result appears to provide quite a reasonable estimate of the value for $\Delta H_f^\circ(g)$ of 1-azabicyclo[3.3.1]nonan-2-one, namely -42.8 kcal/mol. If one examines Equation 4-12, it is clear that the resonance energy is about half that in the amide. We think there is probably somewhat more resonance stabilization. The loss in resonance energy calculated here (≈ 10 kcal/mol) can be compared with 12 kcal/mol of strain due to π torsion in 1-bicyclo[3.3.1]nonene.[63] An interesting potential contribution to strain in *cis*-lactams may arise from dipolar repulsion between the carbonyl and the nitrogen lone pair.[21]

$$\text{(structure)} + \text{(structure)} \xrightarrow{-10.4 \text{ kcal/mol}} \text{(structure)} + \text{(structure)} \quad (4\text{-}12)$$

(-230.6 kJ/mol[63]) (-32.7 kJ/mol)[64]   (-127.5 kJ/mol)   (-179.1 kJ/mol, MNDO)

There are relatively few thermochemical data for amides and lactams. The Pedley–Naylor–Kirby compendium[59] provides heat of formation data for 16 simple amides, but $\Delta H_f^\circ(g)$ values are available for only nine of these amides, which mostly represent homologous series and provide fairly little variation of the amide linkage. The situation is even more sparse for simple lactams; all data are listed in Table 4-5. Only three gas-phase values are found in this table. If one employs homodesmotic[65] Equation 4-13, then a resonance energy of 18.0 kcal/mol is observed for caprolactam, in good agreement with earlier cited estimates for amides. (Six-membered rings are employed in Equation 4-12 because there are no data for the seven-membered ring amine[59]; the approximation should not significantly change the conclusion.) If one employs Trouton's rule ($\Delta H_v = 0.021\ T_b$, in kcal/mol) to estimate $\Delta H_f^\circ(g)$ for pyrrolidone (BP 250–255°C, 742 mm, *CRC Handbook*) using a relationship analogous to Equation 4-13, one comes to the interesting conclusion that the resonance stabilization in the five-membered lactam (using $\Delta H_f^\circ(l)$ data in Table 4-5) is 28.9 kcal/mol. While the approximate ΔH_v does

TABLE 4-5. Published ΔH_f° Values for Various Physical States of Simple Lactams

Lactam	ΔH_f° (kJ/mol)		
	Crystal	Liquid	Gas
2-Pyrrolidone (5)		−286.2 ±0.5	
2-Piperidone (6)	−306.6 ±0.5		−232.1[a]
1-Methyl-2-piperidone		−293.0 ±0.5	
caprolactam (7)	−329.4 ±0.9		−246.2 ±1.3
N-Methylcaprolactam		−306.7 ±0.5	
5-Methylcaprolactam	−364.0 ±1.3		
7-Methylcaprolactam	−362.3 ±1.3		
Enantholactam (8)	−348.5 ±1.3		
1-Methylenantholactam		−325.4 ±1.3	
2-Pyridone	−166.3 ±0.5		−79.7 ±1.4

[a] A value for ΔH°_{subl} (2-piperidone) = 74.5 kJ/mol has been reported [Chickos, J. S. In "Molecular Structure and Energetics", Vol. 2; Liebman, J. F.; and Greenberg, A., Eds.; VCH Publishers: Deerfield Beach, FL, 1987, pp. 67–150].
Source: Reference 59.

not take into account hydrogen bonding in the liquid, this still appears to be surprisingly high and perhaps reflects some experimental error.

$$\text{piperidine} - \text{cycloheptanone} - \text{cyclohexane} \xrightarrow{-18.0 \text{ kcal/mol}} \text{caprolactam} \quad (4\text{-}13)$$

One of the most interesting forays into lactam calorimetry was that summarized in 1949 by Woodward, Neuberger, and Trenner in collaboration with the U.S. National Bureau of Standards.[66] The data are not cited in any compendia[59] to our knowledge, in part due to the unusual source and, in part, due to some confusion between values and structures (at points the thermochemical data for two compounds are switched). It is historically interesting that these calorimetric data also helped in the decision over whether penicillin was β-lactam or an oxazolonethiazolidine.[66] The values are for crystalline compounds; the clever comparison between analogous β-lactams and the ring-opened amino esters in two cases allowed these authors to conclude that there is an additional 6 kcal/mol of strain in the penicillin (40) due to ring fusion compared to that in the monocyclic lactam 38. Their estimate led to an estimate of a 10^4–10^5 increase in hydrolysis rates of penicillins relative to monocyclic β-lactams. Our value (Scheme VII) is somewhat smaller (2–5 kcal/mol), but in the same direction.* Thus, the effect of fusion of the five-membered ring in the penicillins is increase in strain, pre-

---

* Note, however, in connection with 41, that the actual compound obtained was the methyl ester, but to remain consistent with the first reaction, the ethyl ester was estimated based on the liquid phase enthalpy difference[59] between methyl acetate and ethyl acetate.

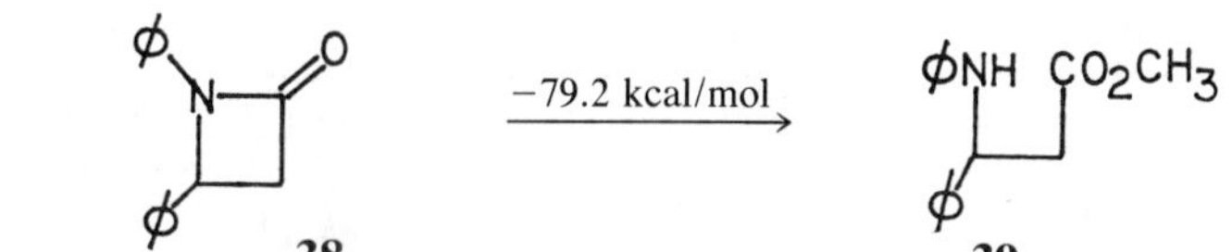

Scheme VII

[Structure 38: β-lactam with N-φ and C-φ substituents]
38
$\Delta H_f^\circ(c) = -3.6$ kcal/mol

$\xrightarrow{-79.2 \text{ kcal/mol}}$

[Structure 39: φNH–CH(φ)–CO$_2$CH$_3$]
39
$\Delta H_f^\circ(c) = -82.8$ kcal/mol

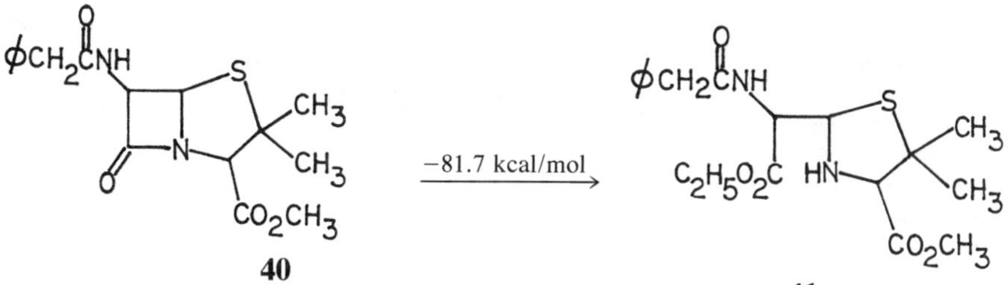

[Structure 40: penicillin methyl ester]
40
$\Delta H_f^\circ(c) = -159.8$ kcal/mol

$\xrightarrow{-81.7 \text{ kcal/mol}}$

[Structure 41: ring-opened product]
41
(-241.5 kcal/mol)
$\Delta H_f^\circ(c) = -233.5$ kcal/mol

sumably coupled with increased reactivity and bioactivity, and we are full circle to the structural work of Sweet and Dahl.[1] The analogous reaction for a simple amide (using gas phase values) is depicted in Equation 4-14 (the value for ethyl formate is estimated from those of methyl formate, methyl acetate, and ethyl acetate[59]).

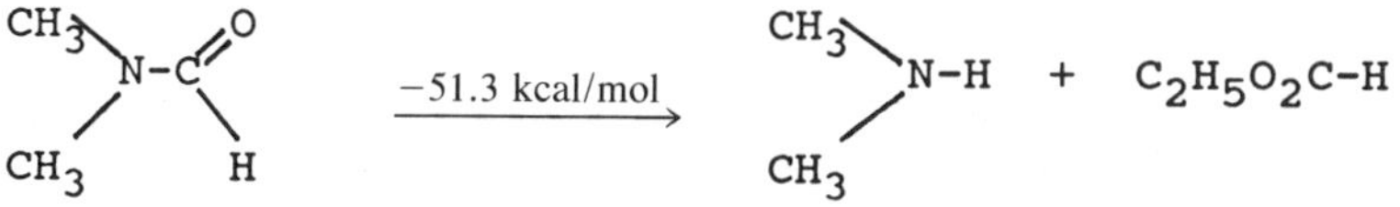

$$(CH_3)_2N-CHO \xrightarrow{-51.3 \text{ kcal/mol}} (CH_3)_2N-H + C_2H_5O_2C-H \quad (4\text{-}14)$$

Based on this value, the extra destabilization in **38** is 27.9 kcal/mol and that in **40** is 30.4 kcal/mol, due to ring strain and presumably decreased resonance stabilization. However, the values cited have enough uncertainty (estimate ±2 kcal/mol) that they cannot be used very quantitatively, and this is why we have not taken into account phase corrections to make Equation 4-14 and Scheme VII "exactly" comparable.

We have employed ab initio molecular orbital calculations using the GAUSSIAN 82 program series and the 3-21G basis set[67] to estimate resonance energies according to Equations 4-15 and 4-16. Selected structural

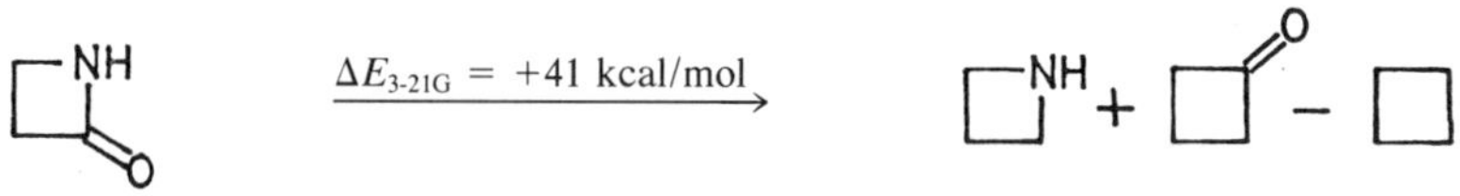

$$\text{[β-lactam]} \xrightarrow{\Delta E_{3\text{-}21G} = +41 \text{ kcal/mol}} \text{[azetidine-NH]} + \text{[cyclobutanone]} - \text{[cyclobutane]} \quad (4\text{-}15)$$

$$\text{[α-lactam]} \xrightarrow{\Delta E_{3\text{-}21G} = +12.7 \text{ kcal/mol}} \text{[aziridine-NH]} + \text{[cyclopropanone]} - \text{[cyclopropane]} \quad (4\text{-}16)$$

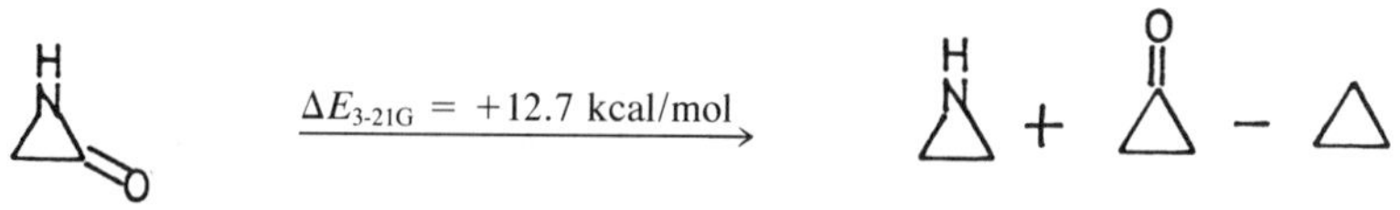

TABLE 4-6. Calculated (ab initio, 3-21G basis set) and Experimental (X-ray) Geometries for β- and α-Lactams

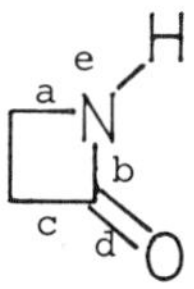

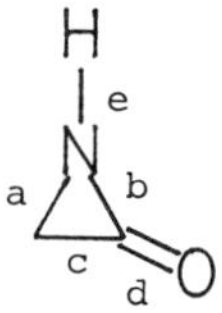

	3-21G Calculated	Experimental[a]	3-21G Calculated	Experimental[b]
a	1.482	1.467	1.491	1.509
b	1.374	1.333	1.321	1.328
c	1.552	1.552	1.509	1.446
d	1.2002	1.226	1.196	1.199
ab	96.5	96.2	65.6	60.9
bc	90.9	91.7	67.0	65.7
bd	132.7	132.4	144.9	147.3
e	0.995	0.85	0.990	—

[a] See Reference 12d.
[b] Wang, A. H.-J.; Paul, I. C.; Talaty, E. R.; Dupuy, A. E., Jr., *J. Chem. Soc. Chem. Commun.* **1972**, 43. The report is for 1,3-diadamantylaziridinone, which is nonplanar at N.

parameters for alpha- and beta-lactams are found in Table 4-6, and there is good agreement with experimental X-ray data. The resonance energy for the four-membered ring seems to be overestimated; the resonance energy for three-membered ring is considerably less. As in the cases of the experimental compounds propiolactam and 1,3-di-*tert*-butyl α-lactam, the former is completely planar including amide H, while the latter is pyramidal at nitrogen for both the alkyl substituent and the amide H.

There are two bridgehead lactams of the type we are concerned with for which there are published X-ray data. 5-Phenyl-1-azabicyclo[3.3.1]nonan-2-one (**22a**)[34] has considerable torsion about the lactam linkage and, as expected, the six-membered ring containing this grouping is in the boat conformation. The values of the distortion parameters are $\chi_N = -48.8°$, $\chi_C = 5.9°$, and $\tau = 200.8°$ (twist of 21°), while the values in *trans*-caprylolactam are $\chi_N = 21.5°$, $\chi_C = -5.5°$, and $\tau = 162.0°$ (twist 18°).[34] The relationship noted by Norskov-Lauritsen and colleagues[19] concerning the alignment of the bisector holds for this molecule; that is, the relationship of the carbonyl axis to the NCOC bisector indicates "acylium ion" character in the strained linkage, since from the data[34] ⟨cd − bd⟩ = +1.1° (see Scheme I).

The twisting in the $\overline{3}.2.2$ lactam **29** is quite significant, and an electron density map seems to locate a lone electron pair near nitrogen.[37b] The extra strain in this molecule relative to **22a** is apparent in its higher carbonyl frequency relative to the corresponding ketone (Table 4-4) and a higher positive value for ⟨cd − bd⟩ (+3.6°). As expected, the N—CO bond length in **29** (1.401 Å) is greater than that in **22a** (1.374 Å), reflecting reduced resonance. However, the carbonyl bond in **29** is longer (1.216 Å vs 1.201 Å). The question of CO bond length is particularly interesting in light of the calculations of Wiberg and Laidig.[21] They predict carbonyl bond lengths of 1.19 Å in the planar conformer and 1.18 Å in the perpendicular conformers. The very slight increase in the planar conformer agrees with resonance intuition, but the lengthening seems too small. We note that X-ray-derived carbonyl bond lengths in valerolactam, caprolactam, and enantholactam (1.243, 1.242, 1.246 Å) are longer than those in propiolactam (1.226 Å)[12d] and 1,3-diadamantyl α-lactam (1.199 Å).[37c] The shortening in the last two molecules may largely be due to hybridization, although reduced resonance (both have pyramidal nitrogen) may play a role. The interesting comparison is with **29**, whose IR frequency (see Tables 4-2 and 4-4) seems to reflect much reduced resonance while the carbonyl bond is only 0.02 Å shorter than those of the cited model lactams. In this light, the reported carbonyl bond length in **22a** seems to be anomalously short. Thus, in seeming agreement with the Wiberg–Laidig study, the bond length changes are not large. Perhaps this simply reflects the resistance to bond stretch of a particularly strong double bond.

7. STRUCTURE, STRAIN, AND REACTIVITY: CONCLUSIONS

The small amount of structural and energetics data on bicyclic bridgehead lactams indicates the extensive work that remains to be performed in this area. The synthesis of a wide variety of bridgehead lactams, their characterization by X-ray and other structural techniques, and the determination of experimental enthalpies of hydrolysis offers the opportunity to construct accurate three-dimensional energy/distortion diagrams, such as that depicted in Figure 4-5.

Another interesting area for exploration of bridgehead lactams is in the stereoelectronic effects on reactivity.[68] Thus, not only can we expect changes in reactivity based on the strain energy of distorted lactam linkages, but the very shape of the linkage may play a significant role.

8. NEW ANTIBIOTICS?

It is difficult enough to try to obtain effective β-lactam antibiotics based on relatively minor changes on the penicillin and cephalosporin nuclei.[69] The possibility of exploring vastly different systems for similar pharmaceutical activity borders on hubris. Nevertheless, bridgehead bicyclic lactams like those discussed here may be worth some brief exploration if only by molecular modeling as a screen. Thus, to be active, a suitable molecule must[70] (a) diffuse through the outer layers of bacteria, (b) diffuse rapidly enough to occupy sites and exceed the rate of destruction by β-lactamases, (c) exhibit suitable structural requirements to be recognized as a substrate by the enzymes of interest—transpepidases, (d) be reactive enough to rapidly alkylate the active serive residue of the transpeptidase, and (e) form a complex long-lived enough to interfere with cell wall construction. This is a very demanding group of criteria. Of further interest is the recent finding that while there is little homology between the targets of penicillin molecules, the D-alanyl-D-alanine peptidases, and the penicillin-destroying β-lactamases, there is very strong homology between the tertiary structures of these proteins.[71-73] This is depicted in Figure 4-9.[71]

Thus, a suitable antibiotic should, in principle, be a very good substrate for the D-alanyl-D-alanine peptidases and a relatively poor substrate for the β-lactamases, even though the three-dimensional structures are similar.

Anticipating accusations of hubris and naiveté, we nevertheless wonder whether biological activity may be observed for molecules such as **42** or related bridgehead bicyclics. There is much potential for affecting strain and reactivity by varying bridging in these compounds.

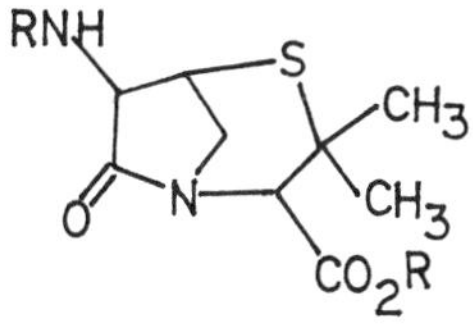

42

From another point of view, it is worthwhile understanding the limitations imposed on synthesis by bridgehead lactam linkages. Thus, Dieckman condensation failed to provide **43**. This compound, obtained in 1% yield through an alternate route, exhibits an NCO band at 1670 cm^{-1}, reflecting the 1-azabicyclo(3.3.1)nonan-2-one embedded in the structure.[74]

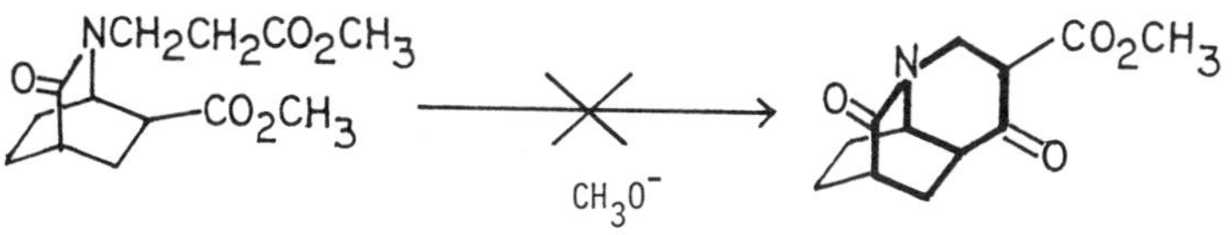

43

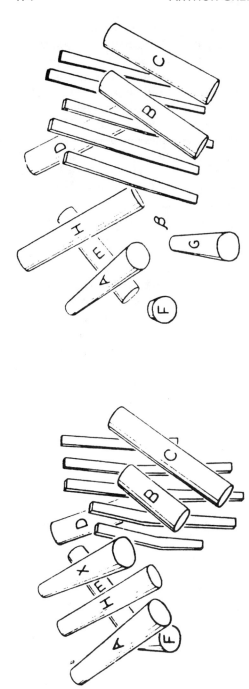

Figure 4-9. Secondary structure elements in *B. licheniformis* β-lactamase (left) and *Streptomyces* R61 DD-peptidase (right) illustrate homology at the tertiary structure level between the enzyme that destroys penicillin activity and the enzyme that is the penicillin target molecule, whose interception inhibits bacterial cell wall synthesis. (After Reference 71, copyright, © American Association for the Advancement of Science.)

9. MODEL SUBSTRATES FOR PROTEASES

The very imaginative work of Brown and co-workers has recently shed new light on the widely accepted mechanism for acid-catalyzed hydrolysis.[37a,75] Thus, investigation of **29** indicated oxygen exchange under acidic conditions[37a] and reinvestigation of more "classical" amides indicated similar behavior under suitable conditions.[75] Moreover, the enhanced reaction of **29** with β-amino alcohols suggested a model for serine proteases,[37b] while enhanced reaction with dicarboxylic acids indicates the potential for this twisted lactam to help model the activity of aspartate proteinases.[76] It is the distortion of the amide linkage that helps to bring this chemistry into focus, and such distortion may take place upon binding of a protein substrate to an enzyme. A rather specific pathway for amide linkage distortion may involve binding with a suitable transition metal ion. In one example (Scheme VIII),

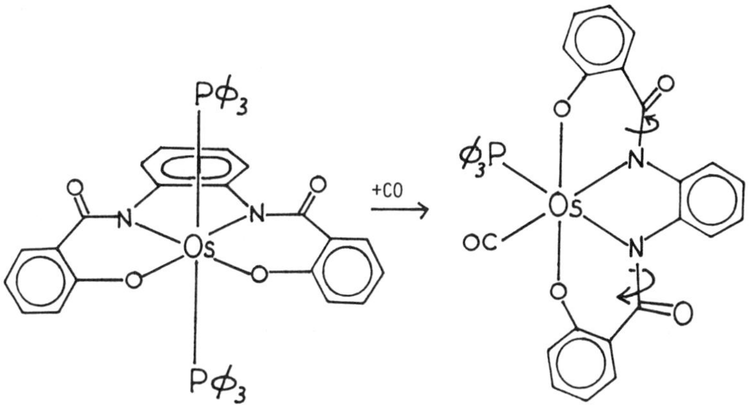

Scheme VIII

exchange of a (higher affinity) carbon monoxide for a triphenylphosphine ligand produces a change from the trans to the cis geometry for the monovalent ligands. The increase in stability associated with this new structure compensates for partial loss of amide resonance, itself mitigated by presumed p-d overlap between N and Os.[77] The distortions of amide linkages in these and related molecules are quite high[77] and rival those of the bridgehead bicyclic lactams covered in this chapter.

REFERENCES

1. Sweet, R. M.; Dahl, L. F. *J. Am. Chem. Soc.* **1970**, *92*, 5489.
2. Lukes, R. *Collect. Czech. Chem. Commun.* **1938**, *10*, 148.
3. (a) Yakhontov, L. N.; Rubsitov, M. V. *J. Gen. Chem. USSR* (Engl. Transl.), **1957**, *27*, 83.

(b) Levkoeva, E. I.; Nikitskaya, E. S.; Yakhontov, L. N. *Khim. Geterot. Soed.* **1971** (3), 378. (c) Levkoeva, E. I.; Nikitskaya, E. S.; Yakhontov, L. N. *Dokl. Akad. Nauk SSR* **1970**, *192*, 342. (d) Kostynovsky, R. G.; Mikhlina, E. E.; Levkoeva, E. I.; Yakhontov, L. N. *Org. Mass Spectrom.* **1970**, *3*, 1023.
4. Pracejus, H. *Chem. Ber.* **1959**, *92*, 988.
5. Pracejus, H. *Chem. Ber.* **1965**, *98*, 2897.
6. Pracejus, H.; Kehlen, M.; Kehlen, H.; Matschiner, H. *Tetrahedron* **1965**, *21*, 2257.
7. Hall, H. K., Jr.; El-Shekeil, A. *Chem. Rev.* **1983**, *83*, 549.
8. Bláha, K.; Malon, P. *Acta Universita Palackinia Olomucensis Facultatis Medicae, Int. Org. Chem. Biochem., Czech. Acad. Sci.* **1980**, *93*, 81.
9. Pauling, L. "The Nature of the Chemical Bond," 3rd ed. Cornell University Press: Ithaca, NY, 1960, pp. 281–282.
10. (a) Corey, R. B.; Pauling, L. *Proc. R. Soc. London B* **1953**, *141*, 10. (b) Marsh, R. E.; Donohue, J. *Adv. Protein Chem.* **1967**, *22*, 234.
11. Ramachandran G. N.; Sasisekharan, V. *Adv. Protein Chem.* **1968**, *23*, 283.
12. (a) Winkler, F. K.; Dunitz, J. D. *J. Mol. Biol.* **1971**, *59*, 169. (b) Dunitz, J. D.; Winkler, F. K. *Acta Crystallogr.* **1975**, *B31*, 251. (c) Chakrabarti, P.; Dunitz, J. D. *Helv. Chim. Acta* **1982**, *65*, 1555. (d) Yang, Q.-C.; Seiler, P.; Dunitz, J. D. *Acta Crystallogr.* **1987**, *C43*, 565.
13. (a) Gutowsky, H. S.; Jonas, J.; Siddall, T. H. *J. Am. Chem. Soc.* **1967**, *89*, 4300. (b) Stewart, W. E.; Siddall, T. H. *Chem. Rev.* **1970**, *70*, 517.
14. Robin, M. B.; Bovey, F. A.; Basch, H. In "The Chemistry of Amides," Zabicky, J., Ed. Wiley-Interscience, London, 1970, pp. 1–72.
15. Boggs, J. E.; Niu, Z. *J. Comput. Chem.* **1985**, *6*, 46. See also: Jasien, P. G.; Stevens, W. J.; Krauss, M. *J. Mol. Struct. (THEOCHEM)* **1986**, *139*, 197.
16. (a) Costain, C. C.; Dowling, J. M. *J. Chem. Phys.* **1960**, *32*, 158. (b) Hirota, E.; Sugisaki, R.; Nielsen, C. J.; Sorensen, G. O., *J. Mol. Spectrosc.* **1974**, *49*, 251. (c) Lee, S. G.; Hwang, K. W.; Bohn, R. K.; Hillig, K. W., II; Kuczkowski, R. L., Poster Presentation at the Austin Meeting on Molecular Structure, Austin, Texas, March, 1988.
17. Ramachandran, G. N. *Biopolymers* **1968**, *6*, 1494.
18. Warshel, A.; Levitt, M.; Lifson, S. *J. Mol. Spectrosc.* **1970**, *33*, 84.
19. Norskov-Lauritsen, L.; Bürgi, H.-B.; Hofmann, P.; Schmidt, H. R. *Helv. Chim. Acta* **1985**, *68*, 76.
20. (a) Mock, W. L. *Bioorg. Chem.* **1976**, *5*, 403. (b) Cieplak, A. S., *J. Am. Chem. Soc.* **1985**, *107*, 271.
21. Wiberg, K. B.; Laidig, K. E. *J. Am. Chem. Soc.* **1987**, *109*, 5935.
22. (a) Muhlebach, A.; Lorenzi, G. P.; Gramlich, V. *Helv. Chim Acta* **1986**, *69*, 395. (b) White, D. N. J.; Guy, M. H. P. *J. Chem. Soc. Perkin Trans. 2* **1975**, 43.
23. Tichý, M.; Dušková, E.; Bláha, K. *Tetrahedron Lett.* **1974**, 237.
24. Smolíkova, J.; Koblicová, Z.; Bláha, K. *Collect. Czech. Chem. Commun.* **1973**, *38*, 532.
25. Ealick, S. E.; Van der Helm, D. *Acta Crystallogr.* **1975**, *B31*, 2676.
26. Uyeo, S.; Fales, H. M.; Highet, R. J.; Wildman, W. C. *J. Am. Chem. Soc.* **1958**, *80*, 2590.
27. Denzer, M.; Ott, H. *J. Org. Chem.* **1969**, *34*, 183.
28. Hall, H. K., Jr.; Johnson, R. C. *J. Org. Chem.* **1972**, *37*, 697.
29. Hall, H. K., Jr.; Ekechuchwu, O. E.; Deutschmann, A., Jr.; Rose, C. *Polym. Bull.* **1980**, *3*, 375.
30. Hall, H. K., Jr.; Shaw, R. G., Jr.; Deutschmann, A. *J. Org. Chem.* **1980**, *45*, 3722.
31. (a) Steliou, K.; Poupart, M.-A. *J. Am. Chem. Soc.* **1983**, *105*, 7130. (b) Greenberg, A.; Zyla, K. Unpublished results. (c) Brehm, R.; Ohnhäuser, D.; Gerlach, H., *Helv. Chim. Acta* **1987**, *70*, 1981.
32. (a) Buchanan, G. L. *J. Chem. Soc. Chem. Commun.* **1981**, 814. (b) Buchanan, G. L. *J. Chem. Soc. Perkin Trans. 1* **1984**, 2669. (c) Buchanan, G. L.; Kitson, D. H.; Mallinson, P. R.; Sim, G. A.; White, D. N. J.; Cox, P. J. *J. Chem. Soc. Perkin Trans. 2* **1983**, 1709.
33. Grigg, R.; Sridharan, V.; Stevenson, P.; Worekun, T. *J. Chem. Soc. Chem. Commun.* **1986**, 1697.

34. Shaaban, M. A.; Ghoneim, K. M.; Khalifa, M. *Pharmazie* **1977**, *32*, 90.
35. Hall, H. K., Jr.; El-Shekeil, A. *J. Org. Chem.* **1980**, *45*, 5325.
36. (a) Brouillette, W. J.; Einspahr, H. M. *J. Org. Chem.* **1984**, *49*, 5113. (b) Coqueret, X.; Bourelle-Wargnier, F.; Chuche, J. *J. Org. Chem.* **1985**, *50*, 910.
37. (a) Somayaji, V.; Brown, R. S. *J. Org. Chem.* **1986**, *51*, 2676. (b) Somayaji, V.; Skorey, K. I.; Brown, R. S.; Ball, R. G. *J. Org. Chem.* **1986**, *51*, 4866. (c) Wang, A. H.-J.; Paul, I. C.; Talaty, E. R.; Dupuy, A. E., Jr. *J. Chem. Soc. Chem. Commun.* **1972**, 43.
38. Blackburn, G. M.; Skaife, C. J.; Kay, I. T. *J. Chem. Res., Miniprint* **1980**, 3650.
39. Williams, G. M.; Lee, B. H. *J. Am. Chem. Soc.* **1986**, *108*, 6431.
40. Treschanke, L.; Rademacher, P. *J. Mol. Struct. (THEOCHEM)* **1985**, *122*, 35.
41. Treschanke, L.; Rademacher, P. *J. Mol. Struct. (THEOCHEM)* **1985**, *122*, 47.
42. Boese, R.; Rademacher, P.; Treschanke, L. *J. Mol. Struct.* **1985**, *131*, 55.
43. Treschanke, L.; Rademacher, P. *J. Mol. Struct. (THEOCHEM)* **1985**, *131*, 61.
44. Cook, D. *Can. J. Chem.* **1961**, *39*, 31.
45. McAlduff, E. J.; Lynch, B. M.; Houk, K. N. *Can. J. Chem.* **1978**, *56*, 495.
46. Tsuchiya, S.; Seno, M. *J. Org. Chem.* **1979**, *44*, 2850.
47. Tsuchiya, S.; Seno, M.; Lwowski, W. *J. Chem. Soc. Chem. Commun.* **1982**, 875.
48. Brown, R. S.; Tse, A. *J. Am. Chem. Soc.* **1980**, *102*, 5222.
49. Stothers, J. B. "Carbon-13 NMR Spectroscopy." Academic Press: New York, 1972.
50. Levy, G. C.; Lichter, R. L.; Nelson, G. L. "Carbon-13 Nuclear Magnetic Resonance Spectroscopy," 2nd ed. Wiley Interscience: New York, 1980.
51. Nakashima, T. T.; Maciel, G. E. *Org. Magnet. Res.* **1972**, *4*, 321.
52. Ermer, O. *Z. Naturforsch.* **1977**, *B32*, 837.
53. Maier, W. F.; Schleyer, P. v. R. *J. Am. Chem. Soc.* **1981**, *103*, 1891.
54. Warner, P.; Peacock, S. *J. Comput. Chem.* **1981**, *3*, 417.
55. Clark, T. "A Handbook of Computational Chemistry." Wiley: New York, 1985, p. 25.
56. Peters, K. S. *Pure Appl. Chem.* **1986**, *58*, 1263.
57. Liebman, J. F.; Greenberg, A. *Biophys. Chem.* **1974**, *1*, 222.
58. Beak, P. *Acc. Chem. Res.* **1977**, *10*, 186.
59. Pedley, J. B.; Naylor, R. D.; Kirby, S. P. "Thermochemical Data of Organic Compounds," 2nd ed. Chapman & Hall: London, 1986.
60. Shiner, C. S.; Vorndam, P. E.; Kass, S. R. *J. Am. Chem. Soc.* **1986**, *108*, 5699.
61. Value for E,E-2,4-hexadiene (42.2 kJ/mol) estimated by subtracting twice $\Delta H_f^\circ[(E$-1,3-pentadiene)-(1,3-butadiene)] from 1,3-butadiene; value for N-methylacetamide (242.7 kJ/mol) obtained by comparison of acetamide, dimethylamine, and methylamine.[59] The $\Delta H_f^\circ(g)$ for N-methylpiperidone (-237.2 kJ/mol) is obtained from comparison of $\Delta H_f^\circ(g)$ of piperidone (Table 4-5), trimethylamine, and dimethylamine. The $\Delta H_f^\circ(g)$ of N,N-dimethylacetamide (-239.0 kJ/mol) is estimated from acetamide, methylamine, and trimethylamine.[59]
62. Liebman, J. F.; Johnson, J. L. Unpublished observation.
63. Lesko, P. M.; Turner, R. B. *J. Am. Chem. Soc.* **1968**, *90*, 6888.
64. Obtained by comparing bicyclo[3.3.1]nonane, bicyclo[2.2.2]-octane and 1-azabicyclo[2.2.2]octane.
65. George, P.; Bock, C. W.; Trachtman, M. In "Molecular Structure and Energetics," Vol. 4; Liebman, J. F.; and Greenberg, A., Eds.; VCH Publishers: Deerfield Beach, FL, 1987, pp. 163–187.
66. Woodward, R. B.; Neuberger, A.; Trenner, N. R. In "The Chemistry of Penicillin," Clarke, H. T.; Johnson, J. R.; and Robinson, R., Eds.; Princeton University Press: Princeton, NJ, 1949, pp. 415–439.
67. Whiteside, R. A.; Frisch, M. J.; Pople J. A., Eds. "The Carnegie–Mellon Quantum Chemistry Archive," 3rd ed. Carnegie–Mellon University: Pittsburgh, 1983.
68. Deslongchamps, P. "Stereoelectronic Effects in Organic Chemistry." Pergamon Press: Oxford, 1983, Chapter 4.
69. Dunn, G. L. *Annu. Rep. Med. Chem.* **1985**, *20*, 127.
70. Boyd, D. B.; Ott, J. L. *J. Antibiot.* **1986**, *39*, 281.

71. Kelly, J. A.; Dideberg, O.; Charlier, P.; Wery, J. P.; Libert, M.; Moews, P. C.; Knox, J. R.; Duez, C.; Fraipont, C.; Joris, B.; Dusart, J.; Frere, J. M.; Ghuysen, J. M. *Science* **1986**, *231*, 1429.
72. Phillips, D. C.; Cordero-Borboa, A.; Sutton, B. J.; Todd, R. J. *Pure Appl. Chem.* **1987**, *59*, 279.
73. Herzberg, O.; Moult, J. *Science* **1987**, *236*, 694.
74. Augustine, R. L.; Bellina, R. F. *J. Org. Chem.* **1969**, *34*, 2141.
75. Ślebocka-Tilk, H.; Brown, R. S.; Olekszyk, J. *J. Am. Chem. Soc.* **1987**, *109*, 4620.
76. Somayaji, V.; Brown, R. S. *J. Am. Chem. Soc.* **1987**, *109*, 4738.
77. Collins, T. J.; Coots, R. J.; Furutani, T. T.; Keech, J. T.; Peake, G. T.; Santarsiero, B. D. *J. Am. Chem. Soc.* **1986**, *108*, 5333.

CHAPTER 5

Polar Effects on the Lability of Carbon–Carbon Bonds

Tsutomu Mitsuhashi

Department of Chemistry, Faculty of Science, The University of Tokyo, Tokyo, Japan

CONTENTS

1. Introduction... 179
2. Polar Effects on Homolytic Cleavage 181
3. Heterolytic Cleavage..................................... 187
4. Polar Effects on Pericyclic Cleavage...................... 216
5. Concluding Remarks 226
Acknowledgment... 226
References 227

1. INTRODUCTION

Carbon–carbon single bonds possess an inherent bond strength as great as 86 kcal/mol, as has been suggested for the bond dissociation energy for ethane[1]; however, the bonds become quite labile under certain structural circumstances. This dual character of bond strength is indispensable to living organisms. Although a large number of modes of bond cleavage are known, the simplest and most fundamental mode is cleavage in thermal unimolecular reactions. In this chapter, the lability of carbon–carbon bonds is discussed, with special emphasis on the reactivities of thermal unimolecular reactions in which the cleavage of carbon–carbon bonds occurs via dipolar transition states. The compounds subjected to the bond cleavage are limited to uncharged, nonradical species; reactions involving intermolecular interactions other than solvation (eg, acid- and metal-catalyzed reactions) are not covered.

The modes of cleavage of carbon–carbon bonds in the thermal unimolecular reactions may be classified thus:

1. Homolytic cleavage to generate a radical pair
2. Heterolytic cleavage to generate an ion pair (or a zwitterion)
3. Pericyclic cleavage accompanied by synchronous bond forming or breaking to yield molecular products

The transition states for homolysis and heterolysis reactions are visualized as two fragments with mutual interaction. According to the frontier orbital treatment, the activated complex for pericyclic reactions can also be divided into two fragments that are interacting at two different positions of each fragment.[2]

The stability of each fragment is directly reflected in the activation barrier (the Hammond postulate).[3] There are many factors affecting the stabilities of the fragments. Electronic effects (especially due to resonance) as well as steric effects of substituents are determining factors. Whereas electron-releasing (donor, eg, MeŌ—) groups stabilize cations and electron-withdrawing (acceptor, eg, —C≡N) groups stabilize anions, both donor and acceptor groups are responsible for the stabilization of radicals due to resonance effects involving dipolar structures.

$$\text{MeÖ}-\overset{+}{\text{C}}\diagup \longleftrightarrow \text{MeÖ}=\text{C}\diagup \qquad \text{N}\equiv\text{C}-\overset{-}{\text{C}}\diagup \longleftrightarrow \overset{-}{\text{N}}=\text{C}=\text{C}\diagup$$

$$\text{MeÖ}-\overset{\cdot}{\text{C}}\diagup \longleftrightarrow \text{MeÖ}\overset{+\cdot}{=}\text{C}\diagup \qquad \text{N}\equiv\text{C}-\overset{\cdot}{\text{C}}\diagup \longleftrightarrow \overset{-}{\text{N}}=\overset{\cdot}{\text{C}}-\overset{+}{\text{C}}\diagup \longleftrightarrow \overset{\cdot}{\text{N}}=\text{C}=\text{C}\diagup$$

$$\text{MeÖ}-\overset{\cdot}{\underset{|}{\text{C}}}-\text{C}\equiv\text{N} \longleftrightarrow \text{MeÖ}\overset{+\cdot}{\underset{|}{-}}\text{C}=\text{C}=\overset{-}{\text{N}}, \text{ etc}$$

It is therefore suspected that geminal substitution by both groups further facilitates homolytic cleavage. This so-called captodative effect is attributable to radical stability arising from the extended dipolar character of a fragment from one attached group to the other through the p orbital of a carbon; this phenomenon is described in Section 2. In contrast, vicinal substitution by these groups induces a dipolar character in the transition state along the carbon–carbon bond cleaved. Strong development of such dipolar character leads ultimately to heterolytic cleavage (Section 3).

It is well known that a system in which the orbital overlapping of two molecules exists can be stabilized through charge-transfer forces such as represented by the resonance between neutral and ionic structures.[4] The extent of mixing of the ionic structure increases with increasing electron-donating power of one molecule and electron-accepting power of the other. This principle can be expected to hold analogously between two fragments in the transition state.

$$\underset{/}{\overset{\backslash}{\text{C}}}-\underset{\backslash}{\overset{/}{\text{C}}}- \longrightarrow \underset{/}{\overset{\backslash}{\text{C}}}\cdot \ \cdot\underset{\backslash}{\overset{/}{\text{C}}}- \longleftrightarrow \underset{/}{\overset{\backslash}{\text{C}}}^{+} \ \ ^{-}\underset{\backslash}{\overset{/}{\text{C}}}-$$

Heterolytic cleavage is naturally the case where the ionic structure is more important than the neutral; however as a general rule, carbon–carbon bond cleavage would be facilitated to the extent of the mixing induced by a charge-transfer interaction in any case. Even the captodative effect may be a phenomenon attributable to an intramolecular charge-transfer interaction restricted within a fragment. In other words, *an electron donor–acceptor arrangement in the reactant that affects the polarity of a transition state can be regarded as one of the important factors in labilizing carbon–carbon bonds.* Thus studies of this topic will provide the basic knowledge needed to understand structural features which are required to achieve the facile cleavage of carbon–carbon bonds under mild conditions.

2. POLAR EFFECTS ON HOMOLYTIC CLEAVAGE

Homolysis of carbon–carbon bonds initially giving singlet radical pairs usually requires drastic conditions such as are employed for the thermal cracking of hydrocarbons. The activation energy, however, can be reduced easily by the introduction of various substituents onto the carbons of a bond to be cleaved. Bulkier groups raise the energy of the ground states because of overcrowding and lower the energy of the transition state through relief of steric strain, and thus accelerate bond cleavage. Substituents involving a π system cause the spin delocalization of radicals.

The frontier orbital consideration, as illustrated in Figure 5-1, constitutes a rational explanation of radical stabilization due to resonance through a donor or an acceptor.[5,6] The three-electron interaction between the singly occupied MO (SOMO) of a radical and the highest occupied MO (HOMO) of a donor makes the SOMO of the perturbed radical higher in energy but the HOMO of the perturbed donor lower in energy, resulting in the net stabilization of the substituted radical (Figure 5-1a). The interaction between the SOMO of an unperturbed radical and the lowest unoccupied MO (LUMO) of an acceptor lower-lying than that of the usual π system such as a vinyl group strongly lowers the SOMO of the substituted radical (Figure 5-1b).

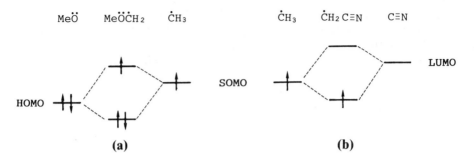

Figure 5-1. Orbital interactions (*a*) between a radical and a donor and (*b*) between a radical and an acceptor.

Geminal substitution by both donor and acceptor groups can induce an extra resonance effect due to dipolar structures because of opposite polarity of the donor and acceptor groups. Such a combination of substituents in a radical is termed "captodative,"[7] and the additional stabilization due to captodative effects can be explained by an analogous perturbation treatment for a pair of a donor-substituted radical and an acceptor as illustrated in Figure 5-2,[6,8] in which the orbital interaction between the SOMO and LUMO is stronger than that shown in Figure 5-1*b* because of a smaller energy difference.

A number of homolysis reactions of carbon–carbon bonds showing the captodative effect have been exemplified. Homolytic ring opening of cyclopropanes has been examined frequently because of facile cleavage due to relief of ring strain.[9,10] The ring opening of cyclopropane proceeds with an activation energy as high as 65 kcal/mol, but the presence of substituents such as CN and Ph reduces the activation energy. Captodative substitution

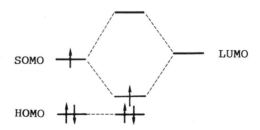

Figure 5-2. An orbital change by captodative substitution.

POLAR EFFECTS ON C—C BOND LABILITY   183

is much more efficient in promoting ring opening, especially in cases of two carbons of a cyclopropane captodatively substituted: the cis → trans isomerizations of compounds **1** and **2** proceed readily in CDCl$_3$ at low temperatures (50°C), and the activation energies (24.1 kcal/mol for **1**, 26.9 kcal/mol for **2**) are considerably lower than that for **3** (31.7 kcal/mol), in which a donor group of one of the captodative centers is replaced by an acceptor.[11] A similar captodative rate enhancement has been observed in the ring opening of methylenecyclopropanes as well (Subsection B of Section 3).[12]

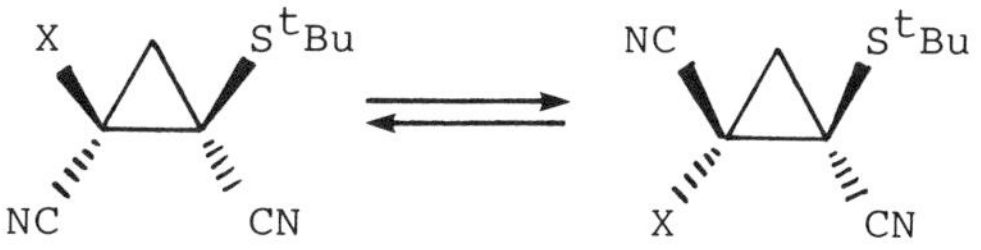

	1	X = MeO
	2	X = PhS
	3	X = CO$_2$Me

Among substituted tetraphenylcyclopropanes **4** through **7**, the lowest barrier for the cis → trans isomerization in benzene is obtained from compound **7** bearing captodative substituents (E_a 30.4 kcal/mol for **4**, **5**, and **6**; 28.2 kcal/mol for **7**).[13]

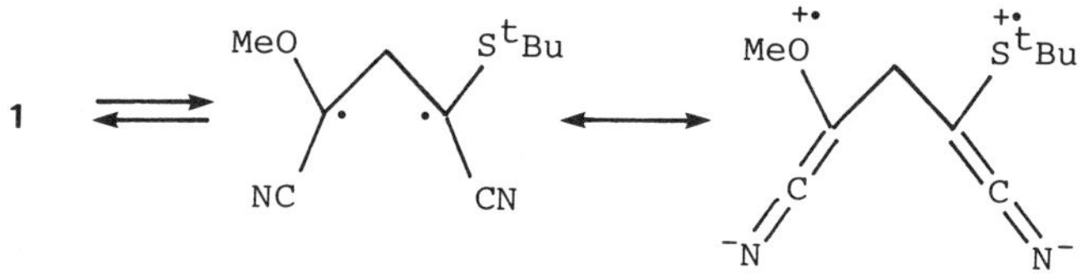

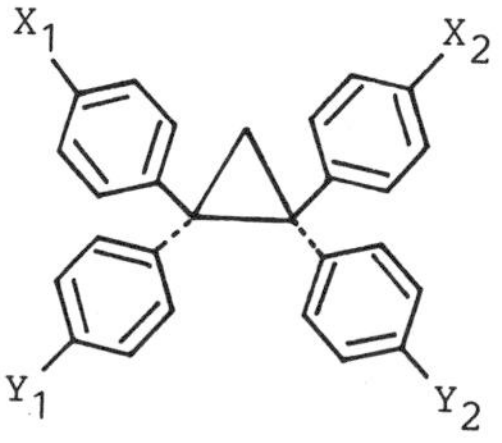

4	X$_1$ = X$_2$ = MeO,	Y$_1$ = Y$_2$ = H
5	X$_1$ = X$_2$ = CN,	Y$_1$ = Y$_2$ = H
6	X$_1$ = CN, X$_2$ = MeO,	Y$_1$ = Y$_2$ = H
7	X$_1$ = X$_2$ = CN,	Y$_1$ = Y$_2$ = MeO

Divergences from the additivity rule of substituent effects may be expressed conveniently by an inequality equation,[14] if the following three rate or equilibrium constants are available:

$$k_{xy} > (k_{xx}k_{yy})^{1/2}$$

in which k_{xy} is the constant for a compound geminally substituted by a pair of donor and acceptor groups, and k_{xx} and k_{yy} are those for the compounds geminally substituted by two identical groups. Actual divergences, however, seem to originate from two factors: one is the synergetic interaction between donor and acceptor groups due to a real captodative effect, and the other is the antagonistic interaction between identical groups, which results in a partial depression of resonance (the actual effect of two identical groups therefore becomes slightly less than the sum of the effect of one group).[15]

The equation above implies that k_{xy} is larger than both the others or at least one of the two, depending on the degree of captodative stabilization. The former case may be expected from the reactions described above, and the latter is seen in Gomberg's triphenylmethyl radical system (**8** → **9**),[16] which is referred to as one of the most facile homolysis reactions ($\Delta H^{\ddagger}_{diss}$ = 16.9 kcal/mol, $\Delta S^{\ddagger}_{diss}$ = −1.5 eu; ΔH_{diss} = 10.8 kcal/mol, ΔS_{diss} = 24.5 eu for the parent compound, X = Y = H, in benzene).[17] All the dissociation constants for the captodatively substituted compounds show intermediate values: $K_{xy}/(K_{xx}K_{yy})^{1/2}$, K_{xy}/K_{xx}, K_{xy}/K_{yy} (in benzene at 25°C) 1.35, 2.05, and 0.89 for X = *t*-Bu, Y = CN; 1.53, 0.93, and 2.53 for X = *t*-Bu, Y = CF$_3$; 1.56, 3.08, and 0.79 for X = MeO, Y = CN. Although the extremely facile bond cleavage should be attributed to steric and isopolar resonance effects, the observation above clearly indicates that the radical allows the mixing of ionic structures to effect further stabilization.

$$Ar^1-\underset{Ph}{\underset{|}{\overset{Ar^2}{\overset{|}{C}}}}-\underset{}{\bigcirc}=C\underset{Ar^1}{\overset{Ar^2}{<}} \xrightleftharpoons{K} 2\ Ar^1-C\underset{Ph}{\overset{Ar^2}{<}}$$

K_{xx}: Ar^1 = Ar^2 = *p*-XC$_6$H$_4$
K_{yy}: Ar^1 = Ar^2 = *p*-YC$_6$H$_4$
K_{xy}: Ar^1 = *p*-XC$_6$H$_4$
  Ar^2 = *p*-YC$_6$H$_4$

8   **9**

There are some distinct examples of captodative stabilization in other noncyclic systems. Evidence for the generation of radical **11** from *meso* and *dl* dimers **10** and **12** has been presented by electron spin resonance (ESR) measurements, stereochemical equilibration, air oxidation, disproportionation reactions, and trapping experiments.[18] The enthalpy of activation is rather low ($\Delta H^{\ddagger}$ = 26.5 kcal/mol, $\Delta S^{\ddagger}$ = 6.0 eu for *meso*; $\Delta H^{\ddagger}$ = 27.3 kcal/mol, $\Delta S^{\ddagger}$ = 6.7 eu for *dl* in chloroform). The dissociation is accelerated greatly with changing from the nonpolar solvent benzene (ΔH_{diss} = 21.5 kcal/mol) to the protic solvent ethanol (ΔH_{diss} = 10.6 kcal/mol). The value of ΔH_{diss} in ethanol is therefore comparable to that for the dissociation of the triphenylmethyl dimers. Steric interactions may be one of the important factors affecting such facile homolysis, since the bond [1.591(4) Å] is significantly longer than the usual (1.54 Å). However, fully substituted crowded ethanes have bond lengths ranging from 1.58 to 1.65 Å,[19] although few of them undergo such facile bond cleavage. The unusually persistent behavior of radical **11** as well as the presence of solvent dependence provides strong support for captodative stabilization through the polar resonance structure **11a**, which can induce further stabilization due to hydrogen bonding with a protic solvent.

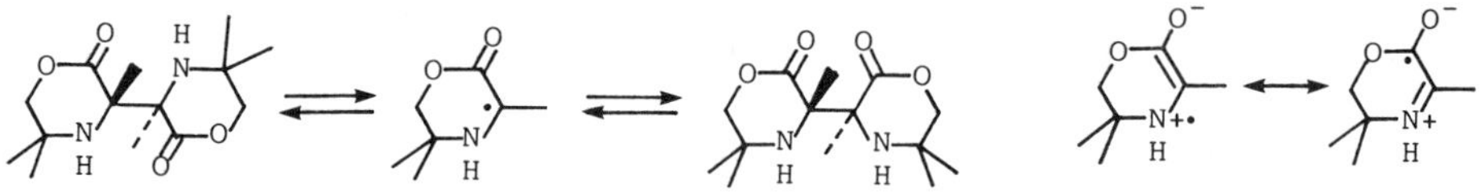

Likewise, *dl* and *meso* dimers **13** (Ar = *o*-nitrophenyl) split off dipolar radical **14**. A very low value of the enthalpy of dissociation (12.2 kcal/mol) in chlorobenzene has been observed.[20]

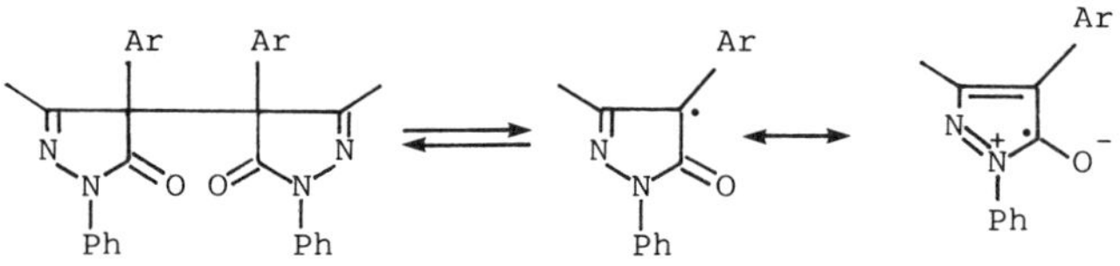

Dimer **15** (or **16**) is in equilibrium with dipolar radical **18** (or **19**), in mesitylene at 40°C.[21] On the other hand, the generation of radical **20** from dimer **17** requires temperatures above 100°C.[22] Facile dissociation in **15** and **16** compared to **17** has been explained in terms of merostabilization (a term equivalent to captodative stabilization).

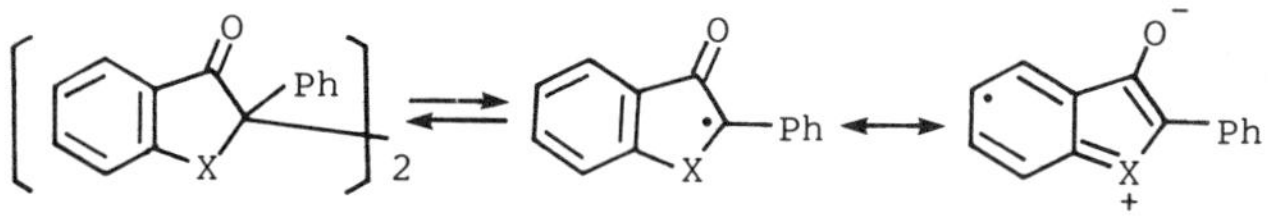

15	**18** X = S
16	**19** X = NH
17	**20** X = CO

It has been confirmed from ESR measurements that the facile interconversion of meso dimer **21** with the *dl* dimer proceeds via captodative radical **22** in chlorobenzene.[23] The temperatures leading to a good ESR signal intensity, which may serve as a qualitative measure of the ease of dissociation, decrease with increasing electron-withdrawing power of substituents on the phenyl group (120–140°C for *p*-MeO, 100–120°C for H, and 70–90°C for *p*-NO$_2$). With *p*-nitro dimers, conversion of pure meso or *dl* solutions into a meso–*dl* equilibrium mixture is complete within 5 minutes at 40°C. The result suggests that both captodative resonance forms **22a** and **22b** are important. On the other hand, higher temperatures (140–150°C) are necessary for a satisfactory ESR measurement of the radical generated from the interconversion of *dl*–meso isomers of **23** bearing only electron-withdrawing groups as geminal substituents.[24]

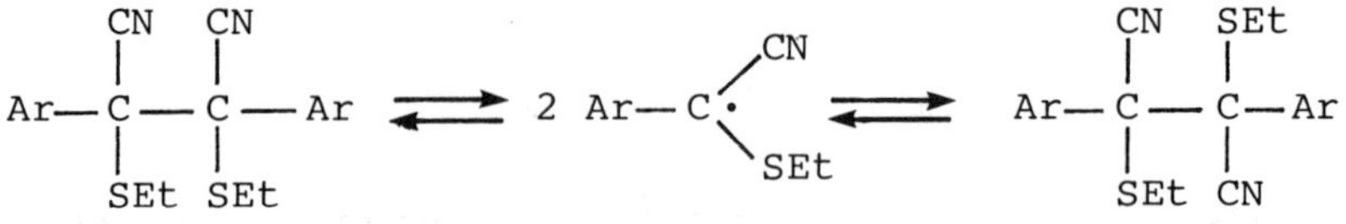

$$\text{Ar}-\underset{\underset{\text{SEt}}{|}}{\overset{\overset{\text{CN}}{|}}{\text{C}}}-\underset{\underset{\text{SEt}}{|}}{\overset{\overset{\text{CN}}{|}}{\text{C}}}-\text{Ar} \rightleftarrows 2\ \text{Ar}-\text{C}\overset{\text{CN}}{\underset{\text{SEt}}{\cdot}} \rightleftarrows \text{Ar}-\underset{\underset{\text{SEt}}{|}}{\overset{\overset{\text{CN}}{|}}{\text{C}}}-\underset{\underset{\text{CN}}{|}}{\overset{\overset{\text{SEt}}{|}}{\text{C}}}-\text{Ar}$$

21 (*meso*)   **22**   **21** (*dl*)

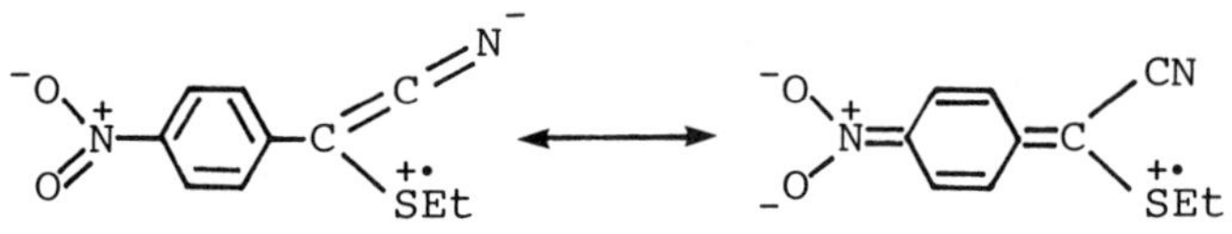

22a ↔ **22b**

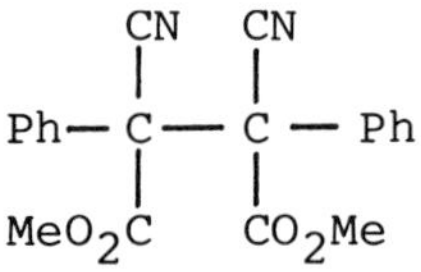

23: Ph–C(CN)(CO$_2$Me)–C(CN)(CO$_2$Me)–Ph

The effect of vicinal donor–acceptor substituents on the rate of homolysis has not been investigated deliberately. The isomerization of tetraarylcyclopropanes **4**, **5**, and **6** exhibits no effect of substitution, and thus $k_6/(k_4 k_5)^{1/2} = 1$. However, the well-known concerted two-bond fission mechanism of the homolysis of peresters **24** via dipolar transition states **25** is suggestive of the possibility of stabilization due to charge transfer along the ruptured bonds.[25] In this case, the cationic character of benzyl groups is evident from a ρ value of -1.2 (in toluene at 56°C) for the Brown–Hammett equation (ie, $\log k/k_0 = \rho\sigma^+$). Although the importance of resonance form **25a** has not been stressed,

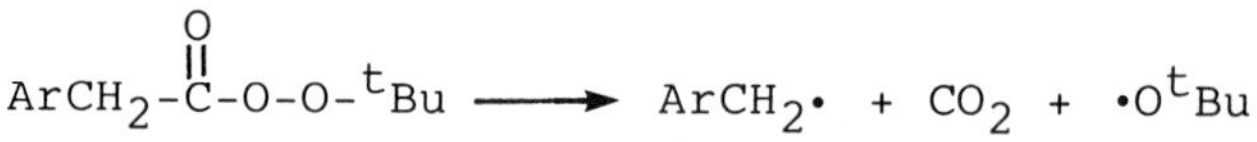

$$\text{ArCH}_2\text{-C(=O)-O-O-}^t\text{Bu} \longrightarrow \text{ArCH}_2\cdot + \text{CO}_2 + \cdot\text{O}^t\text{Bu}$$

24

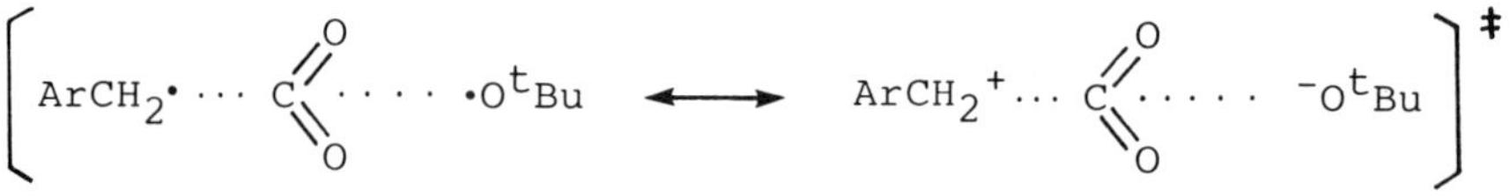

$$[\text{ArCH}_2\cdot \cdots \text{C(=O)(O)} \cdots \cdot\text{O}^t\text{Bu}] \leftrightarrow [\text{ArCH}_2^+ \cdots \text{C(=O)(O)} \cdots {}^-\text{O}^t\text{Bu}]^\ddagger$$

25

$$[\text{ArCH}_2^+ \cdots \text{C(=O)(O}^-\text{)} \cdots \cdot\text{O}^t\text{Bu}]^\ddagger$$

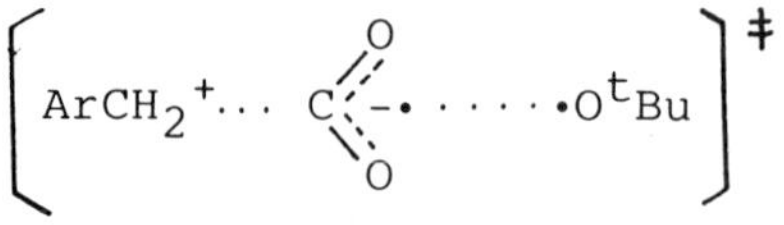

25a

it should be taken into consideration, because the bent CO_2 group would favor the anion radical form over the neutral.

3. HETEROLYTIC CLEAVAGE

A. Generation of Zwitterions

There are numerous examples of zwitterion generation via ring opening. The heterolytic ring opening of cyclopropanes using optically active compounds **26** through **29** has been studied extensively. Heating compound **26** in methanol at 150°C for 1.5 hours results in the partial racemization (15%) of **26** (37% recovery) and the formation of methanolysis produce **30** (55%).[26] Because of an unracemized backward process, the sum of the racemization rate (k_r) and the solvolysis rate (k_s) is regarded as a minimal heterolysis rate $(k_r + k_s = 20 \times 10^{-5}$ s$^{-1})$. Likewise, compound **27** yields methanolysis product **31** (46%) and olefin **32** (36%) after heating at 150°C for 3 days. The reactivity of **27** is slightly higher than that of **26**; $k_r + k_s = 57 \times 10^{-5}$ s^{-1} (calculated by $\Delta H^{\ddagger} = 25.5$ kcal/mol, $\Delta S^{\ddagger} = -14$ eu, and $k_r/k_s > 10^2$).[27] The rate of racemization of **27** is sensitive to solvent polarity, and the following results provide strong evidence for the model in which the dipolar transition state leads to a zwitterion:

	$(k_r)_{rel}$ at 126°C
Benzene	1.0
DMF	5
Methanol	20
AcOH	25
HCO_2H	2×10^4

With **28**, the rate of geometrical isomerization shows a large solvent polarity dependence [eg, $(k_{DMF}/k_{benzene})_{126°}, \approx 3.2 \times 10^4$].[28] In sharp contrast, the isomerization of **29** is insensitive to solvent polarity [$(k_{DMF}/k_{benzene})_{175°}, \approx 0.5$], and it has therefore been assumed that **28** undergoes heterolytic ring opening while **29**, in which the cyano group of **28** is replaced by a phenyl group, produces a diradical intermediate.[29] These data suggest that the heterolytic cleavage of this type of cyclopropane requires two negative groups at a carbon of the anionic site. As substituents at a carbon of the cationic site, phenyl groups are naturally more efficient than alkyl groups. An allyl group is also effective: compound **33** is racemized, presumably via a zwitterion, with a half-life of 75 hours at 140°C in xylene.[30]

Direct attachment of heteroatoms such as oxygen and nitrogen at a carbon of the cyclopropane ring may cause considerable lowering of decomposition temperatures because of positive charge stabilization through the interaction

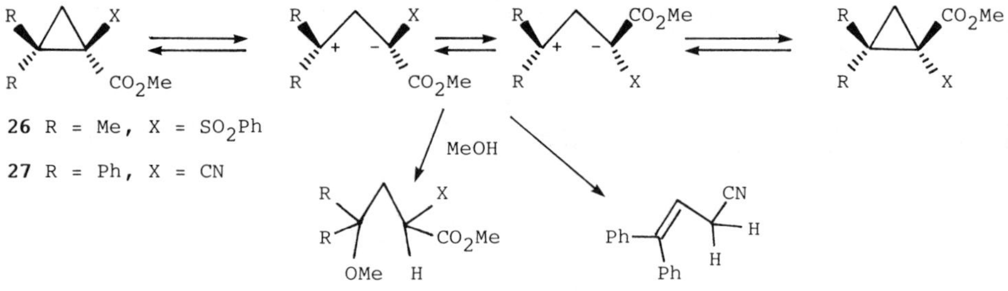

26 R = Me, X = SO$_2$Ph
27 R = Ph, X = CN

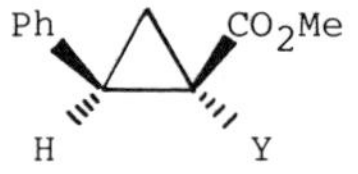

28 Y = CN
29 Y = Ph

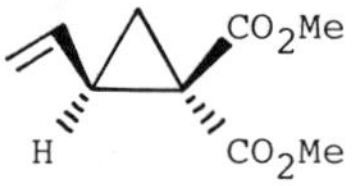

33

with lone pair electrons of heteroatoms. Treatment of indanone (**34**) with the diethyl malonate anion produces cyclopropane (**35**), whose NMR absorptions gradually decrease at room temperature in proportion to an increased amount of those of naphthalenol (**37**). The intermediacy of zwitterion **36** has been proposed.[31]

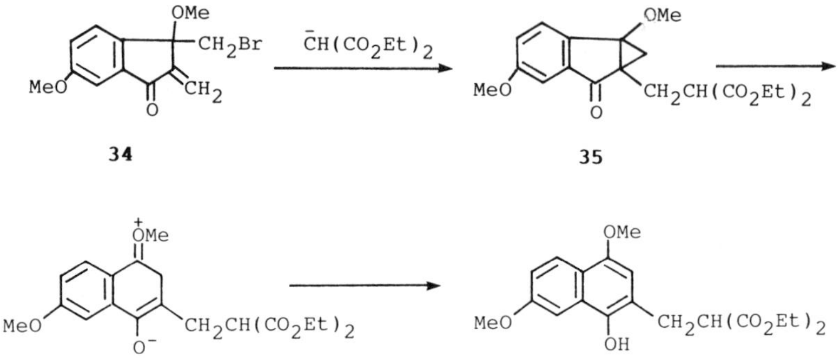

Action of diazoacetonitrile on **38** in the presence of dirhodium tetraacetate at room temperature yields a stable cyclopropane **39**. However, an analogous reaction using diazoacetone affords only a compound (11%) tentatively assigned to enol ether (**42**), which may be formed by way of cyclopropane **40** and zwitterion **41**. The difference in stability between **39** and **40** has been attributed to the inability of a cyano group to act as an internal trap rather than the absence of a cyclopropane–zwitterion equilibrium in **39**.[32]

38

39 R = CN

40 R = COMe

41

42

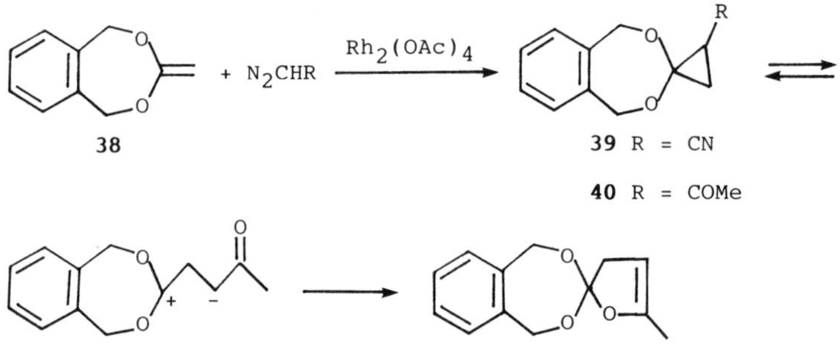

It has been reported that cyclopropenone ketal (**43**) has also the ability to generate 1,1- or 1,3-dipole **44** via ring opening in benzene at 80°C, giving solvolysis product **45** in methanol, and [1 + 2] and [3 + 2] adducts (eg, **46a** and **46b**) with electron-deficient olefins.[33]

43 **44** **45**

46b

46a

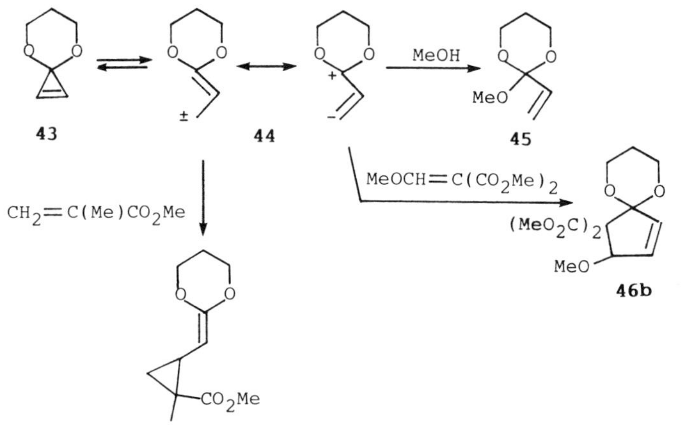

Carbene adduct **47**, obtained from the reaction of 2,3-dimethylindole with :CClF generated at 20°C, in which an amino nitrogen is linked to the cyclopropane ring, appears to be in equilibrium with zwitterion **48**.[34]

47 **48**

X, Y = Cl or F

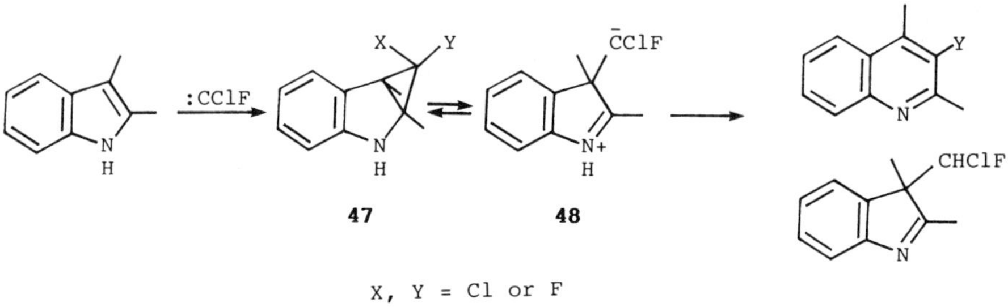

Thermolysis (100°C) of diazoimidazole **49** in benzene derivatives yields an isomeric mixture of arylimidazoles **52**.[35] Electron-donating groups are o,p-directing (eg, anisole, $o:m:p = 67:0:20$), and electron-withdrawing groups are o,m-directing (eg; CF$_3$Ph, $o:m:p = 12:88:0$). It is likely that the reaction proceeds via zwitterions (**51**) arising from ring opening of spironorcaradiene intermediates (**50**). The negative charge of **51** is incorporated into a nonbenzenoid aromatic π system, and the positive charge is stabilized by delocalization through a cyclohexadienyl system. Since a similar result is obtained by photolysis of **49** at lower temperatures ($\approx$25°C), it is evident that ring opening proceeds with low activation energy. Although the product ratios have been assumed to result from highly selective ring opening, it seems difficult to explain the ratios in terms of cation stability.

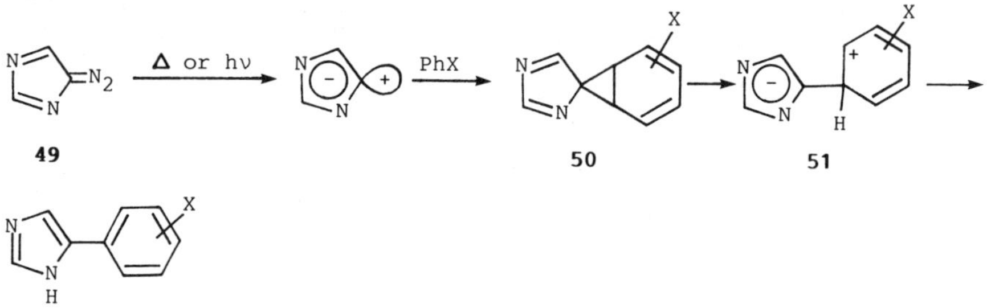

Three-membered rings in which a heteroatom replaces one of cyclopropane carbons are expected to undergo facile heterolytic ring opening.

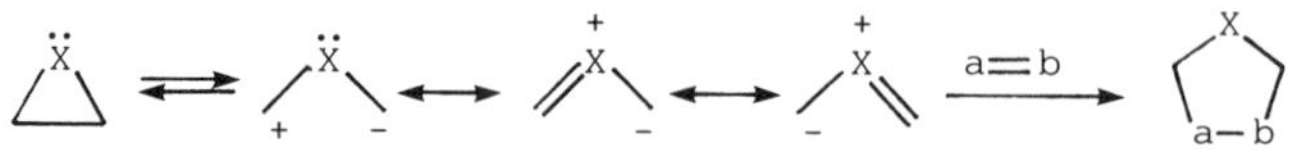

Oxiranes bearing electron-withdrawing groups such as CN and CO$_2$R react with dipolarophiles to give cycloadducts,[36–42] and strong pieces of evidence for the formation of carbonyl ylides have been presented. The trans isomer **53a** of dicyanodiphenyloxirane isomerizes in dioxane at 100°C to cis isomer **53b** until an 85.5:14.5 equilibrium is reached ($\Delta H^\ddagger = 27.2$ kcal/mol, $\Delta S^\ddagger = -8$ eu for the process **53a** → **53b**) and 1,3-dipolar addition occurs at temperatures exceeding 100°C as well, and thus carbonyl ylides have been postulated as the common intermediates involved in both reactions.[37] Oxirane **54** reacts with dimethyl fumarate more rapidly than with benzaldehyde ($k_2/k'_2 = 3.0$ at 125°C). Carbonyl ylide **55**, which is assumed to intervene, can be generated via another unambiguous route showing a very similar reactivity ($k_2/k'_2 = 3.3$).[38]

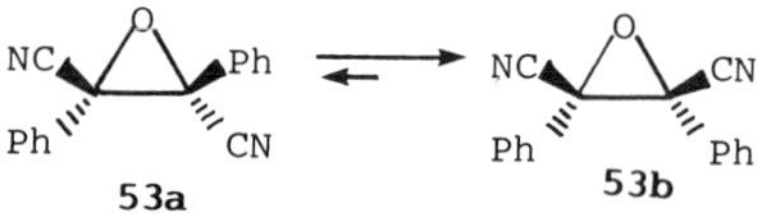

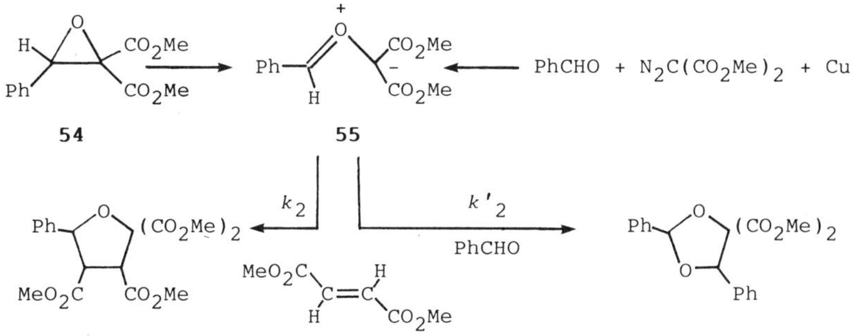

The orbital symmetry rule predicts that thermal conversion of cyclopropyl anions to allyl anions proceeds via conrotatory ring opening so that a cis isomer of cyclopropyl anions would lead to a trans form of allyl anions, and vice versa.[39] Oxiranes as well as aziridines, discussed later, are isoelectronic with cyclopropyl anions. Careful examination of the geometry of cycloadducts obtained from the reactions of cyano-*trans*-stilbene oxide (**56**) and the cis isomer (**58**) with dimethyl fumarate (130°C, 45 hours) indicates that the intermediate carbonyl ylides **57** and **59** result from conrotatory ring opening of the oxiranes.[40] Heterolytic ring opening of oxiranes has now gained general acceptance and has been assumed even in high-temperature thermolyses without solvents[41] (eg, **60** → **61**), as well as in nonpolar solvents (**62** → **63**).[42]

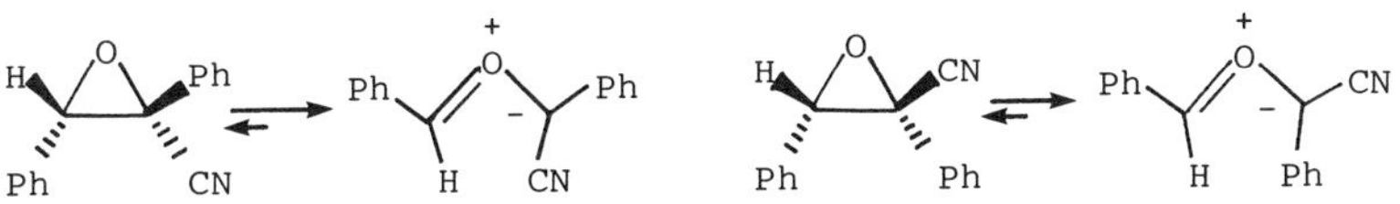

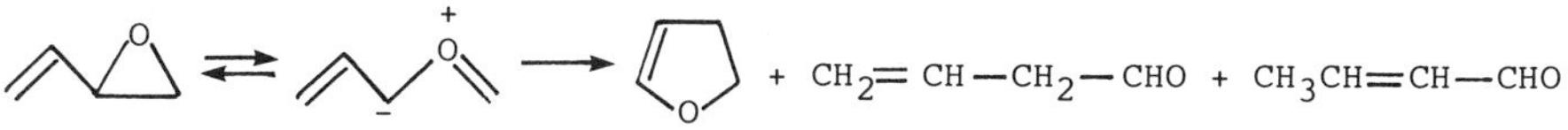

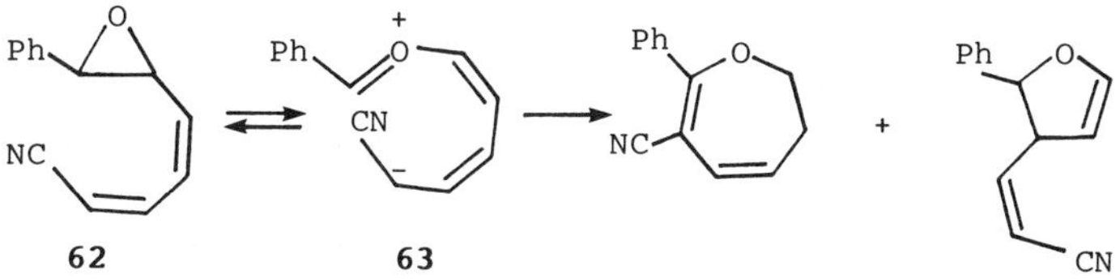

Although a quantitative comparison of reactivities has not yet been presented, a rough survey of reaction conditions suggests that heterolytic ring opening of simple oxiranes would be facilitated by introducing negative groups onto both carbons, as expected from the resonance forms of carbonyl ylides. There are some exceptions due to structural specificity. The strain relief of a fused ring often makes up for the lack of a stabilizing group and causes ring opening against orbital symmetry restraint. Cyclobutene oxide (**64**) affords a purple solution of carbonyl ylide **65** at 100°C in benzene.[43] The ylide can be intercepted by benzonitrile or dimethyl acetylenedicarboxylate.

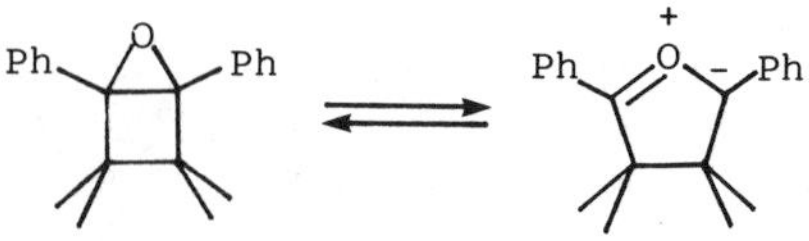

Ring opening is further accelerated when accompanied by aromatization. Conversion of indenone oxide (**66**) into red benzopyrium oxide (**67**) occurs merely upon heating to 80°C in solution.[44] The *cis*-tropone dioxide (**68**) isomerizes rapidly to the trans isomer **69** upon warming at 50°C in acetonitrile (half-life, ≈40 minute), and thus a couple of highly resonance-stabilized ylides (eg, **70** or **71**) have been postulated as plausible intermediates.[45]

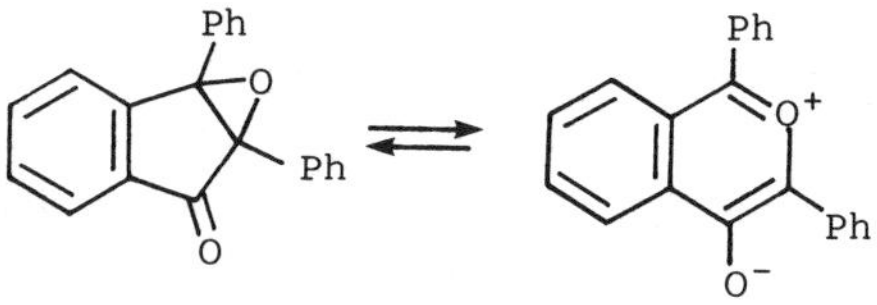

Generation of azomethine ylides from aziridines has been studied more extensively because their 1,3-dipolar cycloadducts have the synthetic potential to produce pyrrolidine-type alkaloids. Aziridine rings bearing phenyl or carboxy ester groups at both carbons undergo carbon–carbon bond scission under moderate conditions.[46] The thermal cis–trans equilibrium of **72** and **73** can be achieved even in the nonpolar solvent CCl_4 at 100°C (cis:trans = 1:4): k(cis–trans) = 3.1 × 10^{-5} s^{-1}; k(trans–cis) = 0.77 × 10^{-5} s^{-1}; $\Delta H^{\ddagger}$(cis–trans) = 26 kcal/mol.[47] Their cycloadditions with a very strong dipolarophile such as tetracyanoethylene (TCNE) proceed by a first-order rate process independent of the TCNE concentration, indicating that the observed rate represents the rate of ring opening itself.

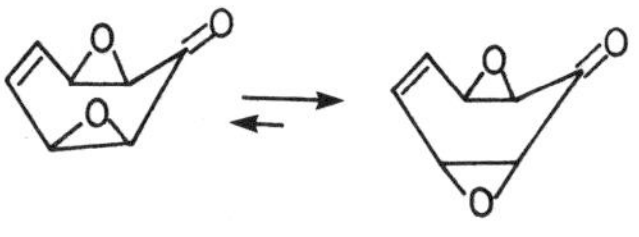

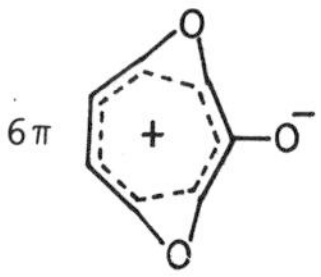

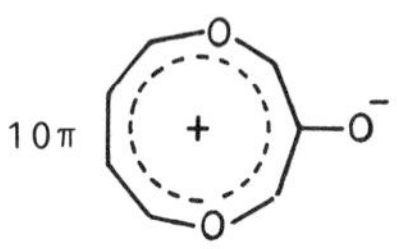

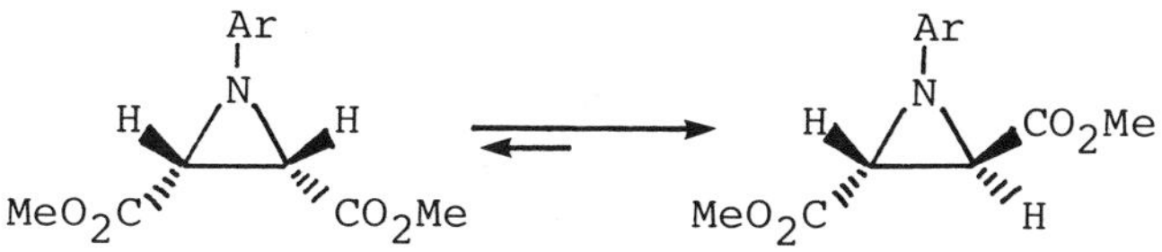

POLAR EFFECTS ON C—C BOND LABILITY 193

[Structures 72 and 73 shown: aziridines with Ar = p-MeOC₆H₄, MeO₂C and CO₂Me substituents in cis (72) and trans (73) configurations]

72 Ar = p-MeOC$_6$H$_4$ **73**

Unequivocal evidence for the generation of azomethine ylides via conrotatory ring opening has been provided by a detailed mechanistic study using these aziridines.[48] Sometimes this mode of ring opening strongly influences the lability of carbon–carbon bonds of aziridines. Although a geometrical difference between **72** and **73** affords little change in the rate of ring opening ($k_{cis} = 4.2 \times 10^{-4}$ s^{-1} and $k_{trans} = 3.8 \times 10^{-4}$ s^{-1} in ethyl acetate at 119°C), trans isomer **74** of the N-t-butyl aziridines undergoes the heterolytic cleavage, giving pyrrole **77**, 40 times more rapidly than cis isomer **75** in methanol at 70°C ($k_{trans} = 1.52 \times 10^{-4}$ s^{-1}; $\Delta H^{\ddagger} = 25.4$ kcal/mol, $\Delta S^{\ddagger} = -1.7$ eu), and such a large difference has been attributed to the nonbonding interaction between the methyl (or benzoyl) and t-butyl groups during conrotatory ring opening of the cis isomer, leading to azomethine ylides **76b** and **76c**.[49]

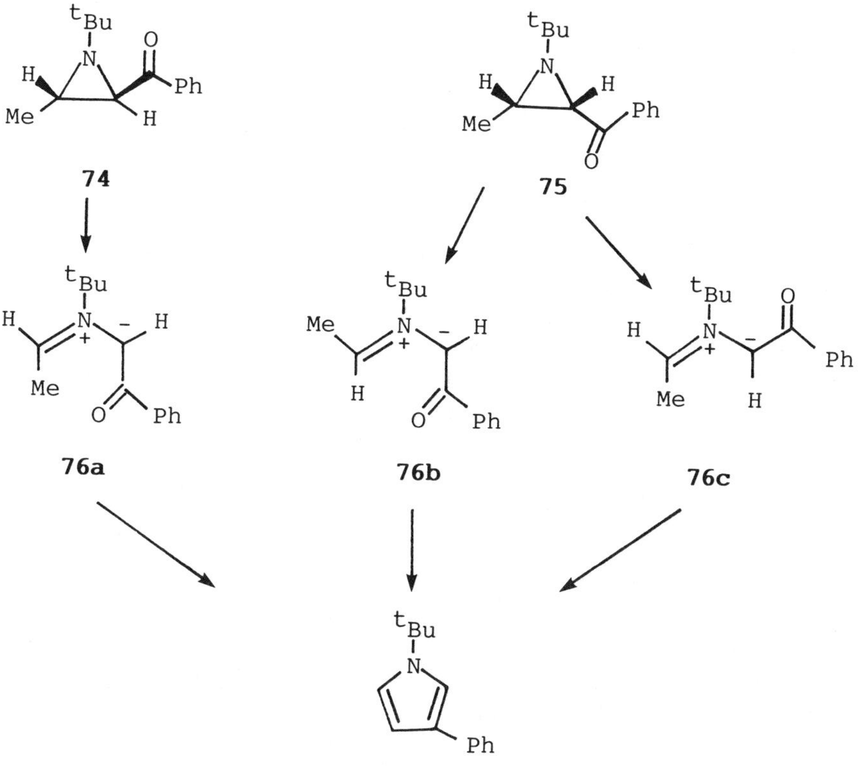

74   **75**

76a   **76b**   **76c**

77

Aziridine **78** constrained in a fused ring is unable to undergo conrotatory ring opening even at 180°C, while photochemical disrotatory ring opening occurs rapidly.[50] However, as already discussed in connection with fused oxirane **66**, ring opening (135°C in toluene) of the nitrogen analogue **79** overcomes orbital symmetry restraints.[51]

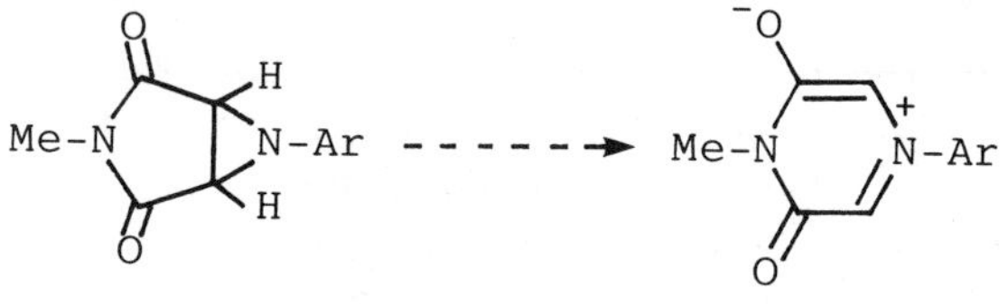

78 Ar = *p*-MeOC$_6$H$_4$

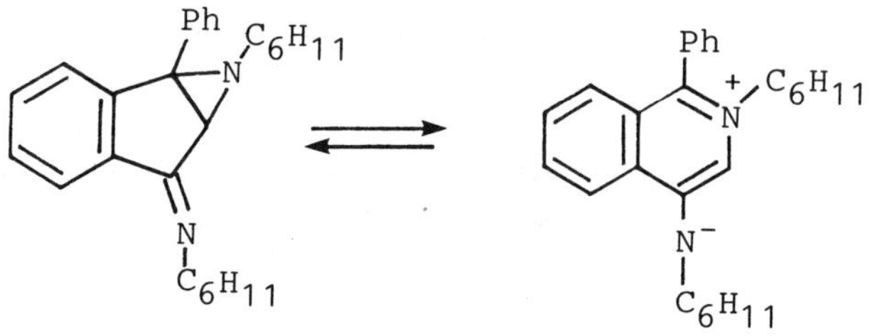

79

Proof that electron-withdrawing groups promote the carbon–carbon bond scission of aziridines has been presented.[52] Thermal decomposition of sulfonylazide **80** in anisole at 155°C affords a mixture of sulfonamides **82** (16%) via carbon–nitrogen bond scission of the intermediate nitrene adduct **81**. On the contrary, decomposition in nitrobenzene leads to the formation of azepine **84** (21%) via azomethine ylide **83** arising from carbon–carbon bond scission of the nitrene adduct in addition to a small amount of the corresponding sulfonamide (5%). It is conceivable that the site of bond scission is changed by the stabilization of the negative charge in the ylide.

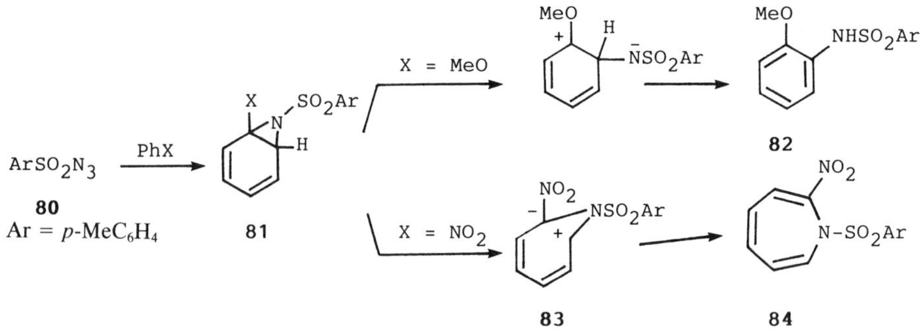

Ring opening of methylenecyclopropanes gives rise to either resonance-stabilized diradicals or zwitterions.[53] The presence of negative charge-stabilizing groups may cause heterolytic bond cleavage. Although strong evidence is lacking, the intervention of zwitterions has been suggested in the following two reactions. Heating a benzene solution of optically active methylenecyclopropane (**85**) at 210°C results in racemization, which presumably proceeds via planar zwitterion **86**.[54] More facile ring opening has been observed in the decomposition of bicyclic dienone (**87**) at 110°C in a benzene solution including isobutene or methanol.[55] The dienone **87** affords isobutene adducts **89a–c**, which are those derived from the intermediate Markovnikov products. With methanol, **87** yields solvolysis product **90**, and the rate of pseudo-first-order kinetics is independent of the methanol concentration ($\Delta H^{\ddagger}$ = 30 kcal/mol, $\Delta S^{\ddagger}$ = 3 eu). The results would support the intermediacy of zwitterion **88**.

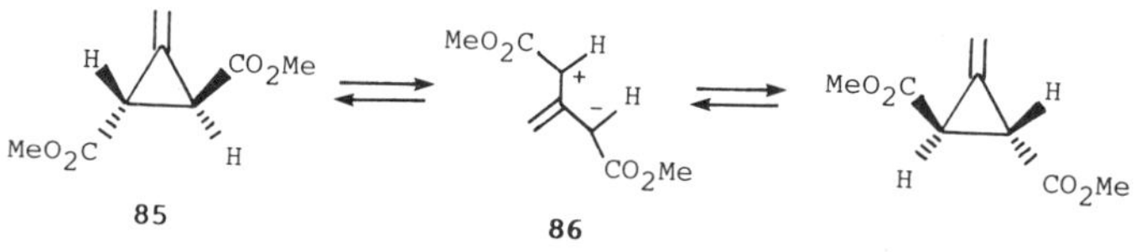

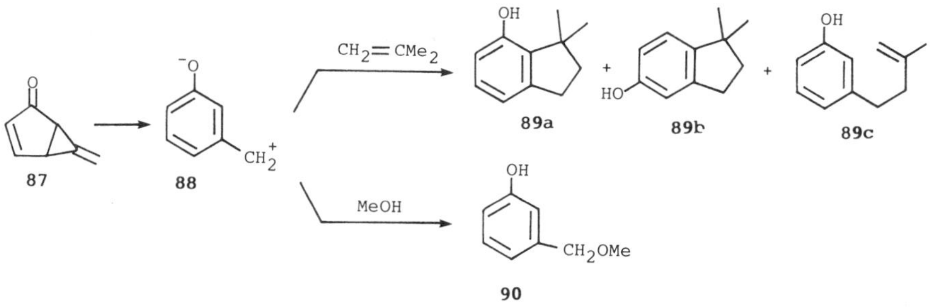

However, there is persuasive evidence supporting the intermediacy of diradicals in the ring opening of a typical methylenecyclopropane system.[12] The rearrangement of ethoxycarbonyl-substituted compound **91** to the isomer **92** proceeds rapidly under mild conditions (X = H, k = 5.62 × 10^{-4} s^{-1} in isooctane at 50°C). Although further introduction of an electron-withdrawing group increases the rate (X = CO$_2$Et, k_{rel} = 1.63), a more pronounced rate enhancement has been effected by an electron-donating group (X = MeO, k_{rel} = 4.35). The result is rationally explained in terms of the captodative effect. If the negative charge is developed strongly enough to induce heterolysis at the carbon-carrying ethoxycarbonyl and phenyl groups, the presence of the electron-donating methoxy group would result in the depression of ring opening.

With cyclopropanones, corresponding to heteromethylenecyclopropanes, zwitterions have been postulated as the common intermediates involved in racemization, cycloaddition, and the Favorskii rearrangement.[56] Whether the zwitterions intervene in cycloadditions of cyclopropanones with dienes to give seven-membered ring products, however, remains a matter of controversy.[57] Cyclopropanones are isoelectronic with cyclopropyl cations and thus disrotatory ring opening is allowed.[38] Although the predicted cycloadduct **97** is actually formed from cyclopropanone **95** (in methanol at room temperature),[58] definitive evidence is difficult to obtain from the product analysis, since direct addition of **95** to a diene yields the same product as well. However, formation of ether **98** may be qualified as evidence for the intervention of zwitterion **96** because an alternative pathway[59] via enol allylic chloride **94**, which is produced from starting substrate **93** before the cyclopropanone formation, seems improbable in this case.

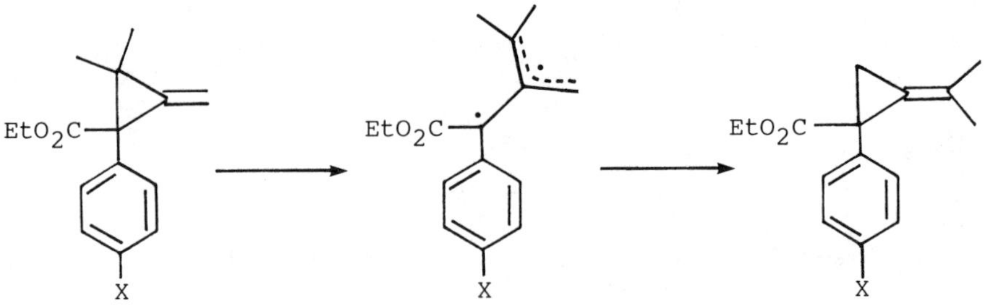

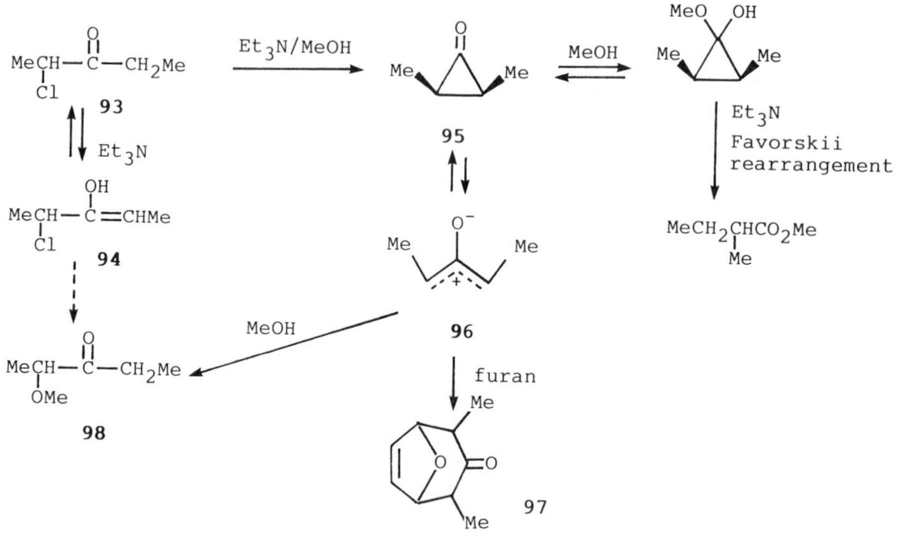

The reaction of α-chloropropanones **99** or **100** with MeONa (0.05 M) in methanol affords Favorskii ester **103** almost quantitatively. However, at very low (10^{-5} M) methoxide concentrations, indanone **104** has been obtained as a by-product (**103**:**104** = 1.6:1), suggesting that cyclopropanone **101** and zwitterion **102** coexist as an equilibrium mixture.[60]

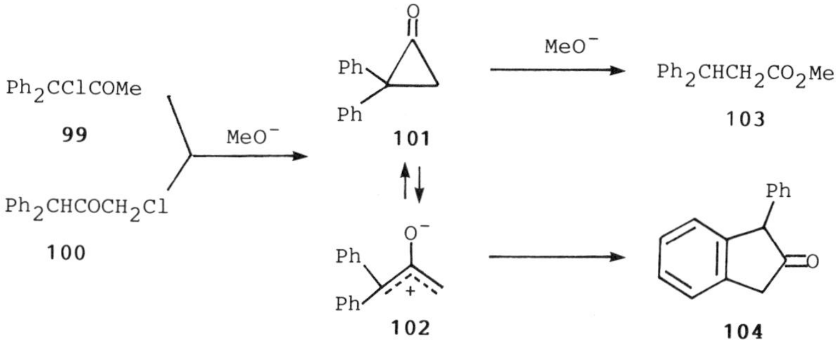

Ring opening of the nitrogen and sulfur analogues leading to the corresponding zwitterions has been suggested without convincing proof.[61,62] Two other examples of heterolytic cleavage of the three-membered ring have been reported. It has been established that azirines **105** undergo ring opening photochemically at the carbon–carbon bond but thermally at the carbon–nitrogen bond. In special cases, however, azirines afford thermal decomposition products that are reasonably assumed to form via zwitterions or carbenes arising from carbon–carbon bond scission.[63] t-Butanolysis of fused episulfone **106** at 0°C has been presumed to proceed via a zwitterion.[64]

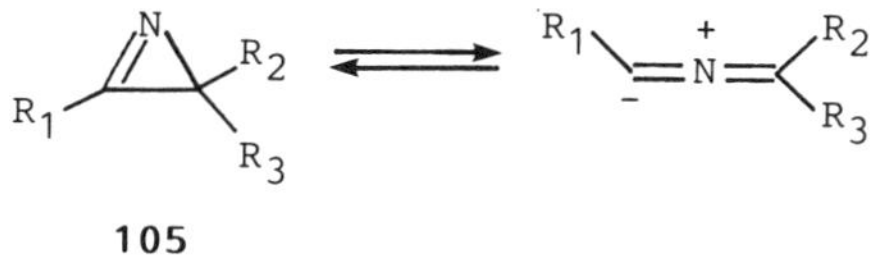

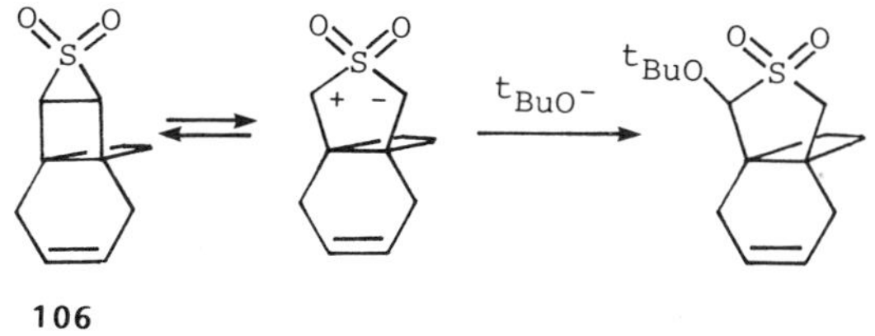

Relief of strain is expected from ring opening of cyclobutane derivatives as well.[9] The presence of electron-withdrawing groups favors generation of zwitterions over that of diradicals. The [2 + 2] adducts of TCNE with dienes bearing two geminal phenyl or cyclopropyl groups (eg, **107**) undergo facile isomerization to the [2 + 4] adducts via zwitterions, which can be trapped by thiophenol.[65] The conversion is complete after 3 hours in acetonitrile at room temperature and faster than in the less polar solvent $CDCl_3$:
$$k(MeCN):k(CDCl_3) = 6.4:1.$$
The ionic character of this type of intermediate has been confirmed by the observation that the reaction of 1,1-diarylbutadienes with TCNE giving [2 + 2] and [2 + 4] adducts exhibits a large ρ (= −5) value in acetonitrile at

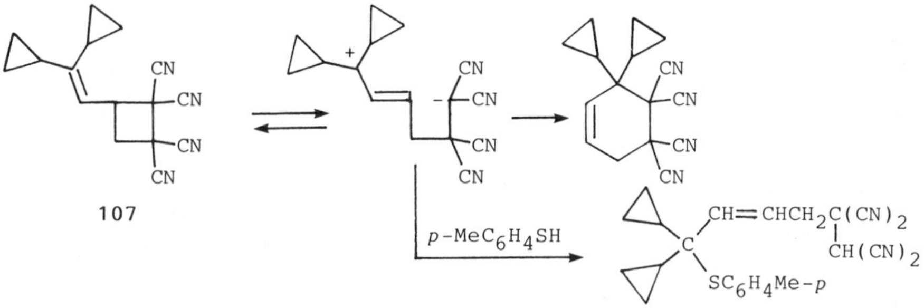

20°C and the [2 + 2] adduct of the *p*-methyl derivative is readily converted to the [2 + 4] adduct via the same intermediate in acetonitrile at 80°C.[66] Similarly, zwitterions are produced from the [2 + 2] adducts **108** of TCNE with enol ethers at room temperature with comparative facility because of high stability of the positive charge and can be trapped by ethanol or acetone.[67]

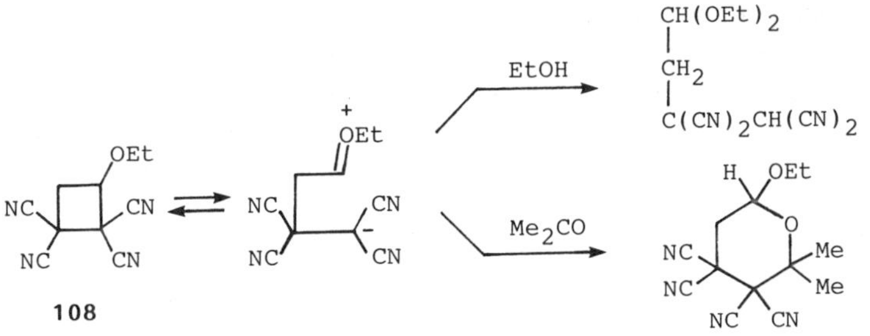

The [2 + 2] adducts of ketenes with allenes undergo heterolytic cleavage if extensive charge delocalization is expected in the corresponding zwitterions (eg, **109** in *o*-dichlorobenzene at ≈130°C).[68]

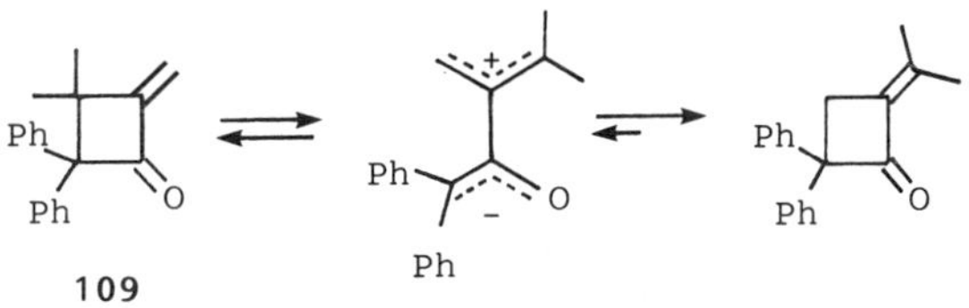

Cyclobutene **110** carrying electron-withdrawing groups at olefinic carbons and an amino group at a saturated carbon is stable in nonpolar solvents but suffers rapid heterolytic ring opening when dissolved in methanol at 20°C, giving a pyrrolizine quantitatively.[69] Analogously, heating **111** in tetrachloroethylene under reflux (121°C) results in the complete conversion into a 2,3-dihydroazepine. Allenic intermediate **112** has been detected.[70]

Generation of a zwitterion through a concerted process in which scission occurs at two positions during charge separation may be advantageous because of extensive charge delocalization, as seen in the isomerizaton of **68**. In this connection, the adducts of cyclobutadiene **113** with TCNE are unique in that the presence of a bicyclobutonium ion in zwitterion **114** induces the heterolytic cleavage not only of four-membered ring **115** but also of more weakly strained five-membered ring **116** in $CDCl_3$ at room temperature.[71]

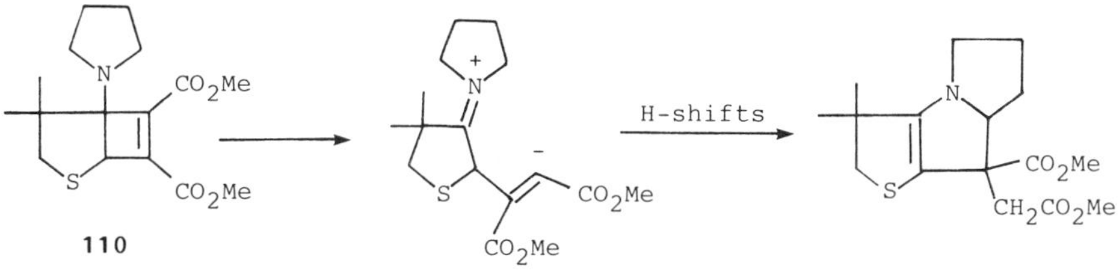

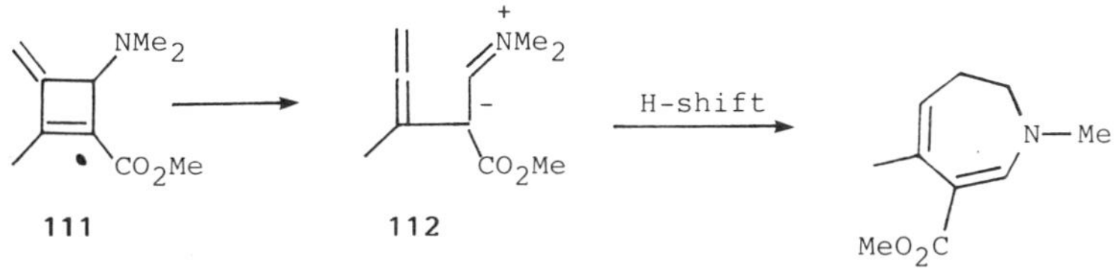

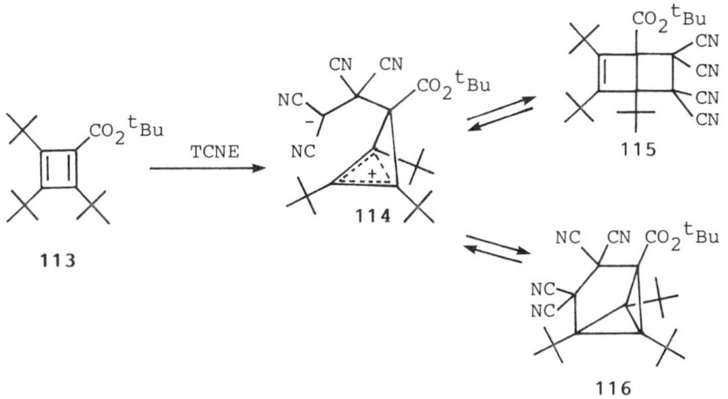

Thermal rearrangement of spiropentanes to methylenecyclobutanes (above 360°C) proceeds via diradical intermediates.[72] However, geminal dicyano substitution not only lowers decomposition temperatures (170°C) but increases the dipolar character of the transition state, since both rearrangements **117** → **118** and **119** → **120** are more than 10 times faster in the polar solvent acetonitrile than in the less polar solvent benzene. Both are stereospecific, but the latter rearrangement in benzene proceeds about 6 times faster than the former, and thus a concerted process has been postulated at least in the latter case, in which disrotatory ring opening of one cyclopropane is induced by the polarization during ring opening of the other cyclopropane carrying the cyano groups.[73]

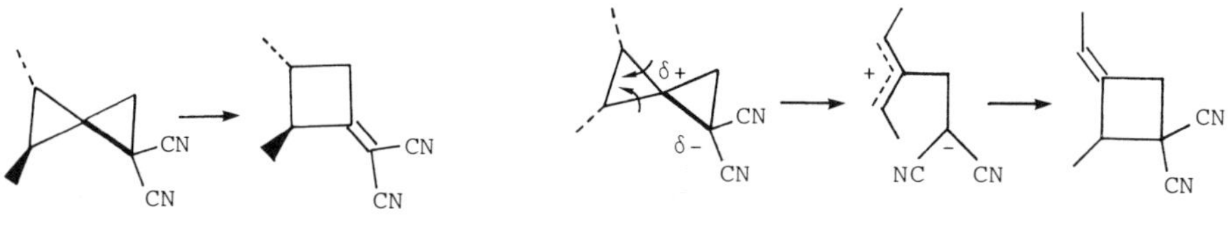

117 → **118**

119 → **120**

Thermal rearrangements of *syn-* and *anti-*azatricycloheptanes **121** and **122** afford dihydroazepine **124** in quantitative yield but differ greatly in reaction temperatures ($k = 4 \times 10^{-5}$ s^{-1} at 121°C for **121** but $k = 3 \times 10^{-5}$ at 350°C for **122**).[74] Compound **121** reacts with *N*-phenylmaleimide with a rate substantially equal to that of the rearrangement to give an isomeric mixture of cycloadducts **125**, suggesting the intervention of zwitterion **123**, whose formation via a concerted π^2s + π^4s process is allowed only in the syn compound.

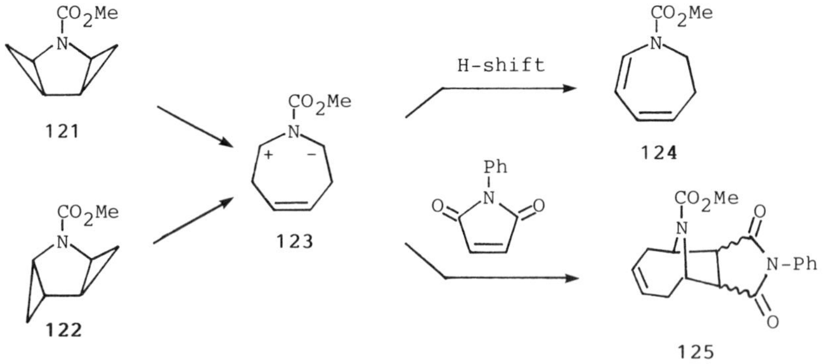

121 → **123** → (H-shift) → **124**

122 → **123**

123 + Ph-maleimide → **125**

A similar concerted mechanism was predicted for the conversion of heteroquadricyclanes **126** and **127** to seven-membered trienes (usually formed as a cycloheptatriene–norcaradiene type of mixture)[74,75] and has gained recent acceptance.[76] These conversions proceed in benzene without difficulty:

126 $\Delta H^{\ddagger} = 27.3$ kcal/mol, $\Delta S^{\ddagger} = 11.1$ eu;
127 $\Delta H^{\ddagger} = 31.9$ kcal/mol, $\Delta S^{\ddagger} = 11.5$ eu (see p. 201)

Substituents strongly affect both the rate and the site of bond scission. For instance, the presence of methoxycarbonyl groups (**128**) in **126** lowers the activation energy by 2.6 kcal/mol and favors exclusive bond scission of the neighboring cyclopropane ring, whereas chloro-substituted compound **129** undergoes bond scission of the opposite cyclopropane ring much more easily (half-life = 10 minutes or $\Delta G^{\ddagger} = 18$ kcal/mol in acetone at −30°C). Zwitterions **130** can be intercepted by dipolarophiles.

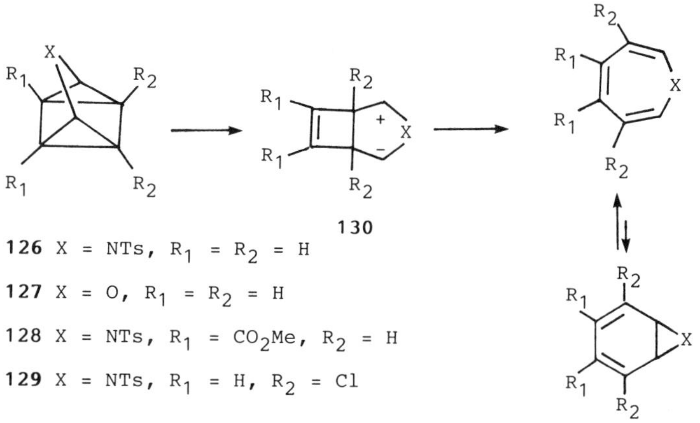

126 X = NTs, $R_1 = R_2 = H$
127 X = O, $R_1 = R_2 = H$
128 X = NTs, $R_1 = CO_2Me$, $R_2 = H$
129 X = NTs, $R_1 = H$, $R_2 = Cl$

Quadricyclanone **131** has a different mode of ring opening. Although the yield is only 6%, tricycloheptenone **133** has been obtained upon heating in refluxing methyl propiolate as a by-product from **131** but not from **134**, and such a difference in reactivity is rationalized by suggesting a mechanism involving zwitterion **132**, whose carboxy ester group is suitably placed at a position capable of stabilizing the negative charge.[77]

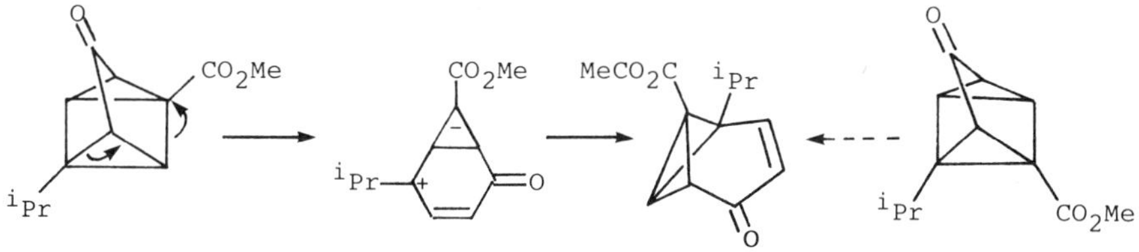

Examples of heterolytic cleavage of five- and six-membered rings are rare, and the possibilities are discussed only in a few cases other than the above-mentioned reaction (**116** → **114**). A plausible one is the conversion of N,N-dimethylaminobenzobarrelene (**135**) into biphenyl **137** via zwitterion **136** in boiling aqueous ethanol.[78] In the presence of D_2O a deuterated product is obtained. Positive charge dispersal ranging from an amino group to two olefinic bonds might be the driving force for the ring opening. Analogously, it seems likely that the interconversion of diastereomeric rugulovasines **138** and **140** in polar solvents proceeds via zwitterion **139** in which the positive

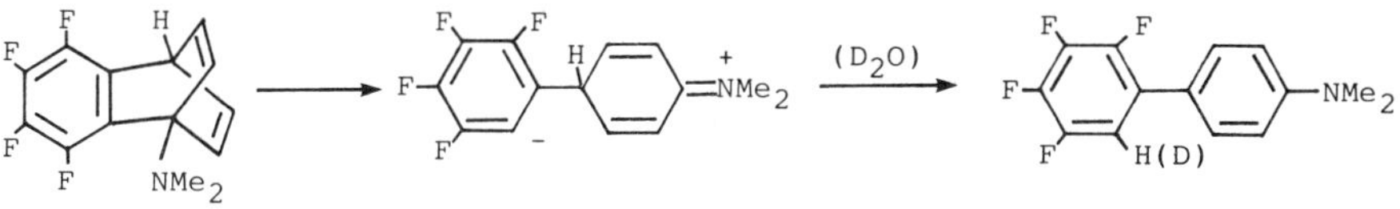

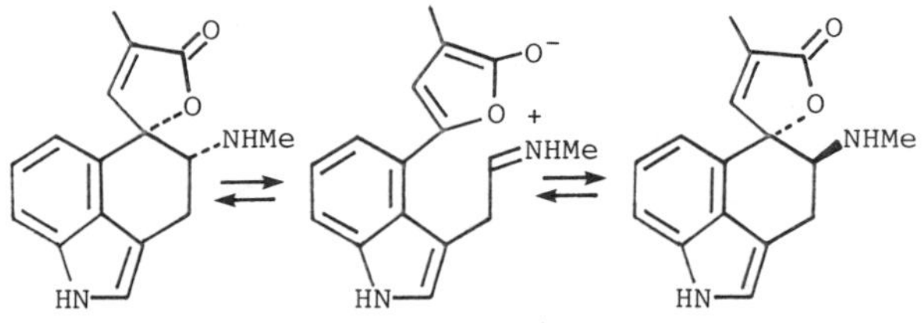

138 ⇌ **139** ⇌ **140**

charge is stabilized by an amino group and the negative charge can be delocalized through a furan ring produced.[79]

The generation of dipolar species **142** from cyclic oligomer **141** (a dimer or trimer) is shown below as the final example of this section.[80] Although it is uncertain whether the bond cleavage proceeds via a zwitterion, the dissociation is facilitated with increasing solvent polarity. On the contrary, a similar compound (**143**) is stable and undergoes no oligomerization.[81] Electronic and steric reasons have been proposed for the differences in reactivity.

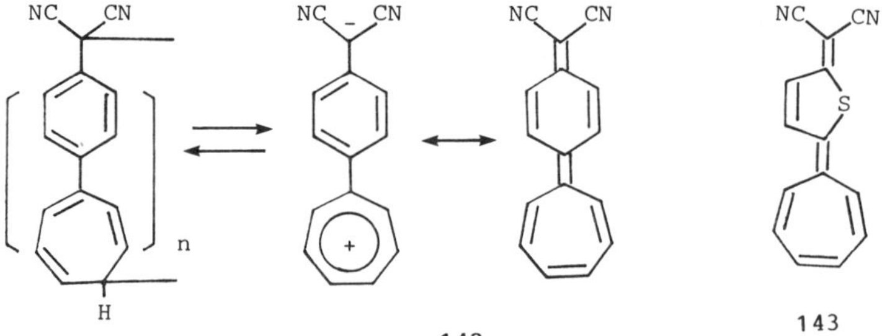

141 ⇌ **142**   **143**

B. Generation of Ion Pairs

The S_N1–$E1$ mechanism is one of the dominant subjects of organic chemistry and has yielded many exciting phenomena as well as important concepts; however, until very recently, its application to the study of the heterolysis of a carbon–carbon bond itself in a common open chain system has evoked little interest because of the lack of efficient leaving groups that terminate in a carbon atom. The relevant works previously reported are limited to those of the reactions involving extrusion of cyanide ion. For instance, triarylmethylnitriles are in equilibrium with triarylmethyl cations and cyanide ion in aqueous solutions.[82] This system favors the covalent form overwhelmingly over the ionic form, which prevents the determination of the exact value of equilibrium constants ($K < 10^{-6}$ mole). Nevertheless, it is clear that the equilibrium is induced by labile carbon–carbon bonds.

$$Ar_3C-CN \rightleftharpoons Ar_3C^+ + {}^-CN$$

The migration of a cyano group from **144** to **146** to 140°C via ion pair **145** has been reported.[83] The cyanide ion can be trapped by the addition of silver nitrate.

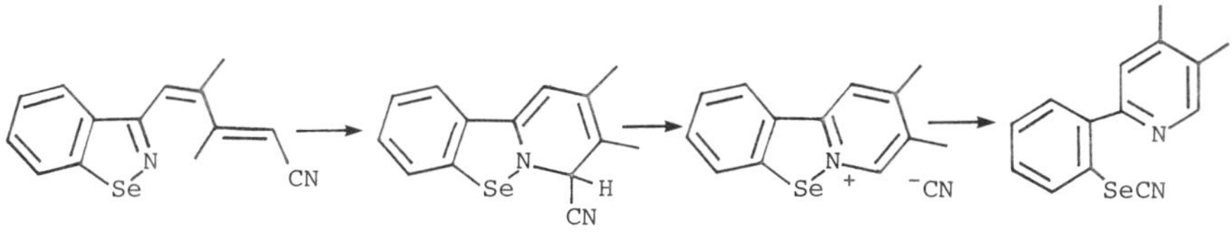

The first clear-cut demonstration of heterolytic cleavage of the bond between two quaternary carbons has been made by using (*p*-nitrophenyl)malononitrile anion **151** as a leaving group anion.[84] The anion reacts with tropylium ion or tris-*p*-methoxyphenylmethyl cation to produce the corresponding covalent product, and the neutral and ionic species coexist at equilibrium in solution. With cyclopropenium ions **149** and **150**, the neutral products **147** and **148** are stable in nonpolar solvents and can be isolated.[85]

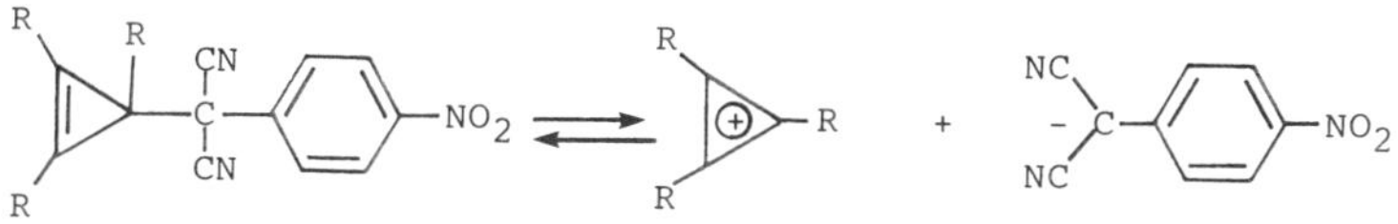

When dissolved in dipolar aprotic solvents, the covalent species rapidly dissociate and come to equilibrium with the generated ions. The equilibrium constant for the heterolysis of **148** increases with increasing solvent polarity (K_{het} at 25°C: eg, CH_2Cl_2, 2.15×10^{-8}; acetone, 1.09×10^{-5}; acetonitrile, 1.28×10^{-4}) and the Born equation holds between the free energies for heterolysis (ΔG) and the reciprocal of the solvent dielectric constants ($1/\varepsilon$). The enthalpy of heterolysis (ΔH_{het}) can be determined by calorimetric measurements of the heat of coordination reaction ($\Delta H_{rxn} \equiv -\Delta H_{het}$) for the cation–anion combinations (Table 5-1). As anticipated, the value of ΔH_{het} or ΔG_{het} is well correlated with that of pK_a for carbanions, in which pK_a is a measure of the anion stability and decreases with increasing its stability or with increasing electron-withdrawing power of substituents in the arylmalononitrile moiety.

Figure 5-3a shows these linear relations along with that obtained from the reaction of 9-substituted fluorenyl anions (pK_a, 8.3 − 24.4) with triphenylmethyl cation ($pK_{R^+} = -6.63$). The assumption that the large discrepancy in intercept results from the difference in the cation stability such as expressed by the pK_{R^+} values of carbocation precursors leads to the derivation of the

TABLE 5-1. Thermodynamic (kcal/mol) and Kinetic (s^{-1}) Data for Heterolysis of Substituted (Cyclopropenyl)arylmalononitriles in Acetonitrile at 25°C (Arnett, ref 85)

		$R^+ =$ trimethylcyclopropenium ion ($pK_{R^+} = 7.4$)				$R^+ =$ triphenylcyclopropenium ion ($pK_{R^+} = 3.1$)	
X	pK_a^a	ΔH_{het}	ΔG_{het}	k_{het}	$(\Delta G^{\ddagger}_{het})^b$	ΔH_{het}	ΔG_{het}
p-NO$_2$			2.76	560	(13.7)		5.39
p-CN			4.37	190	(14.3)		7.00
m-NO$_2$			5.63				8.44
m-CN	1.90	5.78	5.90			10.01	8.78
m-CF$_3$	2.18	6.17	6.55	33	(15.4)	10.63	
m-Cl	2.54	6.47	7.29	27	(15.5)	10.94	
p-Cl	3.14	7.34	7.87	14	(15.9)	11.30	
H	4.24	8.63	9.44	3.8	(16.6)	12.20	
p-Me	4.85	9.57	10.32	2.2	(17.0)	13.40	
p-MeO	5.67	10.91				14.00	

^{a} pK_a of conjugate acids of anions:

$$XC_6H_4C(CN)_2R \rightarrow XC_6H_4C(CN)_2^- + R^+$$

^{b} In MeCN-d_3.

following master equation representing the heat of heterolysis as a function of both cation and anionic stabilities (Figure 5-3b)[85]:

$$\Delta H_{het} = 1.18(pK_a - pK_{R^+}) + 12.0$$

The dynamic NMR method makes it possible to determine the rate for the very fast heterolysis process (Table 5-1). The free energy of activation decreases with decreasing heat of enthalpy for the heterolysis process, reflecting stabilities of the anions produced.

The length of a cleaved carbon–carbon bond may provide information about the nature of such a labile bond in the ground state. It has been confirmed that the carbon–carbon bonds in (trimethylcyclopropenyl)arylmalononitriles [1.588(4) Å and 1.581(3) Å for the p-nitro and p-methoxy compounds, respectively] are longer than usual (1.54 Å). Such elongation may be an important factor in labilizing the bond. However, bond elongation does not reflect product stability, since little difference in bond length appears between the p-nitro and p-methoxy compounds, whereas the p-nitro compound is expected to dissociate about 650 times faster than the p-methoxy compound. The rate of the latter has not actually been measured but can be estimated from the Hammett $\rho\sigma$ plot of the data for other substituents listed in Table 5-1 ($\rho = 2.5$ in MeCN-d_3 at 25°C, $r = 0.99$).

As another example involving extrusion of a nonbenzenoid aromatic cation, heterolysis of compound **152** giving tropylium ion and carboanion **153**

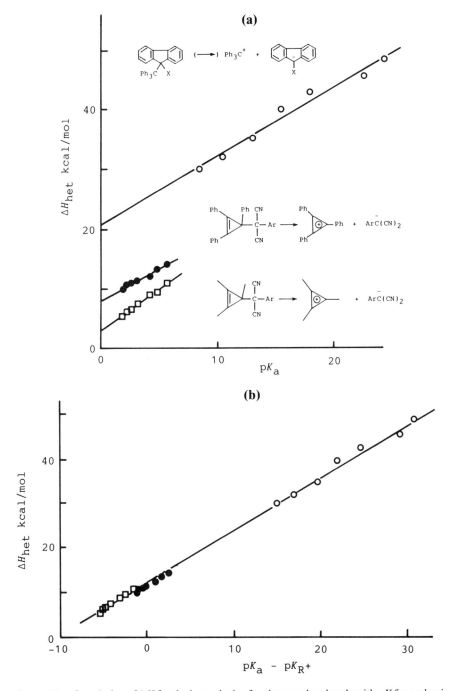

Figure 5-3. Correlation of ΔH for the heterolysis of carbon–carbon bonds with pK for carbanions and carbocations: open circles, solid circles, open squares in (b) correspond to those reactions shown in (a). (From: Arnett, E. M. *J. Am. Chem. Soc.* **1984**, *106*, 6726.)

is unique in that **152** is a pure hydrocarbon. The extent of the dissociation form amounts to about 17% in DMSO at ambient temperature.[86]

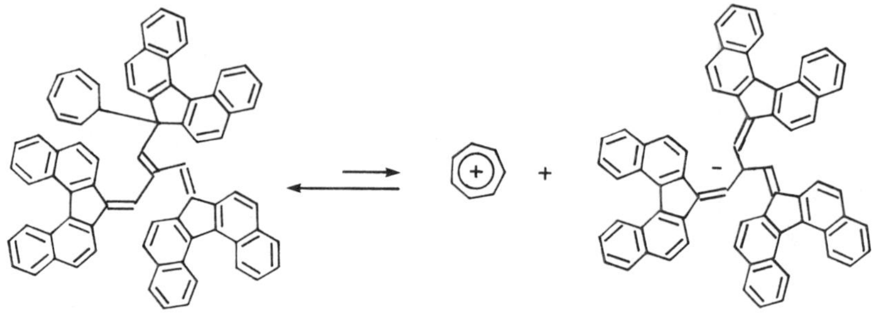

152   **153**

The reaction of 1,3-di-*p*-tolyltriazene **154** with TCNE in methanol at room temperature affords Schiff's base **159** and hydrazone **160** in good yields, formally analogous to the products of olefin metathesis, and a mechanism via cycloadduct **161** was first suggested.[87] However, IR measurements using doubly ^{15}N-labeled triazene **162** provide definitive evidence for a mechanism via adduct **155** followed by rearrangement to azo compound **156**, which quickly undergoes carbon–carbon bond heterolysis to give ionic species **157** and **158**, leading to the final products.[88] Crossover experiments using an arylamine different from the component of the triazene in the presence of acetic acid (10% in benzene) further support this mechanism. Under acidic conditions, triazenes are in equilibrium with amines and diazonium salts. A 1:1 mixture of triazene **163** and TCNE yields Schiff's base **164** (85%) and hydrazone **165** (88%). On the contrary, a 1:1:2 mixture of **163**, *p*-toluidine, and TCNE affords another Schiff's base, **166** (37%), at the expense of **164** (39%), while the yield of **165** is little reduced (86%). Although a possibility of retro-ene type of bond cleavage instead of the heterolytic process cannot be ruled out in this reaction, an analogous reaction of azo intermediate **168**

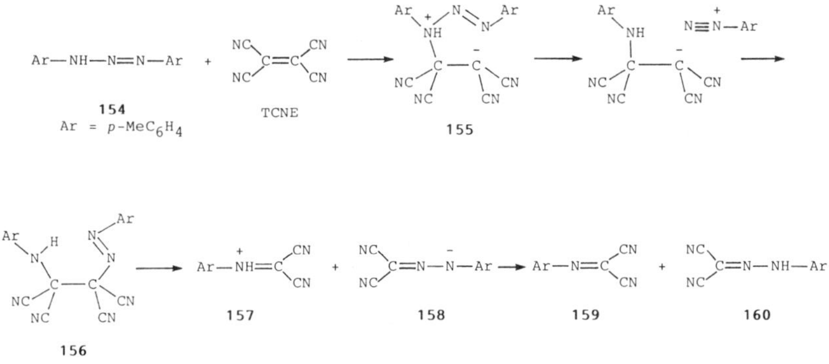

produced by the diazonium coupling to malononitrile **167** furnishes sufficient evidence for the heterolytic cleavage, since the *retro*-ene type of cleavage directly giving final products **160** (94%) and **169** (91%) is sterically impossible in this case. These azo intermediates are too unstable to be isolated. Both cations and anions generated from the azo compounds are highly stabilized by delocalization through the electron-donating amino group and through the diaza allylic system bearing two electron-withdrawing cyano groups, respectively.

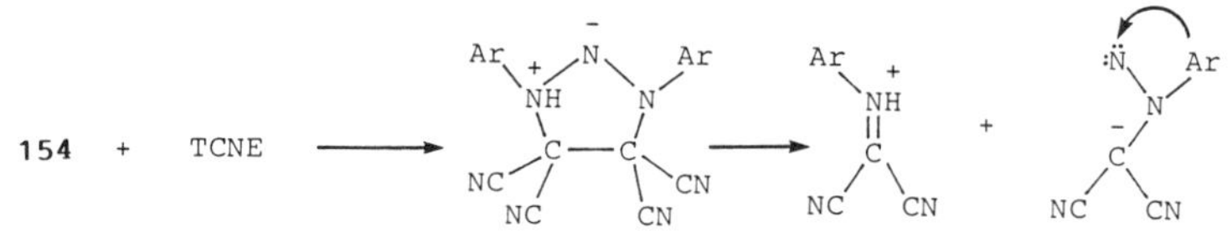

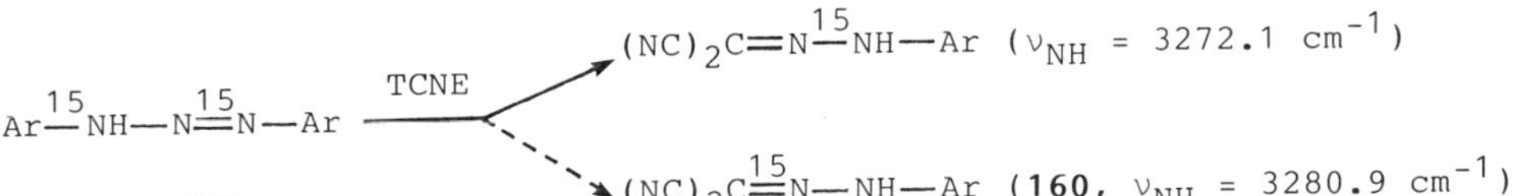

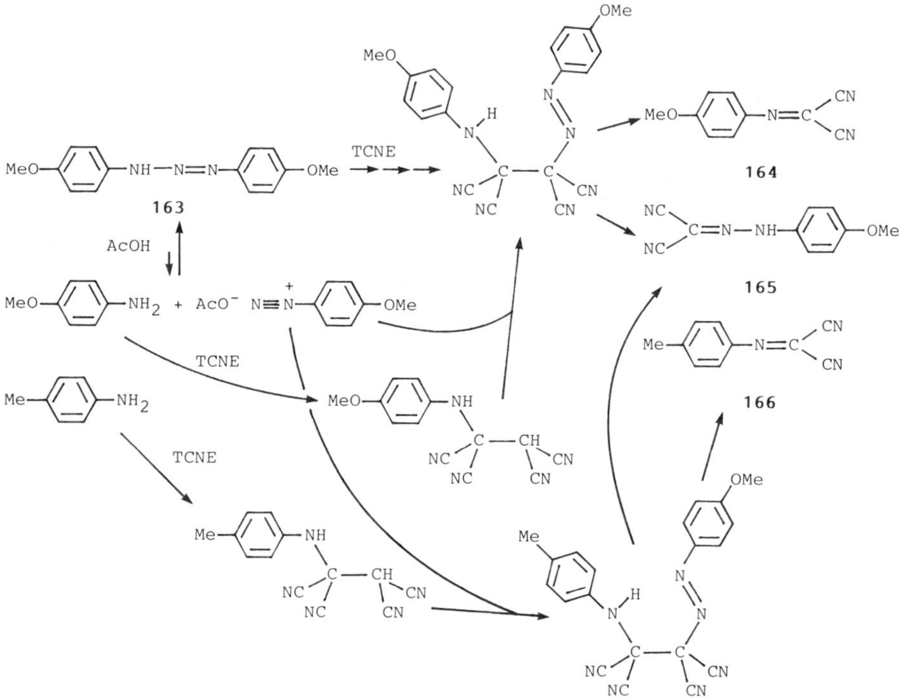

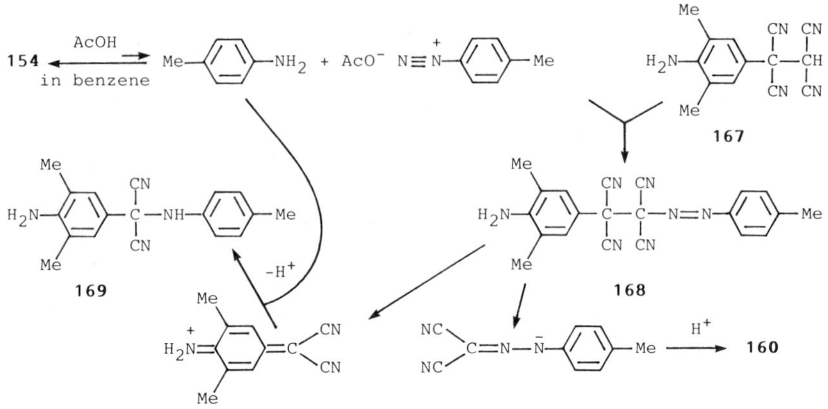

The high stability of the hydrazone anions is important in achieving carbon–carbon bond heterolysis capable of generating unstable carbocations. In fact, it has turned out that arylazodicyanomethanides substituted by an electron-withdrawing nitro or cyano group are able to serve as efficient leaving groups for the generation of *t*-cumyl cation **171**, a typical carbocation in solvolysis reactions, as is evident from the fact that solvolysis of *t*-cumyl chloride is the reference reaction of the Brown–Hammett equation. Heating azo compound **170** gives rise to *t*-cumyl methyl ether **173** in methanol and α-methylstyrene **174** in DMSO and in DMF together with hydrazone **175**. The reaction in pyridine affords *N-t*-cumylpyridinium hydrazonide **176**. Formation of another novel product **177** (ie, a rearranged isomer of **170**) has been observed upon reaction in acetonitrile and in acetone. In any polar solvent used, the total yields of the products derived from cation **171** amount to 93 to 99%. These products are consistent with those expected from a S_N1–E1 mechanism.[89]

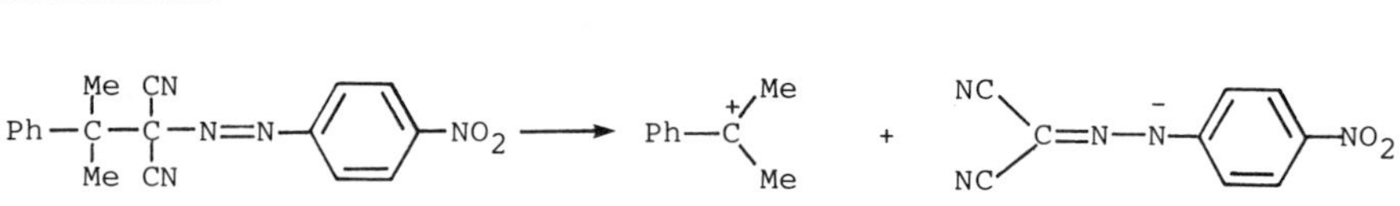

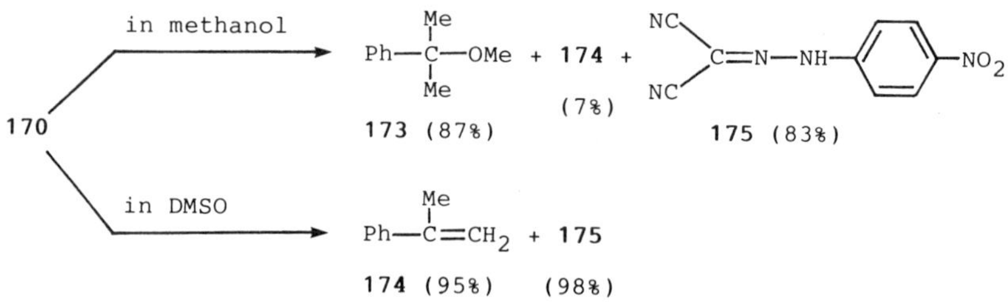

[Ph—C(Me)(Me)—N⁺=⟨C₆H₄⟩ , 172]

176

Me   N=C(CN)₂
Ph—C—N
Me   ⟨C₆H₄—NO₂⟩

177

The rate, determined by the NMR measurements using deuterated solvents, increases with increasing electron-withdrawing power of substituents on the phenyl group. The activation parameters for compound **170** showing the highest reactivity are as follows: $\delta H^{\ddagger} = 24.4$ kcal/mol, $\Delta S^{\ddagger} = 0.7$ eu in methanol; $\Delta H^{\ddagger} = 21.5$ kcal/mol, $\Delta S^{\ddagger} = -3.8$ eu in DMSO.

The solvent effect on the rate of decomposition offers an intriguing field of study. Usually, S_N1–$E1$ reactions of uncharged substrates bearing halides or sulfonates as leaving groups are faster in protic solvents than in dipolar aprotic solvents because of the stabilization of the leaving group anion through hydrogen bonding. For instance, methanolysis of t-cumyl chloride proceeds 90 times faster than its decomposition in DMSO at 31°C ($k = 742 \times 10^{-5}$ s^{-1} in methanol-d_4; $k = 8.30 \times 10^{-5}$ s^{-1} in DMSO-d_6). In sharp contrast, the rate of decomposition of azo compound **170** increases by a factor of 12 on changing the solvent from methanol to DMSO at 31°C ($k = 2.67 \times 10^{-5}$ s^{-1} in methanol-d_4; $k = 32.3 \times 10^{-5}$ s^{-1} in DMSO-d_6). It is noteworthy that the (p-nitrophenyl)azodicyanomethanide moiety departs more easily than chloride in DMSO. The results indicate that hydrazone anion **172** is more stable than chloride ion in aprotic solvents because of extensive charge delocalization, which is spread over the whole molecule, whereas in protic solvents chloride ion is much more stabilized by hydrogen bonding. With anion **172**, charge

$$\text{Ph—C(Me)(Me)—Cl} \xrightarrow{\text{in methanol}} \left[\text{Ph—C}^{\delta+}(\text{Me})(\text{Me})\text{-----Cl}^{\delta-}\text{-----H—OMe}\right]^{\ddagger} \longrightarrow \text{Ph—C}^+(\text{Me})(\text{Me}) + \text{Cl}^-\text{-----H—OMe}$$

dispersal as well as the decrease in nitrogen basicity induced by electron-withdrawing cyano and nitro groups, precludes tight hydrogen bonding as compared with the undissociated state.

There is reliable evidence that the unprecedented solvent effect comes from the nature of leaving group anion **172** but not from that of the bond cleaved. The hydrazone **177**, an isomer of azo compound **170**, undergoes heterolysis of the carbon–nitrogen bond to produce the same ions as does **170**. Although the heterolysis of **177** proceeds much more slowly, its behavior toward differential solvation is quite similar to that of **170**; eg, k_{rel}(DMSO-d_6 versus methanol-d_4) = 3.4 at 60°C.

It is worth noting, from a bioenergetic point of view, that the behavior of anion **172** cited above is closely connected with the function of p-trifluoromethoxyphenylhydrazonomalononitrile (FCCP) and m-chlorophenylhydrazonomalononitrile (CCCP) as efficient uncouplers of oxidative phosphorylation in mitochondrial systems.[90] According to Mitchell's chemiosmotic theory, conversion of ADP to ATP requires a proton gradient across the mitochondrial bilayer membrane created by respiration. However, the uncouplers behave as proton ionophores, carrying protons across the membrane, because of the lipophilic nature of their anionic forms due to extensive charge delocalization, in which the lipophilicity implies the stability of anions without solvation as well as without the aid of counterions, thus quickly dissipating the proton gradient to inhibit the formation of ATP (Figure 5-4).

Polar unimolecular reactions therefore can be classified into two types with respect to the differential solvation:

1. "Hydrogen-bond-susceptible" reactions in which a growing negative charge is stabilized through hydrogen bonding with protic solvents
2. "Hydrogen-bond-insusceptible" reactions, which lack the ability to strengthen hydrogen bonding with protic solvents during charge separation

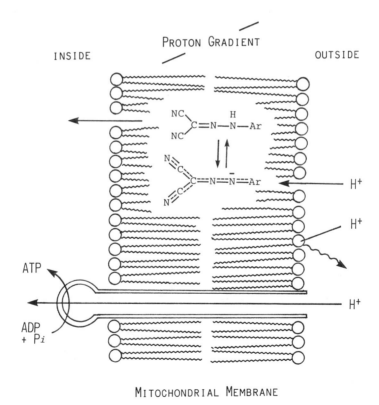

Figure 5-4. Role of uncouplers (FCCP, Ar = p-CF$_3$OC$_6$H$_4$; CCCP, Ar = m-ClC$_6$H$_4$).

The use of the quantity $P_s = RT \ln k_{DMSO}/k_{methanol}$ (kcal/mol), defined as ionizing power of dimethyl sulfoxide relative to that of methanol, is convenient for the classification; that is, we have a polar reaction of type 1 when P_s is less than 0 and type 2 when P_s exceeds 0. Some P_s values for S_N1–$E1$ reactions are as follows: **170** = 1.4 (60°C), **177** = 0.8 (60°C), t-cumyl chloride = −2.7 (31°C), t-butyl chloride = −2.7 (120°C), and **180** = −1.5 kcal/mol (75°C). Hydrogen-bond-insusceptible behavior has been observed not only in the S_N1–$E1$ reactions mentioned above but also in pericyclic reactions, as will be discussed in Section 4.

The heterolysis of azo compound **178** also proceeds easily (k = 70.0 × 10^{-5} s^{-1} in methanol-d_4 and 466 × 10^{-5} s^{-1} in DMSO-d_6 at 60°C; thus P_s = 1.3 kcal/mol) to generate 2-cyclopropyl-2-propyl cation, whose stability is similar to that of t-cumyl cation. The generation of t-butyl cation via carbon–carbon bond heterolysis is important because it is regarded as the most typical carbocation of the purely aliphatic type in S_N1–$E1$ reactions. However, t-butyl cation is produced in solvolysis reactions much more slowly than t-cumyl cation; for instance, the difference in free energy of activation for methanolysis between t-butyl chloride and t-cumyl chloride amounts to about 5 kcal/mol. Although its generation from azo compound **179** is com-

$$\text{R}-\underset{\underset{\text{Me}}{|}}{\overset{\overset{\text{Me}}{|}}{\text{C}}}-\underset{\underset{\text{CN}}{|}}{\overset{\overset{\text{CN}}{|}}{\text{C}}}-\text{N}=\text{N}-\underset{}{\underset{}{\bigcirc}}-\text{NO}_2$$

178 R = c-C_3H_5

179 R = Me

pletely suppressed by the preferential occurrence of carbon–nitrogen bond heterolysis, sooner or later this difficulty will be overcome by using an efficient leaving-group anion of the novel hydrogen-bond-insusceptible type.

TABLE 5-2. Solvent Effects $k \times 10^5$ (s^{-1}) on the Rate for S_N1–$E1$ Reactions

Solvent	**170** (60°C)	**177** (60°C)	PhCMe$_2$Cl (50°C)	t-BuCl (120°C)	**180** (75°C)[a]	
Methanol-d_4	97.2	2.53	6410[b,c]	1360[b,d]	132	(166)[b]
DMSO-d_6	772	8.66	43.2[b,e]	43.7[b,f]	17.2	(18.2)[b]
DMF-d_7	245	5.11	3.72[b,e]	8.71[b,f]	5.57	(4.96)[b]
Pyridine	76.8	2.94	0.512	1.05	2.36	(2.14)
MeCN-d_3	57.6	2.04	6.40[b,e]	6.92[b,f]	6.65	(6.01)[b]
Acetone-d_6	29.6	1.36	0.166	0.603[b,f]	0.94	(0.86)[b]

[a] Rates in parentheses are from Reference 91.
[b] Undeuterated solvents.
[c-e] Calculated from data at other temperatures:
[c] Okamoto, Y.; Inukai, T.; Brown, H. C. *J. Am. Chem. Soc.* **1958,** *80,* 4976.
[d] Biordi, J.; Moelwyn-Hughes, E. A. *J. Chem. Soc.* **1962,** 4291.
[e] Uchida, T.; Marui, S.; Miyagi, Y.; Maruyama, K. *Bull. Chem. Soc. Jpn.* **1974,** *47,* 1549.
[f] Abraham, M. H.; Abraham, R. J. *J. Chem. Soc. Perkin Trans.* 2 **1974,** 47 and references cited therein.

The effects of six representative solvents on the rate of various S_N1–$E1$ reactions are summarized in Table 5-2 and compared in Figure 5-5, in which the hydrogen-bond-susceptible decomposition of **180** is used as a reference reaction, since the most reliable rates without involvement of internal return are expected throughout protic and aprotic solvents; the scale $W_{ion} = RT \ln k_{ion}/k_{ion,methanol}$ (kcal/mol) is introduced, where k_{ion} is the rate of decomposition of **180**.[91] A common feature of the plots is that a straight line composed

MeO—⟨⟩—C(Me)(Me)—CH$_2$OTs   ⟶   Me$_2$C—C(H)(H) [with MeO$^+$-cyclohexadienyl bridge] TsO$^-$ ⟶   Me$_2$C$^+$—CH$_2$—⟨⟩—OMe   TsO$^-$

180

of the points for four aprotic solvents holds with an increase in the order acetone < pyridine < DMF < DMSO irrespective of the degree of negative charge dispersal and thus may be attributed mainly to the difference in cation solvation. In cases of hydrogen-bond-susceptible reactions, all data points are approximately linear. On the other hand, with hydrogen-bond-insusceptible reactions the points for methanol and acetonitrile exhibit downward shifts from the line. Although acetonitrile has little ability to hydrogen bond with anions, its electrophilic property is known and may induce charge transfer in solution from leaving group anions to the sterically less hindered cyano carbon depending on the degree of charge dispersal and nucleophilicity of anions, while a different mode of interaction has been assumed by other workers (the anion is located in front of the methyl hydrogens along the axis of the acetonitrile molecule, but in any mode the increase in charge dispersal leads to the decrease in solvation).[92]

These findings suggest that caution may be needed to prevent the erroneous use of the solvent effect on reaction rates as a criterion for examining the polarity of transition states. In comparing the differential solvation of the reaction of azo compound **170** with that of *t*-cumyl chloride in a very limited series of solvents (eg, among methanol, acetonitrile, and acetone), one will obtain the misleading impression of low slope for **170** and high slope for *t*-cumyl chloride for these three data points (see Figure 5-5). Thus, one would erroneously assign a much less polar transition state for the reaction of **170**. Charge dispersal must be discussed separately from its separation. Even though distinct charge separation is attained in the transition state, charge dispersal, if widely developed, stabilizes charges themselves but results in an accompanying decrease in stabilizing solvation as well. The effect of hydrogen bonding is particularly sensitive to the degree of charge dispersal because it is a highly localized interaction. Dipolar aprotic solvents like DMSO and DMF are ideally suited for the acceleration of hydrogen-bond-

POLAR EFFECTS ON C—C BOND LABILITY 213

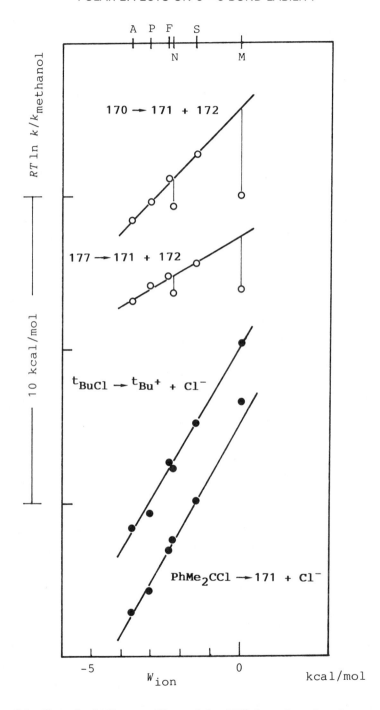

Figure 5-5. Plots of $-\Delta\Delta G^\ddagger$ versus W_{ion}, and the APFS line, where the solvents are: A = acetone, P = pyridine, F = DMF, N = MeCN, S = DMSO, and M = methanol.

insusceptible reactions because of their excellent ability to stabilize cations as well as because of their dispersion forces.

Elongation of the carbon–carbon bond cleaved has been observed in azo compound **170** [1.597(5) Å].[93] Although the bond is slightly longer than that for compound **147** (1.588 Å), **170** undergoes heterolysis much more slowly than **147** ($k_{170}/k_{147} = 1.24 \times 10^{-8}$ in acetonitrile at 25°C). It is inconceivable that the two compounds differ significantly in the steric effect on bond length. Furthermore, the geometry around the carbon–carbon bond is typical of sp^3 hybridizations in any case. Such an enormous difference in reactivity, therefore, strongly supports the view that destabilization due to bond elongation in the ground state is not important as a factor increasing the rate of heterolysis and is overwhelmed by stabilization through resonance effects during charge separation in the transition state. Two empirical rules for the relationship between bond length and reactivity have been proposed on the basis of studies of carbon–oxygen bond heterolysis[94]:

1. "The longer the bond, in a given system, the faster it breaks."
2. "The more reactive the system, the more sensitive is the length of the bond to structural variation."

Obviously carbon–carbon bond heterolysis violates both rules. If the interactions of the n or π orbital of donor-type substituents with the σ^* orbital of the bond and those of the π^* orbital of acceptor-type substituents with σ orbitals of the bond exist in the ground state, they are related directly to cation and anion stabilities, respectively. Bond polarization due to negative groups causes the lowering of both σ and σ^* orbitals, which may be favorable for the former interaction but not for the latter because of the high energy gap except in cases of cleavage occurring in a bond of pseudo-π character in highly strained, small-ring systems. Even though the former interaction related to cation stabilization is effected by electron-withdrawing substituents on the leaving group, it is merely a secondary effect, since these substituents are of primary importance in stabilizing the leaving group anion itself through the latter-type interaction in the transition state.

In a carbon–oxygen bond, it appears difficult to examine whether the σ–π^* interaction induces effective bond elongation, since the n–π^* interaction

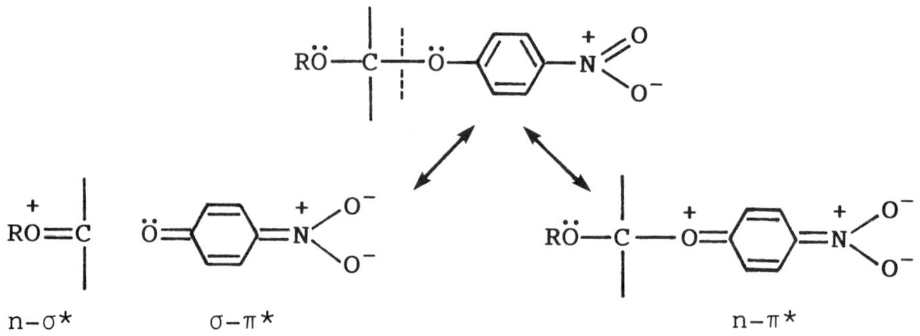

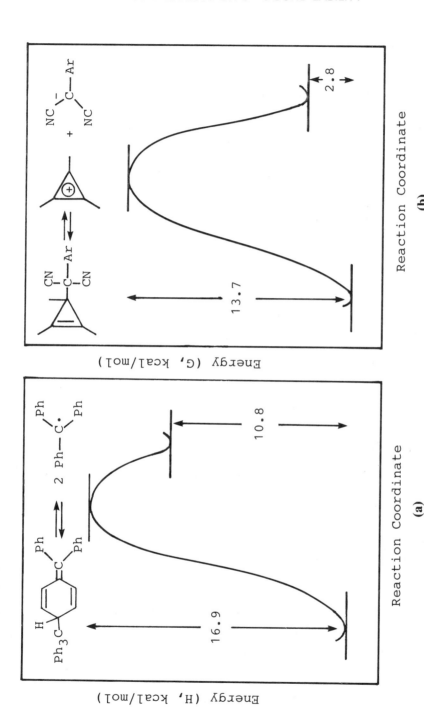

Figure 5-6. Energy profiles of very facile carbon–carbon bond cleavage reactions: (*a*) in benzene[17] and (*b*) in acetonitrile at 25°C (Ar = *p*-NO$_2$C$_6$H$_4$).[85e]

between the mobile lone pair electrons of the oxygen atom and the π system of an acceptor-type substituent attached to the oxygen may behave like a σ–π^* interaction in such a way that it acts as a factor affecting the inductive effect, which strongly causes polarization and thus elongation of the carbon–oxygen bond. On the other hand, studies of the cleavage of quaternary carbon–carbon bonds preclude such a complicated problem.

If one compares the reactivities in a series of analogous compounds in which only the bulkiness of substituents is important as a contribution to the effect on relative rate, a linear free energy relationship would be expected between the reactivity and destabilization due to bond elongation, because the steric congestion in the ground state may correlate with its relief in the transition state. Homolytic reactions generating simple radicals without spin or charge delocalization may exhibit a good correlation.[95] However, there is still a contradictory case even in a simple homolytic reaction, in which the extent of bond elongation does not exactly correlate with that of strain relief.[19,96]

Figure 5-6 shows energy profiles for two very facile carbon–carbon bond cleavage reactions, namely, homolysis of the triphenylmethyl dimer and heterolysis of trimethylcyclopropenylmalononitrile **147**. The most striking characteristic of these reactions is the relatively large energy difference between the activated complex and the product, as well as the low dissociation barrier compared to the usual cleavage reactions generating unstable intermediates. It is noteworthy that the reactivity is still influenced essentially by the product stability despite the very small energy difference between the reactant and the product in the heterolysis of **147**. It is likely that the highly resonance-stabilized transition state is attained only by the almost planar structures around the carbons of the bond cleaved, and thus each fragment has a geometry much closer to the product than to the reactant. The high stability of the product as compared with the transition state may be attributed to the relief of nonbonding interactions between fragments together with full contributions of resonance and solvation. These considerations suggest that a geometrical resemblance between the transition state and the product might be regarded as a propensity intrinsic to homolytic and heterolytic paths irrespective of the energy difference between the two.

4. POLAR EFFECTS ON PERICYCLIC CLEAVAGE

All pericyclic reactions involving carbon–carbon bond cleavage are objects of discussion. However, only some selected topics are described here. The Diels–Alder reaction is the most typical of cycloadditions.[97] It is well known that a combination of donor-type dienes and acceptor-type dienophiles greatly enhances reactivity relative to the parent reaction of butadiene with ethylene. The dipolar nature of the transition state has been established by observing the substituent and solvent effects on rate. For instance, the reaction of 1-arylbutadiene with maleic anhydride in dioxane at 25°C exhibits a ρ value of -0.69.[98] This implies that the reversal of reactions of this type (retro-Diels-Alder reactions) proceeds via a dipolar transition state as well. Compound **181** undergoes very facile fragmentation to give addends **182** and **183** under mild conditions (in dioxane at temperatures 30–50°C). The resulting dienophile **183** can be trapped by cyclopentadiene, which allows measurements of the fragmentation rate ($\Delta H^{\ddagger} = 27.2$ kcal/mol, $\Delta S^{\ddagger} = 11$ eu).[99] In addition to the steric congestion in **181** and the product stability, the charge-transfer contribution to the transition state may be responsible for lowering of the activation energy.

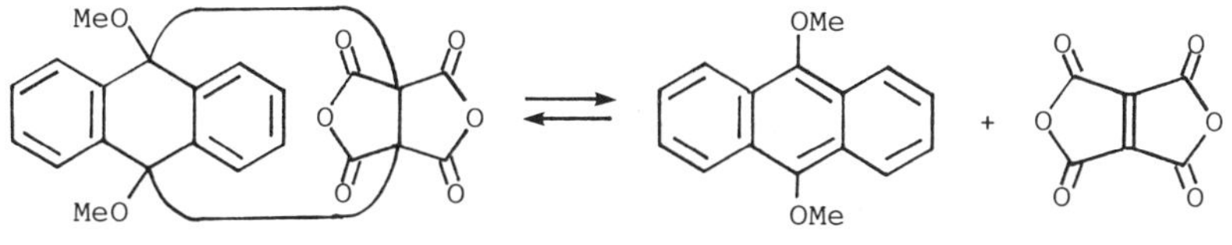

Analogous dipolar transition states are expected for 1,3-dipolar cycloreversions, which often proceed under moderate conditions.[100] For instance, upon heating in refluxing toluene, pyrrolidine **184** produces olefin **185** and azomethine ylide **186**, which undergoes proton transfer leading to benzylidenebenzylamine but can be intercepted readily as the cycloadduct with methyl maleate or fumarate.[101]

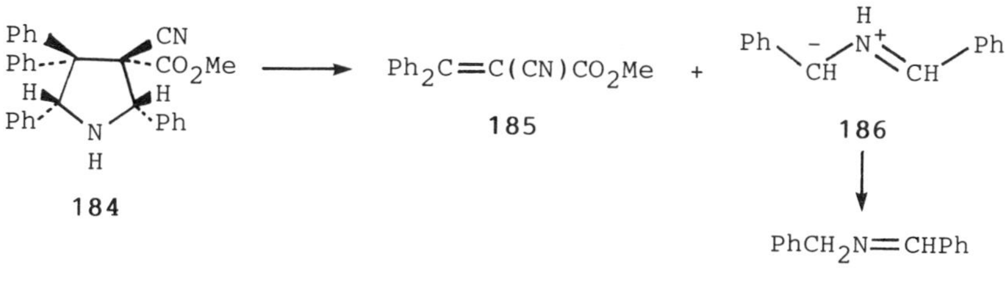

The substituent effects on norcaradiene–cycloheptatriene equilibrium constitute a relevant topic in electrocyclic reactions. Cycloheptatriene itself is quite stable. However, the presence of electron acceptors such as CN in position 7 favors the norcaradiene form.[102] Theoretically, this effect has been explained by the delocalization of electrons existing in the antisymmetric Walsh orbitals of the C_1–C_6 bond over the LUMO of the acceptor, which reduces the intrinsic weakness of the bond.[103] This principle seems useful and has been applied to the prediction of the equilibrium preference of more complicated systems such as substituted semibullvalenes **187**.[104] This aspect has been reviewed clearly.[9]

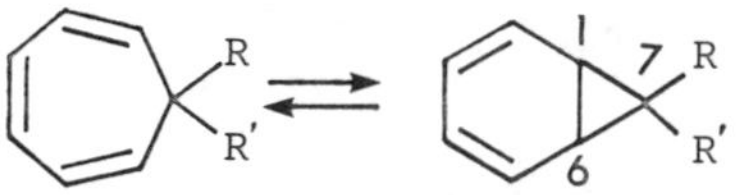

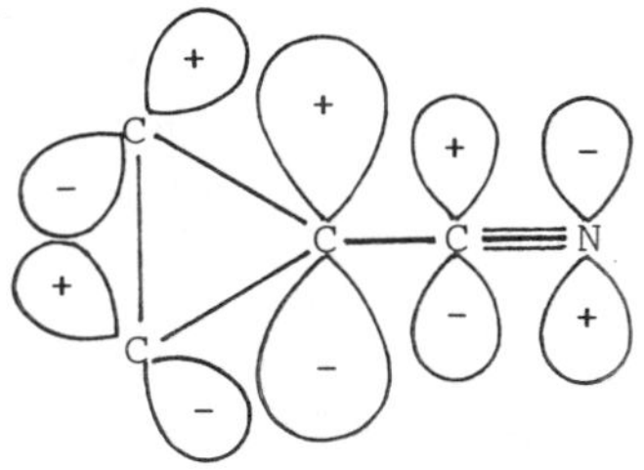

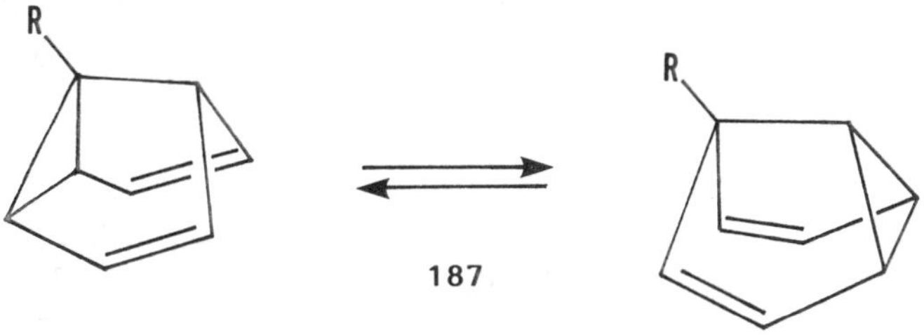

The Cope rearrangement is by far the most important of the sigmatropic reactions involving carbon–carbon bond cleavage. The degenerate rearrangement of 1,5-hexadiene, the simplest substrate, requires elevated temperatures (> 230°C, $\Delta H^{\ddagger}$ = 33.5 kcal/mol, $\Delta S^{\ddagger}$ = −13.8 eu for the 1,1-dideuterated compound).[105] Structural changes can reduce the barrier up to only about 5 kcal/mol in an extreme case such as the semibullvalene system, in which the driving force may be extensive delocalization of the frontier electrons but not charge development.[104]

The charge-transfer type of interaction in the transition state is naturally an important factor affecting the rate of the Cope rearrangements. The electronic structure of the transition state is best visualized as the cyclic interaction between the SOMOs of a pair of allyl radicals[2,106]: the larger the energy difference between the SOMOs, the greater the extent of charge transfer. The energy gap can be made by the introduction of donor- and/or acceptor-type substituents.

In this regard, the substituent effect on the rate of rearrangement of 2-aryl-1,5-hexadienes is interesting (k = 1.5 × 10^{-5} s^{-1} for the unsubstituted compound in cyclohexane at 164°C).[107] Although only four points are available and the rate change is small, the plot of log k/k_0 versus σ is concave upward (Figure 5-7). It is therefore tempting to presume that the dipolar nature of the transition state changes from **188** to **189** with an increase of electron-withdrawing power of the substituent.

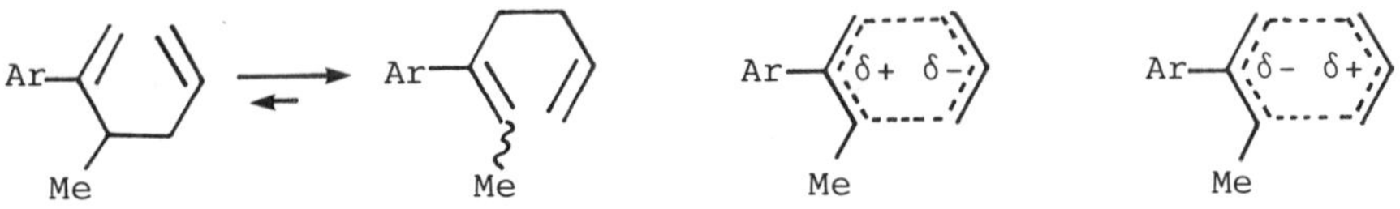

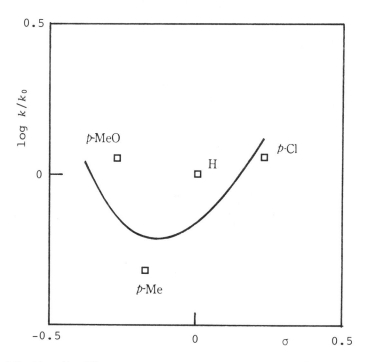

Figure 5-7. Plot of log k/k_0 against σ for the Cope rearrangement of 2-aryl-1,5-hexadienes.

The presence of a strong donor in position 2 and a strong acceptor in position 5, however, produces a zwitterion as a result of bond formation prior to carbon–carbon bond cleavage because of the high stabilization of charges.[108] Zwitterion **190** can be trapped by benzaldehyde or acrylonitrile as cycloadducts.

1,5-Hexadienes bearing electron-withdrawing groups such as CN and CO_2R on the saturated carbon (position 3 or 4) rearrange much more easily than the parent compound,[109] suggesting the dipolar nature of the transition states. The first recognized Cope rearrangement involved such a compound (**191**; at 260°C, 20 minutes, 67%).[110]

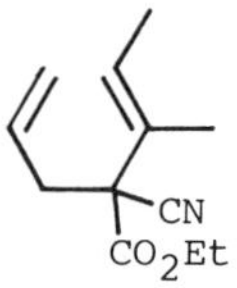

191

Compounds **192** bearing an arylazo moiety together with two cyano groups at a saturated carbon rearrange with remarkable ease to hydrazones **193** (usually 90–98% yields) under mild conditions.[111] The rate increases with

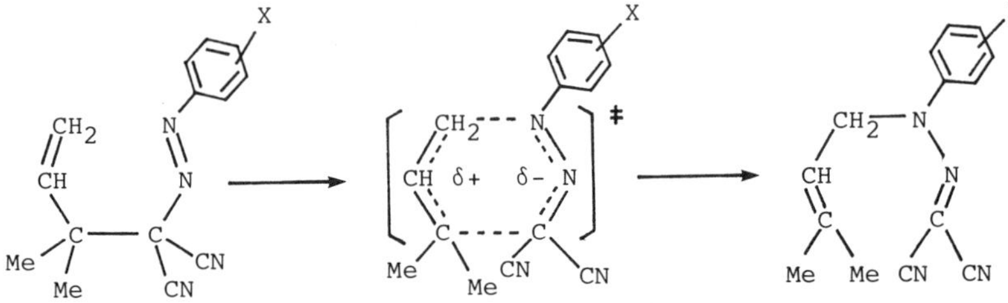

192 → **193**

increasing electron-withdrawing power of the substituent on the phenyl group. The half-life varies from 7.7 hours (p-MeO; $\Delta H^{\ddagger} = 22.8$ kcal/mol, $\Delta S^{\ddagger} = -11.2$ eu) to 6 minutes (p-NO$_2$; $\Delta H^{\ddagger} = 20.0$ kcal/mol, $\Delta S^{\ddagger} = -11.2$ eu) in o-dichlorobenzene at 60°C. Contrary to the rearrangement of 2-aryl-1,5-hexadienes, the plot of log (k/k_0) versus σ exhibits a good linear correlation with a ρ value of 1.64 (Figure 5-8). The large ρ value suggests considerable charge development in the transition states. The possibility of a mechanism via an ion pair, however, is exclusively ruled out by a comparison of the reactivity of azo compound **194** (X = p-NO$_2$ in **192**) with that of azo compound **170** in methanol:

1. Heating **194** affords the rearranged product nearly quantitatively, but there is no methanolysis product arising from the expected carbocation.
2. The rate ratio k_{194}/k_{170} (3.4 at 60°C, 6.7 at 31°C) is in conflict with the relative stability of carbocations expected from the ethanolysis rate for the chlorides [k_{rel}(CH$_2$=CHCMe$_2$Cl versus PhCMe$_2$Cl) = 0.065 at 45°C].[112]
3. The rearrangement exhibits a large negative $\Delta S^{\ddagger}$ value (-11.2 eu) compared to the methanolysis of **170** (0.7 eu), suggesting a cyclic structure of the transition state.
4. A good linear free energy relationship between the rates of the hetero-

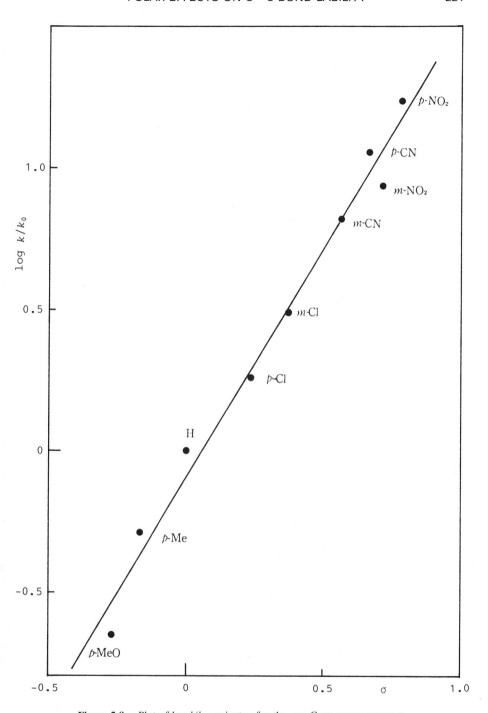

Figure 5-8. Plot of log k/k_0 against σ for the azo-Cope rearrangement.

lysis and the rearrangement measured for four derivatives (p-NO$_2$, p-CN, m-NO$_2$, and m-CN) of the azo compounds at 60°C holds with:

$$\log k_{\text{rear}} = 0.53 \log k_{\text{het}} - 0.90$$

where $r = 0.995$ (Figure 5-9), which indicates that the rearrangement is more facile but less sensitive to substituent changes than the heterolysis. Consequently, the growing charges in the rearrangement appear to be stabilized by an interaction analogous to anchimeric assistance.

5. The ortho effect on the rate of rearrangement also shows a sharp contrast to the rate of heterolysis. With **192**, the first o-chloro substitution causes a rate enhancement due to the electron-withdrawing power of the chloro group, which however is canceled by a steric effect upon the introduction of a second o-chloro group, whereas such an introduction of a second o-chloro group results in a further increase in the rate of methanolysis of **195**. The observation of the steric retardation indicates the existence of carbon–nitrogen bond forming in the transition state for the rearrangement.

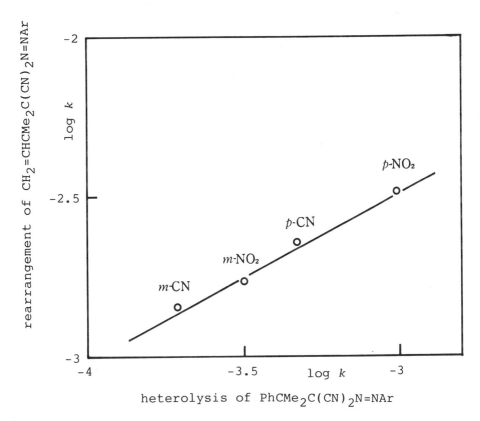

Figure 5-9. Comparison of rates of rearrangement and heterolysis in methanol-d_3 at 60°C.

These findings, which are consistent with a concerted process via a dipolar transition state, definitely exclude the intervention of ionic intermediates.

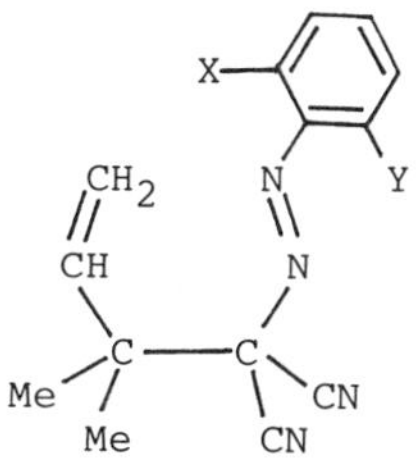

		k_{rel}	
X	Y	192[a]	195[b]
H	H	1	1
H	Cl	4.4	11
Cl	Cl	0.9	31

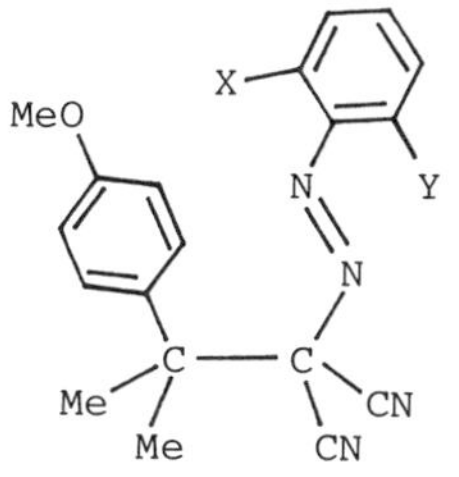

[a] In o-dichlorobenzene at 60°C.
[b] In (1:1) benzene-methanol-d_4 at 31°C.

192

195

Further evidence for the dipolar transition state comes from the solvent effect on rate. The rates for the *m*-chloro compound **196** (X = *m*-Cl in **192**) in a variety of solvents at 60°C are summarized in Table 5-3. The rate increases with increasing solvent polarity (eg, the relative rate CCl$_4$:methanol:DMSO = 1:4.8:19). Clearly this rearrangement exhibits the hydrogen-bond-insusceptible nature (P_s = 0.9 kcal/mol at 60°C, see Subsection B of Section 3). The anionic part in the transition state corresponds to the hydrazone anion that is resistant to hydrogen bonding with a protic solvent as discussed above. However, we can assume that the hydrogen-bond insusceptibility in this case derives from the mode of pericyclic interactions for the rearrangements of this sort, since the cationic part in the transition state interacts with both termini of the anionic part where a hydrogen bond would otherwise be formed.

This assumption is supported by a comparison of the reactivities of the rearrangements of thionbenzoates **197** and **199** to the corresponding

TABLE 5-3. Solvent Effects on the Rate of Rearrangement of *m*-Chlorophenylazo(α,α-dimethylallyl)malononitrile at 60°C

Solvent	$k \times 10^5$ (s^{-1})
DMSO-d_6	251
DMF-d_7	148
MeCN-d_3	92.8
Pyridine	86.4
Methanol-d_4	62.9
Acetone-d_6	59.8
o-Cl$_2$C$_6$H$_4$	34.6
CDCl$_3$	32.3
Benzene	31.9
CCl$_4$	13.2

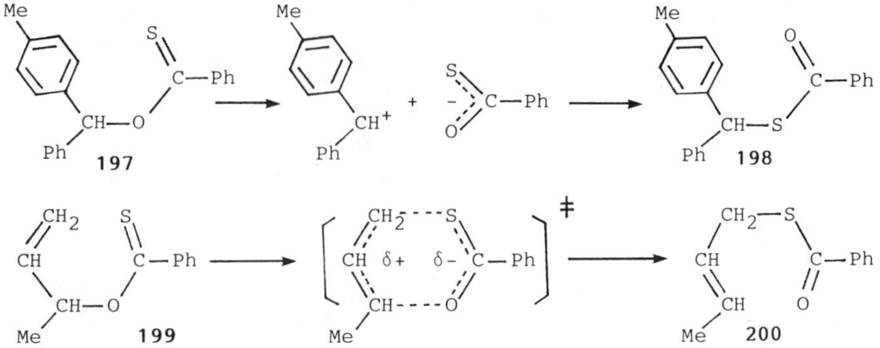

thiolbenzoates.[113] Conversion of **197** to **198** proceeds via an ion pair, as is evidenced by the fact that it competes with solvolysis in protic solvents, and exhibits a negative P_s value (-1.5 kcal/mol at 60°C). On the other hand, **199** yields **200** exclusively in any solvent, and the P_s value turns positive (0.5 kcal/mol at 60°C). Figure 5-10 shows the solvent effects on the free energy of activation as compared with W_{ion}. The plots for the rearrangements of **196** and **199** are typical of the hydrogen-bond-insusceptible reactions, and the plot for the rearrangement of **197** shows a trend analogous to the hydrogen-bond-susceptible $S_N 1$–$E1$ reactions except that the point for acetonitrile deviates upward. The results clearly point to the validity of the assumption above, and thus the rate for the azo Cope rearrangement of **196** shown in Table 5-3 may serve as a measure of solvent ionizing power (eg, in the form of log $k/k_{methanol}$ or its free energy equivalent) other than anion solvation throughout polar and nonpolar solvents.

The compound **201**, bearing two geminal cyano groups at the saturated carbon, rearranges with an enthalpy of activation of 25.0 kcal/mol ($\Delta S^{\ddagger} = -11$ eu), 8.5 kcal/mol less than that for 1,5-hexadiene itself.[109] The solvent effect on the rate has been examined.[114,115] The rearrangement behaves as a hydrogen-bond-insusceptible reaction via a slightly dipolar transition state; the rate is increased by a factor of 1.5 on changing the solvent from methanol to DMSO at 90°C ($P_s = 0.3$ kcal/mol).[115]

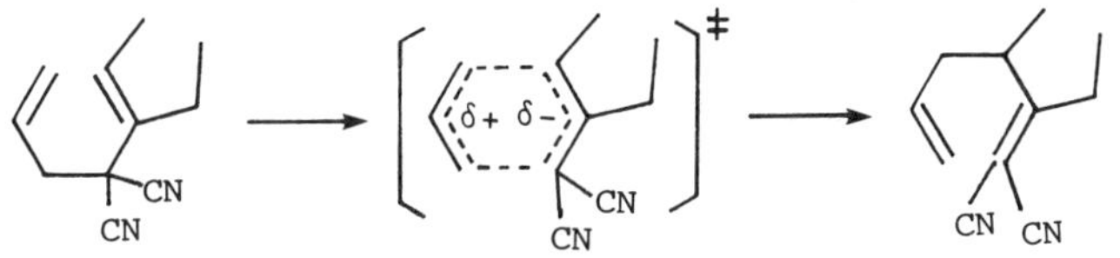

The presence of donor and acceptor groups on the different allylic moieties sometimes leads to the formation of an ion pair. On heating diene **202** at 80°C, a rearranged product **203** is produced with a half-life of 2 hours. Al-

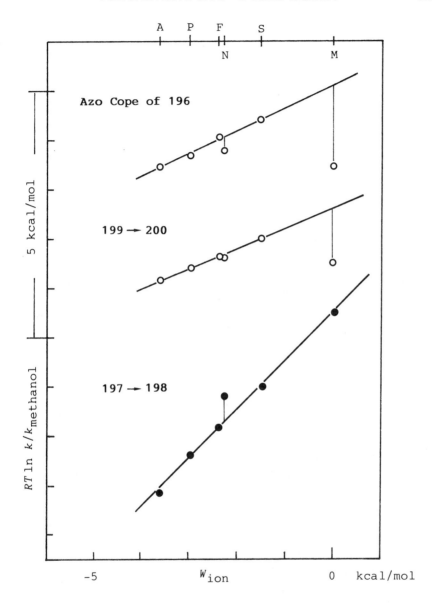

Figure 5-10. Plots of $-\Delta\Delta G^{\ddagger}$ versus W_{ion}, and the APFS line, where the solvents are: A = acetone, P = pyridine, F = DMF, N = MeCN, S = DMSO, and M = methanol.

though the intermediate carbocation cannot be trapped by thiophenoxide, it has been detected as a reduction product **204** by sodium borohydride, together with dinitrile **205** arising from the anionic intermediate.[116] This rearrangement, therefore, constitutes another example of carbon–carbon bond heterolysis in an open-chain system.

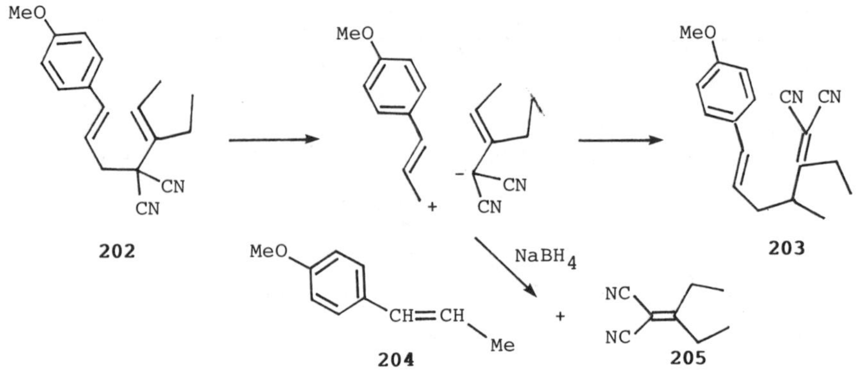

5. CONCLUDING REMARKS

An attempt has been made to outline interactions of the electron donor–acceptor type during carbon–carbon bond cleavage in thermal unimolecular reactions. Homolytic and pericyclic reactions are popular modes of cleavage in which neutral reactants yield neutral products, and thus the isopolar nature of the transition states seems favorable. A suitable combination of donor–acceptor groups, however, can induce polarization of the transition states to some extent and sometimes results in a considerable lowering of the barrier to bond cleavage.

Carbon–carbon bonds are likely to be regarded as reluctant to undergo heterolytic cleavage. In fact, the most common method of generating carbocations via unimolecular reactions is the heterolysis of carbon–heteroatom bonds, so that the term "leaving group" has been defined for convenience as "the part of a substrate to be cleaved not containing the carbon" in a textbook on advanced organic chemistry.[117] It should be noted that this view is not based on a common feature of carbon–carbon bonds but has been advanced merely as a result of failure to focus efforts on the development of efficient carbon-terminated leaving groups. Such leaving groups are obtainable by introducing three strong acceptor groups onto the carbon, and if extensive charge delocalization is expected during bond cleavage, they are able to depart in aprotic solvents more rapidly than usual leaving groups.

Since a carbon–carbon bond is capable of carrying six functional groups, numerous interesting donor–acceptor arrangements are possible in any mode of cleavage and many of them remain unexplored.

ACKNOWLEDGMENT

The author is grateful to Dr. Midori Goto and Dr. Gaku Yamamoto for helpful discussions and collaboration.

REFERENCES

1. McMillen, D. F.; Golden, D. M. *Annu. Rev. Phys. Chem.* **1982**, *33*, 509.
2. Fukui, K. *Acc. Chem. Res.* **1971**, *4*, 62.
3. Hammond, G. S. *J. Am. Chem. Soc.* **1955**, *77*, 334. Fărcasiu, D. *J. Chem. Educ.* **1975**, *52*, 76.
4. Mulliken, R. S.; Person, W. B. "Molecular Complexes." Wiley: New York, 1969.
5. Fleming, I. "Frontier Orbitals and Organic Chemical Reactions." Wiley: New York, 1976, pp. 182ff.
6. Viehe, H. G.; Janousek, Z.; Merényi, R.; Stella, L. *Acc. Chem. Res.* **1985**, *18*, 148.
7. Stella, L.; Janousek, Z.; Merényi, R.; Viehe, H. G. *Angew. Chem. Int. Ed. Engl.* **1978**, *17*, 691. Viehe, H. G.; Merényi, R.; Stella, L.; Janousek, Z. *Angew. Chem. Int. Chem. Engl.* **1979**, *18*, 917.
8. Crans, D.; Clark, T.; Schleyer, P. v. R. *Tetrahedron Lett.* **1980**, *21*, 3681.
9. Greenberg, A.; Liebman, J. F. "Strained Organic Molecules." Academic Press: New York, 1978.
10. Berson, J. A. In "Rearrangements in Ground and Excited States," de Mayo, P., Ed., Academic Press: New York, 1980, Essay 5, pp. 311–390.
11. Merényi, R.; De Mesmaeker, A.; Viehe, H. G. *Tetrahedron Lett.* **1983**, *24*, 2765. De Mesmaeker, A.; Vertommen, L.; Merényi, R.; Viehe, H. G. *Tetrahedron Lett.* **1982**, *23*, 69.
12. Creary, X.; M.-Mohammadi, M. E. *J. Org. Chem.* **1986**, *51*, 2664.
13. Arnold, D. R.; Wayner, D. D. M.; Yoshida, M. *Can. J. Chem.* **1982**, *60*, 2313.
14. Mitsuhashi, T.; Simamura, O. *J. Chem. Soc. B* **1970**, 705. Mitsuhashi, T.; Simamura, O.; Tezuka, Y. *J. Chem. Soc. Chem. Commun.* **1970**, 1300. Mitsuhashi, T.; Miyadera, H.; Simamura, O. *J. Chem. Soc. Chem. Commun.* **1970**, 1301.
15. Sylvander, L.; Stella, L.; Korth, H.-G.; Sustmann, R. *Tetrahedron Lett.* **1985**, *26*, 749.
16. Neumann, W. P.; Uzick, W.; Zarkadis, A. K. *J. Am. Chem. Soc.* **1986**, *108*, 3762.
17. D'yachkovskiĭ, F. S.; Bubnov, N. N.; Shilov, A. E. *Dokl. Akad. Nauk SSSR* **1958**, *122*, 629. Nelsen, S. F.; Landis, R. T., II. *J. Am. Chem. Soc.* **1973**, *95*, 8707.
18. Koch, T. H.; Olesen, J. A.; DeNiro, J. *J. Am. Chem. Soc.* **1975**, *97*, 7285. Haltiwanger, R. C.; Koch, T. H.; Olesen, J. A.; Kim, C. S.; Kim, N. K. *J. Am. Chem. Soc.* **1977**, *99*, 6327. Bennett, R. W.; Wharry, D. L.; Koch, T. H. *J. Am. Chem. Soc.* **1980**, *102*, 2345. Himmelsbach, R. J.; Barone, A. D.; Kleyer, D. L.; Koch, T. H. *J. Org. Chem.* **1983**, *48*, 2989.
19. Rüchardt, C.; Beckhaus, H.-D. *Angew. Chem. Int. Ed. Engl.* **1985**, *24*, 529.
20. Hüttel, R.; Rosner, M.; Wagner, D. *Chem. Ber.* **1973**, *106*, 2767.
21. Baldock, R. W.; Hudson, P.; Katritzky, A. R.; Soti, F. *J. Chem. Soc. Perkin Trans. 1* **1974**, 1422. See, however, Katritzky, A. R.; Zerner, M. C.; Karelson, M. M. *J. Am. Chem. Soc.* **1986**, *108*, 7213.
22. Beringer, F. M.; Galton, S. A.; Huang, S. J. *Tetrahedron* **1963**, *19*, 809.
23. Stella, L.; Pochat, F.; Merényi, R. *Nouv. J. Chim.* **1981**, *5*, 55.
24. de Jongh, H. A. P.; de Jonge, C. R. H. I.; Mijs, W. J. *J. Org. Chem.* **1971**, *36*, 3160.
25. Bartlett, P. D.; Rüchardt, C. *J. Am. Chem. Soc.* **1960**, *82*, 1756. Koenig, T. In "Free Radicals," Vol. 1; Kochi, J. K., Ed.; Wiley: New York, 1973, Chapter 3. Under photolysis conditions, a two-step mechanism via an acyloxy radical has been demonstrated: Falvey, D. E.; Schuster, G. B. *J. Am. Chem. Soc.* **1986**, *108*, 7419. For the electron affinity of CO_2, see: Yoshioka, Y.; Schaefer, H. F., III; Jordan, K. D. *J. Chem. Phys.* **1981**, *75*, 1040.
26. Cram, D. J.; Ratajczak, A. *J. Am. Chem. Soc.* **1968**, *90*, 2198.
27. Yankee, E. W.; Cram, D. J. *J. Am. Chem. Soc.* **1970**, *92*, 6328. Yankee, E. W.; Badea, F. D.; Howe, N. E.; Cram, D. J. *J. Am. Chem. Soc.* **1973**, *95*, 4210.
28. Yankee, E. W.; Cram, D. J. *J. Am. Chem. Soc.* **1970**, *92*, 6329, 6331. Yankee, E. W.;

Spencer, B.; Howe, N. E.; Cram, D. J. *J. Am. Chem. Soc.* **1973**, *95*, 4220. Howe, N. E.; Yankee, E. W.; Cram, D. J. *J. Am. Chem. Soc.* **1973**, *95*, 4230.
29. Chmurny, A. B.; Cram, D. J. *J. Am. Chem. Soc.* **1973**, *95*, 4237.
30. Danishefsky, S.; Rovnyak, G. *J. Chem. Soc. Chem. Commun.* **1972**, 821.
31. Lahousse, H.; Martens, H. J.; Hoornaert, G. J. *J. Org. Chem.* **1980**, *45*, 3451.
32. Dowd, P.; Kaufman, C.; Paik, Y. H. *Tetrahedron Lett.* **1985**, *26*, 2283.
33. Boger, D. L.; Brotherton, C. E. *J. Am. Chem. Soc.* **1986**, *108*, 6695, 6713.
34. Botta, M.; De Angelis, F.; Gambacorta, A. *Gazz. Chim. Ital.* **1983**, *113*, 129.
35. Magee, W. L.; Shechter, H. *J. Am. Chem. Soc.* **1977**, *99*, 633.
36. Linn, W. J. *J. Am. Chem. Soc.* **1965**, *87*, 3665.
37. Hamberger, H.; Huisgen, R. *J. Chem. Soc. Chem. Commun.* **1971**, 1190.
38. Huisgen, R.; de March, P. *J. Am. Chem. Soc.* **1982**, *104*, 4953.
39. Woodward, R. B.; Hoffmann, R. *J. Am. Chem. Soc.* **1965**, *87*, 395.
40. Dahmen, A.; Hamberger, H.; Huisgen, R.; Markowsky, V. *J. Chem. Soc. Chem. Commun.* **1971**, 1192.
41. Paladini, J.-C.; Crawford, R. J. *Can. J. Chem.* **1974**, *52*, 2098. Vukov, V.; Crawford, R. J. *Can. J. Chem.* **1975**, *53*, 1367. Crawford, R. J.; Lutener, S. B.; Cockcroft, R. D. *Can. J. Chem.* **1976**, *54*, 3364. Paladini, J. C.; Chuche, J. *Tetrahedron Lett.* **1971**, 4383. Eberbach, W.; Hädicke, E.; Trostmann, U. *Tetrahedron Lett.* **1981**, *22*, 4953.
42. Eberbach, W.; Trostmann, U. *Chem. Ber.* **1981**, *114*, 2979.
43. Arnold, D. R.; Karnischky, L. A. *J. Am. Chem. Soc.* **1970**, *92*, 1404.
44. Ullman, E. F.; Milks, J. E. *J. Am. Chem. Soc.* **1962**, *84*, 1315. Ullman, E. F. *J. Am. Chem. Soc.* **1963**, *85*, 3529. Ullman, E. F.; Milks, J. E. *J. Am. Chem. Soc.* **1964**, *86*, 3814.
45. Prinzbach, H.; Seppelt, W.; Fritz, H. *Angew. Chem. Int. Ed. Engl.* **1977**, *16*, 198.
46. Heine, H. W.; Peavy, R. *Tetrahedron Lett.* **1965**, 3123. Heine, H. W.; Peavy, R.; Durbetaki, A. J. *J. Org. Chem.* **1966**, *31*, 3924. Padwa, A.; Hamilton, L. *Tetrahedron Lett.* **1965**, 4363. DeShong, P.; Kell, D. A.; Sidler, D. R. *J. Org. Chem.* **1985**, *50*, 2309.
47. Huisgen, R.; Scheer, W.; Huber, H. *J. Am. Chem. Soc.* **1967**, *89*, 1753.
48. Huisgen, R.; Scheer, W.; Szeimies, G.; Huber, H. *Tetrahedron Lett.* **1966**, 397. Huisgen, R.; Scheer, W.; Mäder, H. *Angew. Chem.* **1969**, *81*, 619. Hall, J. H.; Huisgen, R. *J. Chem. Soc. Chem. Commun.* **1971**, 1187. Hall, J. H.; Huisgen, R.; Ross, C. H.; Scheer, W. *J. Chem. Soc. Chem. Commun.* **1971**, 1188. Huisgen, R.; Mäder, H. *J. Am. Chem. Soc.* **1971**, *93*, 1777. Hermann, H.; Huisgen, R.; Mäder, H. *J. Am. Chem. Soc.* **1971**, *93*, 1779.
49. Padwa, A.; Dean, D.; Oine, T. *J. Am. Chem. Soc.* **1975**, *97*, 2822.
50. Huisgen, R.; Mäder, H. *Angew. Chem.* **1969**, *81*, 621.
51. Lown, J. W.; Matsumoto, K. *J. Chem. Soc. Chem. Commun.* **1970**, 692. Lown, J. W.; Matsumoto, K. *Can. J. Chem.* **1971**, *49*, 3443. Lown, J. W.; Matsumoto, K. *J. Org. Chem.* **1971**, *36*, 1405.
52. Ayyangar, N. R.; Phatak, M. V.; Purchit, A. K.; Tilak, B. D. *Chem. Ind. (London)* **1979**, 853.
53. Trost, B. M. *Angew. Chem. Int. Ed. Engl.* **1986**, *25*, 1.
54. Ullman, E. F. *J. Am. Chem. Soc.* **1960**, *82*, 505.
55. Matlin, A. R.; Inglin, T. A.; Berson, J. A. *J. Am. Chem. Soc.* **1982**, *104*, 4954. Inglin, T. A.; Berson, J. A. *J. Am. Chem. Soc.* **1986**, *108*, 3394.
56. Turro, N. J. *Acc. Chem. Res.* **1969**, *2*, 25. Turro, N. J.; Edelson, S. S.; Williams, J. R.; Darling, T. R.; Hammond, W. B. *J. Am. Chem. Soc.* **1969**, *91*, 2283. Edelson, S. S.; Turro, N. J. *J. Am. Chem. Soc.* **1970**, *92*, 2770. Sclove, D. B.; Pazos, J. F.; Camp, R. L.; Greene, F. D. *J. Am. Chem. Soc.* **1970**, *92*, 7488. Bowers, K. G.; Mann, J. *Tetrahedron Lett.* **1985**, *26*, 4411.
57. Hoffmann, H. M. R. *Angew. Chem. Soc. Int. Ed. Engl.* **1984**, *23*, 1.
58. Föhlisch, B.; Gottstein, W.; Kaiser, R.; Wanner, I. *Tetrahedron Lett.* **1980**, *21*, 3005. Föhlisch, B.; Gehrlach, E.; Stezowski, J. J.; Kollat, P.; Martin, E.; Gottstein, W. *Chem. Ber.* **1986**, *119*, 1661.
59. Bordwell, F. G.; Carlson, M. W. *J. Am. Chem. Soc.* **1970**, *92*, 3377.

60. Bordwell, F. G.; Scamehorn, R. G. *J. Am. Chem. Soc.* **1971,** *93,* 3410. Bordwell, F. G.; Strong, J. G. *J. Org. Chem.* **1973,** *38,* 579.
61. De Kimpe, N.; Schamp, N. *Tetrahedron Lett.* **1974,** 3779.
62. Block, E.; Penn, R. E.; Ennis, M. D.; Owens, T. A.; Yu, S.-L. *J. Am. Chem. Soc.* **1978,** *100,* 7436.
63. Weyler, W., Jr.; Pearce, D. S.; Moore, H. W. *J. Am. Chem. Soc.* **1973,** *95,* 2603. Wendling, L. A.; Bergman, R. G. *J. Org. Chem.* **1976,** *41,* 831. Padwa, A.; Akiba, M.; Cohen, L. A.; Gringrich, H. L.; Kamigata, N. *J. Am. Chem. Soc.* **1982,** *104,* 286.
64. Paquette, L. A.; Meisinger, R. H.; Wingard, R. E., Jr. *J. Am. Chem. Soc.* **1973,** *95,* 2230.
65. Shimizu, N.; Nishida, S. *J. Chem. Soc. Chem. Commun.* **1978,** 931. Kataoka, F.; Shimizu, N.; Nishida, S. *J. Am. Chem. Soc.* **1980,** *102,* 711.
66. Drexler, J.; Lindermayer, R.; Hassan, M. A.; Sauer, J. *Tetrahedron Lett.* **1985,** *26,* 2559.
67. Schug, R.; Huisgen, R. *J. Chem. Soc. Chem. Commun.* **1975,** 60. Huisgen, R. *Acc. Chem. Res.* **1977,** *10,* 117, 199. Herges, R.; Ugi, I. *Chem. Ber.* **1986,** *119,* 829.
68. Brook, P. R.; Harrison, J. M.; Hunt, K. *J. Chem. Soc. Chem. Commun.* **1973,** 733.
69. Visser, G. W.; Trompenaars, W. P.; Reinhoudt, D. N. *Tetrahedron Lett.* **1982,** *23,* 1217. Troxler, F.; Weber, H. P.; Jaunin, A.; Loosli, H.-R. *Helv. Chim. Acta* **1974,** *57,* 750.
70. Klop, W.; Brandsma, L. *J. Chem. Soc. Chem. Commun.* **1983,** 988.
71. Regitz, M.; Eisenbarth, P. *Chem. Ber.* **1984,** *117,* 1991.
72. Flowers, M. C.; Frey, H. M. *J. Chem. Soc.* **1961,** 5550.
73. Gajewski, J. J.; Burka, L. T. *J. Am. Chem. Soc.* **1971,** *93,* 4952. The relative rate of rearrangement in benzene was calculated from the logarithmic ratio of unreacted spiropentanes (log 0.25/log 0.8) after one hour heating at 170°C.
74. Tanny, S. R.; Fowler, F. W. *J. Am. Chem. Soc.* **1973,** *95,* 7320.
75. Haselbach, E.; Martin, H.-D. *Helv. Chim. Acta* **1974,** *57,* 472.
76. Prinzbach, H.; Bingmann, H.; Markert, J.; Fischer, G.; Knothe, L.; Eberbach, W.; Brokatzky-Geiger, J. *Chem. Ber.* **1986,** *119,* 589. Prinzbach, H.; Bingmann, H.; Fritz, H.; Markert, J.; Knothe, L.; Eberbach, W.; Brokatzky-Geiger, J.; Sekutowski, J. C.; Krüger, C. *Chem. Ber.* **1986,** *119,* 616.
77. Figeys, H. P.; Destrebecq, M.; Van Lommen, G. *Tetrahedron Lett.* **1980,** *21,* 2369.
78. Heaney, H.; Ley, S. V. *J. Chem. Soc. Chem. Commun.* **1970,** 1184.
79. Yamatodani, S.; Asahi, Y.; Matsukura, A.; Ohmomo, S.; Abe, M. *Agric. Biol. Chem.* **1970,** *34,* 485. Cole, R. J.; Kirksey, J. W.; Clardy, J.; Eickman, N.; Weinreb, S. M.; Singh, P.; Kim, D. *Tetrahedron Lett.* **1976,** 3849. Rebek, J., Jr.; Shue, Y. K. *J. Am. Chem. Soc.* **1980,** *102,* 5426.
80. Takahashi, K.; Hirata, N.; Takase, K. *Tetrahedron Lett.* **1970,** 1285.
81. Takahashi, K.; Takase, K.; Sakae, T. *Chem. Lett.* **1980,** 1485.
82. Ritchie, C. D.; Skinner, G. A.; Badding, V. G. *J. Am. Chem. Soc.* **1967,** *89,* 2063.
83. Bryce, M. R.; Reynolds, C. D.; Hanson, P.; Vernon, J. M. *J. Chem. Soc. Perkin Trans. 1* **1981,** 607.
84. Arnett, E. M.; Troughton, E. B. *Tetrahedron Lett.* **1983,** *24,* 3299.
85. (a) Arnett, E. M.; Troughton, E. B.; McPhail, A. T.; Molter, K. E. *J. Am. Chem. Soc.* **1983,** *105,* 6172. (b) Troughton, E. B.; Molter, K. E.; Arnett, E. M. *J. Am. Chem. Soc.* **1984,** *106,* 6726. (c) Arnett, E. M.; Chawla, B.; Molter, K.; Amarnath, K.; Healy, M. *J. Am. Chem. Soc.* **1985,** *107,* 5288. (d) Arnett, E. M.; Molter, K. E. *Acc. Chem. Res.* **1985,** *18,* 339. (e) Arnett, E. M.; Molter, K. *J. Phys. Chem.* **1986,** *90,* 383.
86. Okamoto, K.; Kitagawa, T.; Takeuchi, K.; Komatsu, K.; Takahashi, K. *J. Chem. Soc. Chem. Commun.* **1985,** 173.
87. Camaggi, C. M.; Leardini, R.; Chatgilialoglu, C. *J. Org. Chem.* **1977,** *42,* 2611. Vernin, G.; Metzger, J. *Synthesis* **1978,** 921.
88. Mitsuhashi, T.; Matsumura, C.; Koga, Y. *J. Chem. Soc. Chem. Commun.* **1986,** 257. Mitsuhashi, T. *J. Chem. Soc. Perkin Trans. 2,* **1986,** 1495.
89. Mitsuhashi, T. *J. Am. Chem. Soc.* **1986,** *108,* 2394.
90. Heytler, P. G.; Prichard, W. W. *Biochem. Biophys. Res. Commun.* **1962,** *7,* 272.

Mitchell, P. *Biochem. Soc. Trans.* **1976**, *4*, 399. Hinkle, P. C.; McCarty, R. E. *Sci. Am.* **1978**, *238*, 104.
91. Smith, S. G.; Fainberg, A. H.; Winstein, S. *J. Am. Chem. Soc.* **1961**, *83*, 618.
92. Yamdagni, R.; Kebarle, P. *J. Am. Chem. Soc.* **1972**, *94*, 2940. Kebarle, P.; Davidson, W. R.; French, M.; Cumming, J. B.; McMahon, T. B. *Faraday Discuss. Chem. Soc.* **1977**, *64*, 220.
93. Mitsuhashi, T.; Yamamoto, G.; Goto, M. Paper presented at the 53rd Meeting of the Japan Chemical Society, Nagoya; October 16–19, 1986; Abstract no. 1F04.
94. Edwards, M. R.; Jones, P. G.; Kirby, A. J. *J. Am. Chem. Soc.* **1986**, *108*, 7067.
95. Winiker, R.; Beckhaus, H.-D.; Rüchardt, C. *Chem. Ber.* **1980**, *113*, 3456.
96. Flamm-ter Meer, M. A.; Beckhaus, H.-D.; Peters, K.; von Schnering, H.-G.; Rüchardt, C. *Chem. Ber.* **1985**, *118*, 4665.
97. Sauer, J.; Sustmann, R. *Angew. Chem. Int. Ed. Engl.* **1980**, *19*, 779.
98. DeWitt, E. J.; Lester, C. T.; Ropp, G. A. *J. Am. Chem. Soc.* **1956**, *78*, 2101.
99. Sauer, J.; Schröder, B.; Wiemer, R. *Chem. Ber.* **1967**, *100*, 306.
100. Bianchi, G.; De Micheli, C.; Gandolfi, R. *Angew. Chem. Int. Ed. Engl.* **1979**, *18*, 721.
101. Sinbandhit, S.; Hamelin, J. *J. Chem. Soc. Chem. Commun.* **1977**, 768.
102. Maier, G. *Angew. Chem.* **1967**, *79*, 446. Simonetta, M. *Acc. Chem. Res.* **1974**, *7*, 345.
103. Hoffmann, R. *Tetrahedron Lett.* **1970**, 2907. Hoffmann, R.; Stohrer, W.-D. *J. Am. Chem. Soc.* **1971**, *93*, 6941. Günther, H. *Tetrahedron Lett.* **1970**, 5173.
104. Cheng, A. K.; Anet, F. A. L.; Mioduski, J.; Meinwald, J. *J. Am. Chem. Soc.* **1974**, *96*, 2887. Gompper, R.; Schwarzensteiner, M.-L.; Wagner, H.-U. *Tetrahedron Lett.* **1985**, *26*, 611. Paquette, L. A.; Volz, W. E.; *J. Am. Chem. Soc.* **1976**, *98*, 2910.
105. Doering, W. v. E.; Toscano, V. G.; Beasley, G. H. *Tetrahedron,* **1971**, *27*, 5299.
106. Doering, W. v. E.; Roth, W. R. *Tetrahedron,* **1962**, *18*, 67. Hoffmann, R.; Woodward, R. B. *J. Am. Chem. Soc.* **1965**, *87*, 4389.
107. Marvell, E. N.; Li, T. H.-C. *J. Am. Chem. Soc.* **1978**, *100*, 883.
108. Gompper, R.; Ulrich, W.-R. *Angew. Chem. Int. Ed. Engl.* **1976**, *15*, 299, 301.
109. Foster, E. G.; Cope, A. C.; Daniels, F. *J. Am. Chem. Soc.* **1947**, *69*, 1893. Rhoads, S. J.; Raulins, N. R. In "Organic Reactions", Vol. 22; Dauben, W. T., Ed.; Wiley: New York, 1975, Chapter 1, pp. 1–252.
110. Cope, A. C.; Hardy, E. M. *J. Am. Chem. Soc.* **1940**, *62*, 441.
111. Mitsuhashi, T. *J. Am. Chem. Soc.* **1986**, *108*, 2400.
112. Vernon, C. A. *J. Chem. Soc.* **1954**, 4462. Smith, S. G.; Goon, D. J. W. *J. Org. Chem.* **1969**, *34*, 3127.
113. Smith, S. G.; *Tetrahedron Lett.* **1962**, 979. Smith, S. G.; Goon, D. J. W. *J. Org. Chem.* **1969**, *34*, 3127. Smith, S. G. *J. Am. Chem. Soc.* **1961**, *83*, 4285.
114. Wigfield, D. C.; Feiner, S. *Can. J. Chem.* **1970**, *48*, 855.
115. Mitsuhashi, T.; Yamamoto, G. To be published.
116. Wigfield, D. C.; Feiner, S.; Taymaz, K. *Tetrahedron Lett.* **1972**, 891. Wigfield, D. C.; Feiner, S.; Malbacho, G.; Taymaz, K. *Tetrahedron,* **1974**, *30*, 2949.
117. March, J. "Advanced Organic Chemistry," 3rd ed. Wiley: New York, 1985, p. 179.

CHAPTER 6

Resonance Structure Contributions Derived from the Experimental Geometry of Molecules

Tadeusz M. Krygowski

Department of Chemistry, Warsaw University, Warsaw, Poland

In memory of my esteemed teacher, Prof. Dr. W. Kemula (d. 1985).

CONTENTS

1. Introduction... 231
2. Factors Influencing the Observed Geometry of Chemical Species in the Crystalline State Obtained by X-Ray Diffractometry.. 234
3. Stabilization Energy and Resonance (Canonical) Structure Contributions Derived from Experimental Bond Lengths 236
4. Applications... 245
5. Concluding Remarks 251
References 253

1. INTRODUCTION

In the past the chemical and physicochemical properties of chemical species were often rationalized in terms of the so-called resonance theory,[1] which is a rough simplification of the more rigorous valence bond theory.[2] In this approach the molecular structure of a chemical compound is presented in the form of a set of "resonance" or "canonical" structures, which contribute various weights to the real structure of the molecule, its geometry, and its physicochemical properties. Few attempts have been made to estimate these contributions in order to get to know more quantitatively about the "weights" of these canonical structures. Such knowledge, however, could be useful in interpretation or even prediction of the chemical and physicochemical properties of chemical species. Two elementary examples will serve for illustration.

Consider p-disubstituted derivatives of benzene with substituents differing strongly one from another: pD − Ph − A, where D stands for the electron-releasing substituent (donor) and A for the electron-accepting one (acceptor). A simple example is p-nitroaniline, which is usually represented by three main canonical structures: two benzenoid ones, B_1 and B_2, and a quinoid one, Q, as indicated in Scheme I. Certainly, for a more detailed

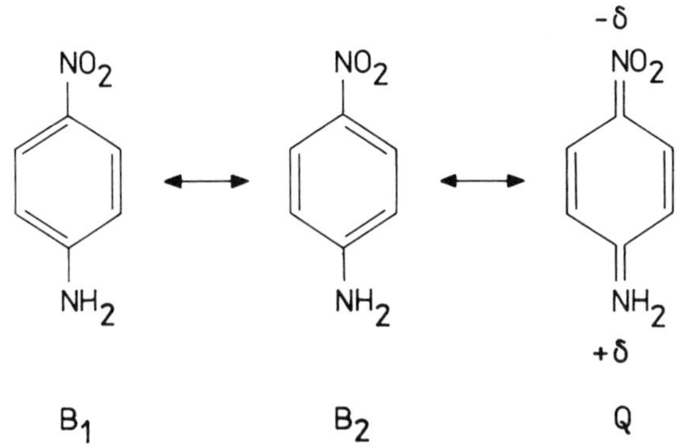

Scheme I

analysis some other structures should be taken into account as well.[3] Nevertheless, even the contributions of these simple structures may help one to understand many properties of p-nitroaniline, such as the dipole moment, its decreased basicity relative to the predictions of the Hammett equation,[4] and other related phenomena.

Another example is provided by electron donor–acceptor (EDA) complexes. Consider an EDA complex of p-phenylenediamine (PDA) and chloranil. Interactions between these two species may be illustrated by Scheme II. The greater the charge transfer from a donor molecule D (PDA this time) to an acceptor A (chloranil), the higher the contribution of the right-hand-side structures in Scheme II. In consequence, the lower the percent contribution of Q_1 for the acceptor species and the higher the percent contribution of Q_2 for the donor species. Thus in the series of EDA systems with varying

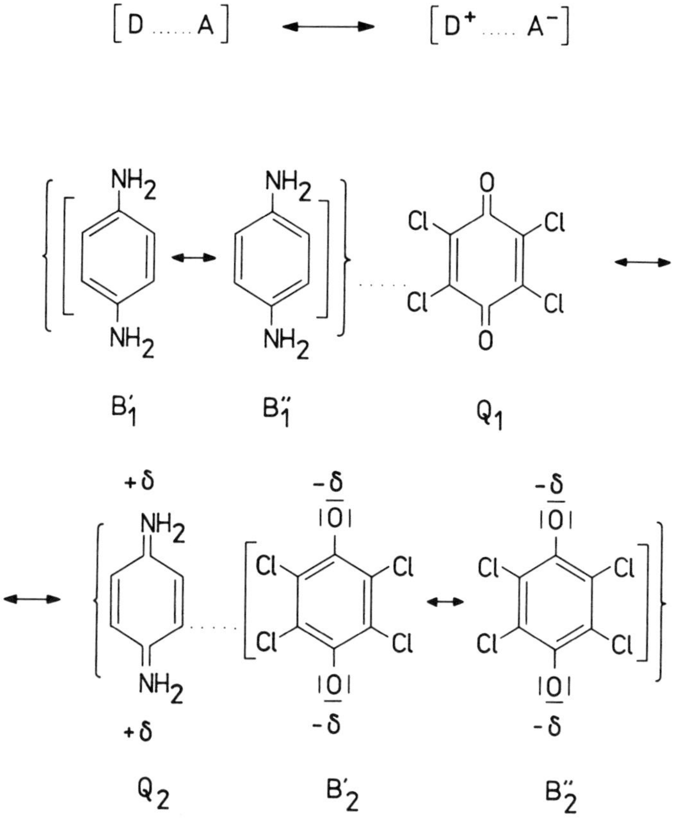

Scheme II

$$[D \cdots A] \longleftrightarrow [D^+ \cdots A^-]$$

[Scheme II depicting resonance structures B_1', B_1'', Q_1, Q_2, B_2', B_2'']

donor or acceptor, the charge transfer in the EDA system may be qualitatively estimated by determining the percentage of Q_1 or Q_2, respectively. This kind of analysis may be carried out for more complex systems as well.

In the two examples above, as a result of intra- or intermolecular charge transfer, respectively, an increase in the percentage of Q or B is expected. This in turn must be visualized in the geometry of the chemical species in question. An increase in the percentage of Q is accompanied by lengthening of the a and c bonds and respective shortening of the b bonds. Obviously, C—X and C—Y bond lengths (cf. Scheme III) are also affected by interactions of this type. The reverse is true of an increase in the percentage of B.

This chapter shows how variation in the geometry of chemical species may be translated into the language of resonance theory—that is, how resonance (canonical) structure contributions may be calculated directly from experimental geometry. First, however, we present briefly the factors that affect the accuracy of experimental geometry.

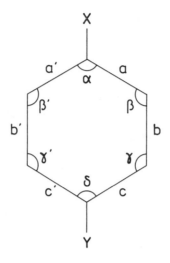

Scheme III

2. FACTORS INFLUENCING THE OBSERVED GEOMETRY OF CHEMICAL SPECIES IN THE CRYSTALLINE STATE OBTAINED BY X-RAY DIFFRACTOMETRY

Since this chapter is devoted to models based solely on molecular geometry, the problem of accuracy and precision is of great importance. Only highly accurate estimations of geometry will lead to reliable results and conclusions.

The accuracy of the various experimental techniques that afford the geometry of chemical species was analyzed in the second volume of this series.[6] Since most information about molecular geometry of chemical species is provided by X-ray diffractometric measurements, a more detailed analysis of such data is given here.

1. First of all it should be noted that the so-called bond length estimated by X-ray diffractometry is not exactly an interatomic distance but a distance between centroids of electron clouds of atoms involved in bond formation. For convenience, these parameters are called bond lengths throughout this chapter, but it is well known that electron clouds of the atomic domains of molecules do not have spherical symmetry.[7] A comparison of CC bond lengths measured by neutron diffraction, with appropriate data measured by X-ray diffraction, leads us to note that short bonds are observed as shorter by as much as by 2–3 pm (= 0.02–0.03 Å). [The picometer (pm) notation is used throughout the chapter: 1 pm = 10^{-2} Å.]

2. Molecules in the crystal lattice undergo thermal motion, which in turn leads to changes in the geometry,[8,9] chiefly in bond lengths. They may be as large as in the case of benzene, where due to libration of the ring around the C_6 axis, the observed CC bond lengths become shorter by 1.4 pm.[10] Usually this effect is smaller, however, and may be taken into account by the use of THMB programs,[11] which calculate corrections for thermal motion for bond lengths. For monosubstituted benzene derivatives, three kinds of motion of the ring should be taken into consideration[12]: (a) yawing, (b) pitching, and (c) rolling. Yawing and pitching cause the largest motion of p- and both m-atoms, leading to a shortening of bonds in the order c > b > a. Rolling leads to a shortening of a and c bonds. In comparison with p-disubstituted derivatives, the librations of monosubstituted derivatives of benzene results in greater changes of the bond lengths. The corrections for libration are usually in the 0.3–0.8 pm range. The observed bond lengths are shorter by this value than after the correction.

3. In some cases orientational disorder may interfere considerably with the observed geometry of molecules in crystals. If molecules in the crystal lattice coexist in two or more orientations in translationally equivalent positions, the observed geometry is a superposition of them. In consequence, it is difficult to estimate the real geometry for this kind of system. As an example, let us look at dimers of carboxylic acids, which may exist in the crystal lattice in two orientations, as shown in Scheme IV.[13,14] An even more

Scheme IV

complex situation may occur[15] if one takes into account noncoplanarity (but still parallelism) of both carboxyl groups.[16] As a result, partial or sometimes even full equalization of the C=O and C—OH bond lengths is observed.

As a conclusion to this short section, we pose a serious warning: one should be very careful when taking various structural data (geometries) from various sources and comparing data obtained by the use of various techniques that differ in precision and accuracy. To these warnings, one should add those from the chapter by Hargittai and Hargittai.[6]

How far may molecular geometry be affected by the crystal lattice forces of symmetry lower than the molecule itself? This difficult problem is still open, and discussion of it would need a separate chapter. Nevertheless, one should take it into account, particularly for very subtle analyses of molecular geometry.

3. STABILIZATION ENERGY AND RESONANCE (CANONICAL) STRUCTURE CONTRIBUTIONS DERIVED FROM EXPERIMENTAL BOND LENGTHS

The geometries of chemical species have long been used to discuss their stabilities. Elvidge and Jackman,[17] as well as Sondheimer,[18] have postulated averaging bond lengths as a typical feature of stable molecules with π-electron systems (aromatic stability). Subsequently, a few attempts were made to translate this rule into a more quantitative form.[19,20] Recently the idea of using a harmonic oscillator model for calculating stabilization energy directly from the geometry of molecules has been successfully applied to chemical species with π-electron systems.[21–23]

A. Principles of the HOSE Model

The main idea of the model called harmonic oscillator stabilization energy (HOSE)[21–23] is based on the following assumptions and factors.

1. Every year a number of relatively accurate sets of structural data are published, and it seemed useful to offer a simple "translation" of this information into the language of organic chemistry.
2. As a qualitative way of interpreting reactivity and most physicochemical properties of organic species, the resonance theory[24] is still frequently applied. It is understood as a model theory to which the presented empirical model is approaching.
3. Deformations of bond lengths (due to inter- or intramolecular interactions in π-electron systems) from some values taken as references, may be approximately described in terms of harmonic potentials. Acyclic polyenes or their heteroanalogues have been chosen as reference systems.
4. Other deformations (eg, angular) are of less importance and are not taken into consideration in this model.

Thus as reference bond lengths for π-electron hydrocarbons the $R_{C=C}$ and R_{C-C} in 1,3-butadiene[25] were chosen. The choice of these reference bond lengths is in line with the generally accepted definition of resonance energy,[26–28] which is a difference in energy between the real molecule and its reference structure with localized single and double π bonds. However, it should be noted here that replacement of these reference bond lengths by more extreme ones (eg, $R_{C=C}$ = 131.4 pm from ethene[29] and R_{C-C} = 151.0 pm from ethane,[30] measured by X-ray diffraction and electron diffraction, respectively) does not change the qualitative picture obtained by the use of the HOSE model.[31] The following formula for HOSE was proposed[21]:

$$\text{HOSE}_i = -E_{\text{def}} = 301.15 \left[\sum_{r=1}^{n_1} (R_r' - R_0^s)^2 k_r + \sum_{r=1}^{n_2} (R_r'' - R_0^d)^2 k_r \right] \quad (6\text{-}1)$$

where R'_r and R''_r stand for the lengths of π bonds in the real molecule, whereas n_1 and n_2 are the numbers of corresponding formal single and double bonds in the ith resonance structure, respectively. In the process of deformation, the n_1 bonds (which correspond to single bonds in the ith resonance structure) are lengthened, whereas the n_2 bonds (which correspond to double bonds) are shortened to the bond lengths R_0^s and R_0^d, respectively (ie, to single and double bond lengths in 1,3-butadiene). The constant, 301.15, in Equation 6-1, allows us to get HOSE values in kilojoules per mole if those for R'_s are in angstrom units and for k in (dynes per centimeter) $\times$ 10^5. The force constants k_r in Equation 6-1 are estimated empirically. For the first-row elements in the periodic table, the formula[21]

$$k_r = a + bR_r \quad (6\text{-}2)$$

is used.

Using k_r and R_r for pure single and double bonds, constants a and b are calculated. They are gathered in Table 6-1, together with reference bond lengths R_0^s and R_0^d for bonds CC, CN, and CO. For longer bonds (eg, C—S), another empirical formula has been successfully applied[32]:

$$\log k_r = 2.15 - 6.60 \log R_r \quad (6\text{-}3)$$

When the exact geometry of a molecule with a π electron is known, by applying its bond lengths, the data of Table 6-1 and Equations 6-1 and 6-3, one can calculate the value of $HOSE_i$, that is, the approximate energy by which the real molecule (with its own geometry) is more stable than the ith resonance structure built up of bonds with lengths R_0^s and R_0^d. In other words, one can say that this ith resonance structure is less stable by energy equal to $HOSE_i$ than the real molecule. Molecules with π systems must be described by at least a few and sometimes many resonance structures. Each resonance structure may have a different calculated HOSE value. Consequently, the total number of HOSE values that may be calculated for a given molecule is equal to the number of resonance structures we want to take into

TABLE 6-1. Constants Used in the HOSE Model[a]

Bond (Ref.)	R_0^s (Å)	R_0^d (Å)	a ($\times 10^5$ dyn/cm)	b ($\times 10^{13}$ dyn/cm^2)
CC (23)	1.467	1.349	44.39	−26.02
CN (23)	1.474	1.274	43.18	−25.73
CO (23)	1.428	1.209	52.35	−32.88
CS (32)	1.820	1.610	— (Eq. 6-3)	—

[a] References 23 and 32 contain details of the choice of R_0^s and R_0^d.

account in our study. Following chemical intuition and ideas of valence bond theory, two assumption have been made[23]:

1. All the most important resonance structures must be taken into consideration in calculating HOSE$_{tot}$ for a given molecule:

$$\text{HOSE}_{tot} = \sum_{i=1}^{N} C_i \, \text{HOSE}_i \qquad (6\text{-}4)$$

where summation runs over all resonance structures.

2. The contribution of the ith resonance structure, C_i, in a description of the real molecule is inversely proportional to its destabilization energy (ie, the energy by which the ith resonance structure is less stable than the real molecule):

$$C_i = \frac{(\text{HOSE}_i)^{-1}}{\sum_{j=1}^{N} (\text{HOSE}_j)^{-1}} \qquad (6\text{-}5)$$

where N is the number of resonance structures, taken into consideration.

B. Comparison of the Results Obtained by the HOSE Model and by Other Models or Theories

The HOSE model was tested by comparing its results with those obtained by other, independent theories and models. When the HOSE$_{tot}$ (hereafter termed HOSE) values were calculated for 22 alternant and 14 nonalternant unsaturated hydrocarbons and then plotted[23] against the Hess and Schaad[28] resonance energy RE, the graphs presented in Figures 6-1 and 6-2 were obtained. The regression line for alternant hydrocarbons has a very high correlation coefficient ($r = 0.991$), whereas for nonalternant species the r value is significantly worse ($r = 0.937$). The reason for this discrepancy is quite obvious. In the first case the structural data (bond lengths) were determined for unsubstituted species. Most of the nonalternant hydrocarbons, however, were studied as substituted species, since in this way they became stable enough to carry out the diffraction measurements. Naturally, an increase in the stability of nonalternant systems due to substitution is reflected in their geometry. Hence the regression line for nonalternant hydrocarbons in Figure 6-2 looks worse than that for alternant ones. The differences in HOSE values calculated for a hydrocarbon skeleton only may be quite large for two differently substituted species. HOSE values for the compounds shown in Scheme V—fulvene (**1**), 6-N,N-dimethylaminofulvene (**2**), and two independent molecules in the crystal of 6-N,N-dimethyl-2-formylfulvene (**3**)—are as follows: −0.75 and 15.8 kJ/mol and 26.7 and 30.0 kJ/mol, respec-

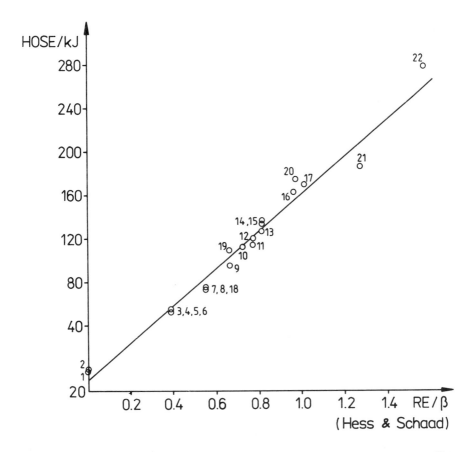

Figure 6-1. Plot of HOSE values (kJ/mol) versus resonance energy by Hess and Schaad[22] (β units) for alternant hydrocarbons. Abbreviations in parentheses describe the data on which points 1–22 were based: (ED) = electron diffraction, (ND) = neutron diffraction, (X-ray) = X-ray diffraction; at room temperature (RT) if not otherwise stated. Data points assigned as follows: 1 = ethylene (ED), 2 = 1,3-butadiene (ED), 3–6 = benzene (ED, ND at 138 and 218 K, X-ray at 270 K), 7 and 8 = naphthalene (X-ray at 123 K and ED), 9 = perdeuteroanthracene (ND), 10 = biphenyl (X-ray), 11 and 12 = phenanthrene (ND and X-ray), 13–15 = pyrene (ND and X-ray RT and 113 K), 16 = chrysene (X-ray), 17 = triphenylene (ND), 18 = naphthalene (ED), 19 = anthracene (ED), 20 = perylene (X-ray), 21 = coronene (ED), and 22 = ovalene (ED). Points 4, 5, and 11–16 corrected for libration. For details, compare Reference 23.

tively. It is immediately clear that substitution increases the stability of these systems very significantly, which in turn is reflected in their geometry. On the other hand, Hess and Schaad[28] RE values were calculated for unsubstituted species. Very recent microwave measurements of the geometry of methylenecyclopropene[33] allowed the application of the HOSE model for this system, and comparison of the HOSE value with the RE of Hess and Schaad. The comparison was not too successful, yielding 5.69 kJ and

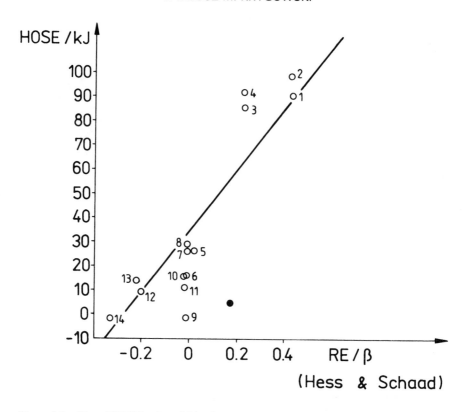

Figure 6-2. Plot of HOSE values (kJ/mol) versus resonance energy by Hess and Schaad[22] for nonalternant hydrocarbons. Data points are assigned as follows: 1 and 2 = benz|f|azulene (X-ray), 3 = azulene (ED), 4 = *cis,cis*-1,3-distyrylazulene (X-ray), 5 = 1,2-dimethyl-3-dicyanomethylene cyclopropene (X-ray), 6 = 6-dimethylaminofulvene (X-ray), 7 and 8 = 6-dimethylamino-2-formylfulvene (two independent molecules in asymmetric unit, X-ray), 9 = fulvene (ED), 10 = dicyanomethylene cycloheptatriene (X-ray), 11 = 2-dicyanomethylene-1,3-dimethyl cycloheptatriene (X-ray), 12 = 1,1'-bicycloheptatrienylidene, 13 = 1,2-di-*tert*-butyl-3,4,5,6-tetramethylbenzocyclobutadiene (X-ray), and 14 = octachlorofulvene. The solid point indicates a very recent microwave measurement of triafulvene.[33] Abbreviation as in Figure 6-1. Data for points 10 and 12 corrected for libration. For details compare Reference 23.

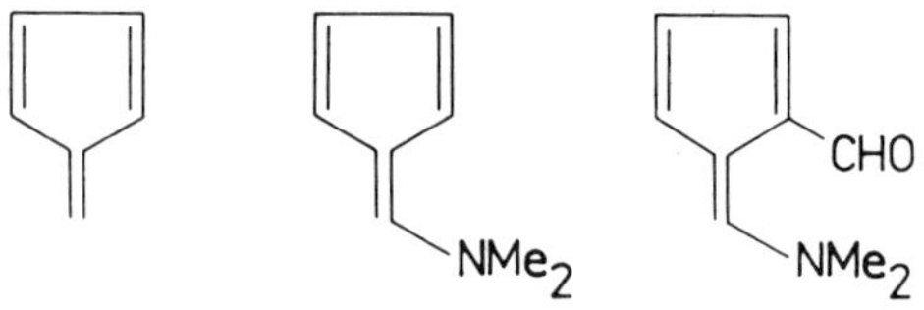

Scheme V

0.019 β, respectively. In Figure 6-2, the point marked by the solid circle deviates considerably from the line—even more than for the fulvene point (9). This effect is caused by the additional loss of energy due to strain, which is evidently taken into account by experimental geometry data used in calculating the HOSE value but not within the HMO calculation used in Hess and Schaad estimation of the resonance energy. This point was not used in calculation of the regression mentioned above.

Similarly, when the percentage contributions of canonical structures calculated by the HOSE model are compared with corresponding quantities obtained by the Randić theory,[34] the regression for 106 data points* reads[23]:

% Kekulé structures (HOSE)
$$= 0.998 \; [\% \text{ Kekulé structures (Randić)}] + 0.135 \quad (6\text{-}6)$$

with correlation coefficient $r = 0.985$. The slope is negligibly different from 1.0, whereas the intercept differs from zero only by 0.135! This shows that both methods of estimating C_i in Equation 6-6 are highly equivalent in spite of quite different theoretical backgrounds for the two models. The Randić theory[34] originates from quantum chemistry, whereas the HOSE model is purely empirical and applies the experimental geometry of molecules.

C. Formal Errors in Calculating $HOSE_i$ and C_i

The accuracy of structural data (ie, bond lengths needed for our purposes) is often not clearly estimated. However, all crystallographic data, such as X-ray and neutron diffraction measurements, are accompanied by estimated standard deviations (esd or σ), that is, measures of precision for bond lengths. Certainly this precision affects the precision of the calculated $HOSE_i$, HOSE, and C_i values. The following formulas are useful for calculating the σ' values[35]:

$$\sigma(HOSE_i) = \left[\sum \left(\frac{d}{dR_r} HOSE_i \right)^2 \sigma^2(R_r)^2 \right]^{1/2} \quad (6\text{-}7a)$$

and

$$\sigma(C_i) = \left[\left(\frac{d}{d \, HOSE_i} C_i \right)^2 \sigma^2 (HOSE_i)^2 \right]^{1/2} \quad (6\text{-}7b)$$

---

* The number of Kekulé resonance structures taken into account for each of the nine benzenoid hydrocarbons is given in parentheses: benzene (2), naphthalene (3), anthracene (4), phenanthrene (5), pyrene (6), chrysene (7), triphenylene (9), coronene (20), and ovalene (50); all together 106 Kekulé structures were taken into calculation in Equation 6-6.

where $\sigma(\text{HOSE}_i)$, $\sigma(C_i)$, and $\sigma(R_r)$ are the estimated standard deviations of stabilization energy of the ith resonance structure, of its contribution to the description of the real molecule, and of the rth bond length, respectively. From these formulas we find that when applying structural data (ie, bond lengths) with precision $\sigma(R_r) \leq 0.6$ pm, the error in precision of C_i, ie, $\sigma(C_i)$, is less than 2%.

D. Dependence of HOSE Model Results on Number of Canonical Structures Taken into Account

It is clear from Equation 6-5 that the numerical values of C_i depend not only on the values of $(\text{HOSE}_i)^{-1}$ and $\Sigma_{j=1}^{N} (\text{HOSE}_j)^{-1}$ but also on N, the number of canonical structures taken into account. Additionally, the size of the π-electron system (ie, of the chemical species or of their fragments) plays an important role in the application of the HOSE model. In many cases in order to interpret chemical and physicochemical properties of molecules it is sufficient to consider only a part of the π-electron system. Since the HOSE model is empirical, its use is reasonable for a series of species in which the π-electron system is topologically the same. For these corresponding parts of the systems in question, the same number of canonical structures N should be taken into account. Obviously, in choosing the kind of canonical structures to be used in the analysis, chemical knowledge and intuition are necessary.

In the case of alternant and nonalternant hydrocarbons,[13] only nonexcited resonance structure of the whole π-electron system must be used to reach good agreement with the Hess and Schaad[28] resonance energies and the Randić[34] Kekulé structure contributions. In a discussion of the chemical properties of complex derivatives of thiourea, Karolak-Wojciechowska[32] used only a part of the system and one nonexcited and three monoexcited resonance structures (II, III and IV), as shown in Scheme VI. The chemical reactivity of these systems was rationalized by analysis of the percentage contributions C_I through C_{IV} for the thiourea fragment as a function of its construction into N-phosphoryl-N,N'-disubstituted ureas and their conformation.[32]

In the case of an approximate estimation of charge transfer in EDA complexes and salts of 7,7,8,8-tetracyanoquinodimethane (TCNQ),[36] four schemes of calculating percentages of quinoid (Q) and benzenoid (B_1 and B_2) were used. First, only a and b bonds (Scheme VII) were applied for calculating the percentages of Q and B_1 and B_2 (Scheme VIIIa). Then the π-electron system was extended over the a, b, and c bonds and again the percentages of Q and the two B's were calculated (Scheme VIIIb). Finally, a, b, c, and d bonds were involved in calculating the percentage of Q and eight structure contributions of the B type (Scheme VIIIc). Obviously the quantitative distributions of the percentage of Q for the first two cases were different from

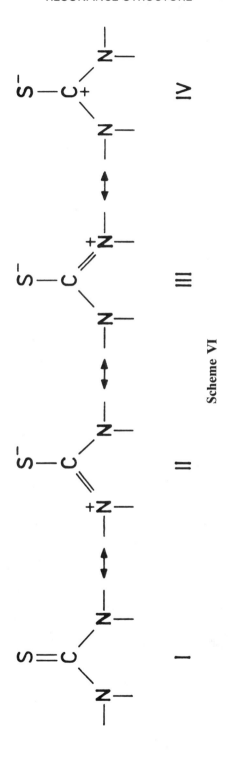

Scheme VI

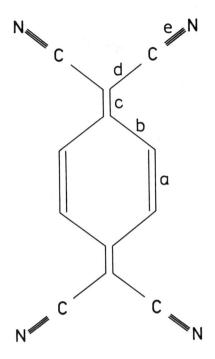

Scheme VII

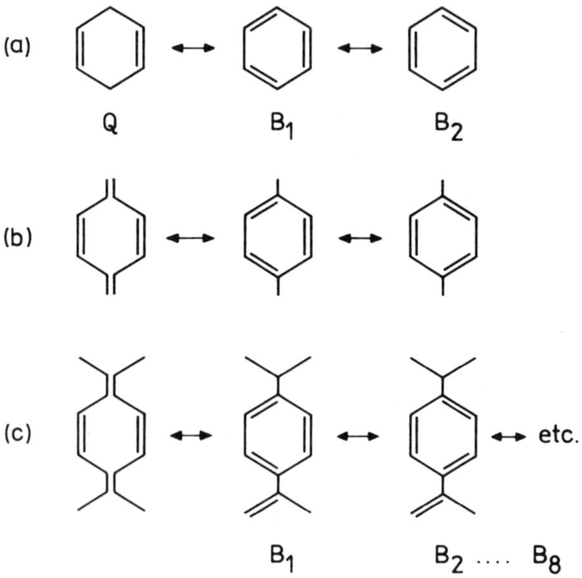

(a) Q ↔ B₁ ↔ B₂

(b)

(c) B₁ ↔ B₂ .... B₈ ↔ etc.

Scheme VIII

that in the last case, but intercorrelations between Q_1, Q_2, and Q_3 were always acceptable with the correlation coefficient $r \geq 0.92$.

In conclusion it may be suggested that the HOSE model works better when (a) the number of structures taken into consideration is not too large, (b) a series of compounds (or their π-electron fragments) taken into analysis contains a π-electron system of the same topology of σ cores, and (c) the number of resonance structures is the same for all the systems studied.

4. APPLICATIONS

The HOSE model as well as other simple uses of the experimental geometry of molecules with π-electron systems seem to be very useful tools in many problems involving the analysis of the structure of molecules for their properties in comparison with other systems of similar properties and structure.

A. Calculation of the Charge Transfer in EDA Complexes and Slats of TCNQ

Flandrois and Chasseau[37] applied the geometry of TCNQ in the approximate estimation of the charge transfer in its EDA complexes and salts. They assumed that for a radical anion, TCNQ$^-$, b, c, and d bonds (Scheme VII) are of the same length. Hence for TCNQ$^-$, the charge at it, q, is -1, and b $-$ c = c $-$ d = 0, whereas for the neutral TCNQ, $q = 0$, b $-$ c = 6.9 pm, and c $-$ d = -6.2 pm.[38] From these data one can calculate q for any TCNQ species, provided the b, a, and d bond lengths are known. Other investigators[39,40] came to similar conclusions. A more detailed analysis of bond lengths for TCNQ$^-$: M$^+$ (where M$^+$ stands for metal cation) in salts with univalent inorganic cations, M$^+$ (1:1), reveals that the assumption b = c = d fails by 1.0–2.0 pm; hence estimating the values of q_{TCNQ} is burdened with a relatively high ($\approx$10–30%) error. The situation can be improved by applying the HOSE model.[36] Calculation of the percentage of Q for neutral TCNQ[38] ($q = 0$) gives 91.28%, whereas it was 50.59 for a singly charged TCNQ species ($q = -1$) in (TCNQ$^-$, Na$^+$) salt,[40] which was solved with very high precision (esd for bond lengths taken into calculation ≤ 0.3 pm). As a consequence, one obtains:

$$q_k = -2.243 - 0.0246\% \ Q_k \tag{6-8}$$

In this approach only the a, b, and c bond lengths have been used. If more bonds are applied, the result is qualitatively the same. Thus, having known a, b, and c lengths, % Q may be calculated and by using Equation 6-8, one can estimate q_k, provided precise geometry of TCNQ is known. The estimated standard deviations for q_k calculated from electrical neutrality for q_k

obtained by the HOSE model[36] and by the Flandrois–Chasseau method[37] are 0.18 e⁻ and 0.28 e⁻, respectively.

B. Relation of HOSE Model to Simpler Applications of Ring Geometry for Substituted Derivatives of Benzene

It has been found recently that for monosubstituted and symmetrically *p*-disubstituted derivatives of benzene, the Walsh–Bent[41,42] rule is revealed by way of regression[43]:

$$\Delta = A\alpha + B \qquad (6\text{-}9)$$

where $\Delta = b - a$ (Scheme III), whereas α is an angle at ipso carbon. Correlation coefficients for those two sets of data are $r = 0.913$ (mono) and 0.979 (para), respectively. Figure 6-3 presents this relation for *p*-disubstituted species of high precision of measurements, the geometry of which was chiefly estimated by electron diffraction. The Walsh–Bent rule states[41,42]: "If a group X_1 attached to carbon is replaced by a more electronegative

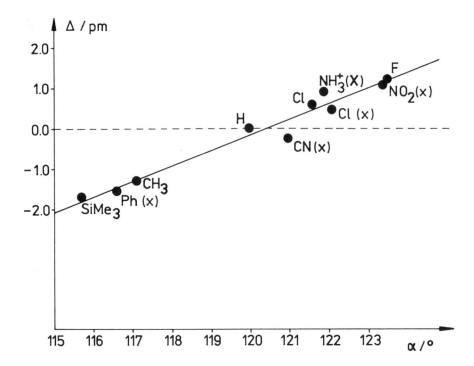

Figure 6-3. Plot of Δ versus α for 10 symmetrically *p*-disubstituted derivatives of benzene. X-ray data of very high precision esd for bond lengths of 0.5 pm or less indicated by X; other points were derived from electron diffraction data. For details, compare Reference 23.

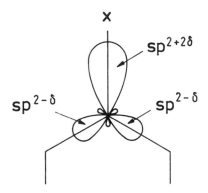

Figure 6-4. Scheme presenting action of the Walsh–Bent rule on sp^2-hybridized orbitals at the substituted carbon atom in a benzene ring.

group X_2, then the carbon valency toward X_2 has more p character than it had toward X_1." This implies a decrease in the p character of the other two hybrid orbitals of the carbon atom and leads to[44] an increase in the α value and a decrease in the bond lengths of the adjacent CC bonds, a (Scheme III). The opposite is true of the electron-repelling substituent. Figure 6-4 illustrates these relations.

Since the electronegativity of the substituent acts no further than at the second carbon,[45,46] and in subtraction $\Delta = b - a$, some systematic errors may be canceled out, the Δ values plotted against α give an excellent picture of fulfillment of the Walsh–Bent rule. If in p-substituted derivatives of benzene, pXPhY, X, and Y, have qualitatively different π-electron properties, there may appear π-electron cooperative effects (eg, conjugation or intramolecular charge transfer). In these cases Δ becomes a very sensitive parameter, since these interactions make a bonds longer and b bonds shorter, thus increasing the quinoid character of the geometry. This in turn is revealed in a strong decrease of Δ values.

All this is illustrated in Figure 6-5, in which points with coordinates (Δ, α) for NO_2 and NEt_2 in p-nitro-N,N-diethylaniline[47] deviate very significantly from the Δ–α line for pXPhX, which may be used as reference line for systems without any π-electron cooperative effect. This is evidently due to the strong through-resonance effect; hence the $\delta\Delta(NEt_2)$ and $\delta\Delta(NO_2)$ values, the distances from the line, may be useful as quantitative parameters describing this effect directly from the geometry of the molecule. One might ask why the HOSE model is not used in the case of pXPhY systems. The answer is very simple. In general, empirical methods should be applied in the way that entails the lowest possible number of empirical parameters. Since for each C—X, C—Y bond new parameters for R_s°, R_d°, and a, b in Equation 6-2 are needed, it is better to use a simpler way, which does not require parametrization. In other words, within the pXPhY series the similarity

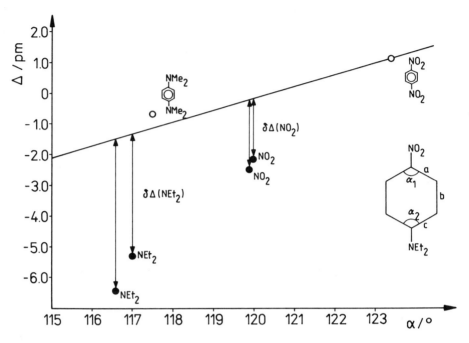

Figure 6-5. Plot of Δ versus α for symmetrically p-substituted benzene derivatives (the line, as in Figure 6-3) with two examples as open points: p-dinitrobenzene and p-N,N,N',N'-tetramethylphenylenediamine. Solid points are: Δ and α values for nitro group (Δ = b − a) and NEt$_2$ (Δ = b − c) in N,N-diethyl-p-nitroaniline[47] (two independent molecules in an asymmetric unit of the crystal cell). Shift down from the line δΔ describes quantitatively (ie, in picometers) a π-electron cooperative effect between the NO$_2$ and NEt$_2$ groups.

between molecules is too low to use the HOSE model safely. Certainly, one could use only *a* and *b* bonds (Scheme III) in the ring when calculating HOSE model results. However, this would be equivalent to the direct use of Δ = b − a or Δ = b − c. It was found for p-substituted derivatives of benzoic acid[48] and nitrobenzene[47] that ΣΔ = (b − a) + (b − c) values plotted against % Q structure give an excellent linear dependence, as shown in Figure 6-6, for which the correlation coefficient $r > 0.98$ ($n = 16$ and 12, respectively). Thus, for some systems it is more convenient to use direct Δ values or ΣΔ values than the more complex HOSE model, since the final results are of comparable importance.

Similar results are obtained for EDA complexes and salts. Figure 6-7 plots Δ versus α for N,N,N'N'-tetramethyl-p-phenylenediamine (TMPD) and its EDA complexes and salts. Points 1 and 2 stand for TMPD itself and its weak complex with 1,2,4,5-tetracyanobenzene, respectively. Other points represent stronger EDA complexes or salts. The more the point is shifted down from the line for pXPhX (taken from Reference 43), the more pronounced is the quinoid structure exhibited by the species is question. The δΔ value (ie, the distance from the line, in picometers, is an approximate measure of

RESONANCE STRUCTURE

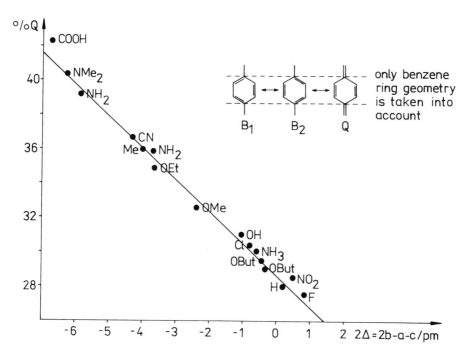

Figure 6-6. Plot of % Q versus $2\Delta = 2b - a - c$ for 16 p-substituted derivatives of benzoic acids. Precision of bond length determination: esd ≤ 0.5 pm.

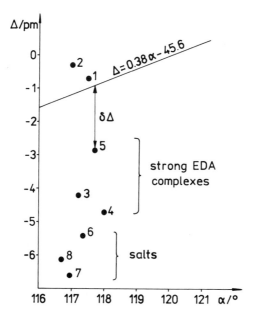

Figure 6-7. Plot of Δ versus α for neutral N,N,N',N'-tetramethylphenylenediamine (1), its weak electron donor–acceptor complex (2), strong complexes (3–5), and salts (6–8).

the contribution of the quinoid structure. In fact, $\delta\Delta$ values plotted against % Q calculated by the HOSE model give a very good linear regression[46]:

$$\% Q = 24.6 - 7.6\Delta \tag{6-10}$$

with the correlation coefficient $r = 0.980$ for $n = 8$. Again it is apparent that such simple quantities as Δ, $\delta\Delta$, or $\Sigma\Delta$ may be very useful parameters in structural analysis, and to some degree they may replace the more complex HOSE model.

Looking at Figure 6-7, one can easily find that in spite of a great variation in bond lengths (ie, $\delta\Delta$ values), the ipso angles at substituted carbons (ie, 1 and 4) are rather immobile, their variation being in the range of 1–1.5°. This finding is in line with an earlier observation[12] that α angle does not depend much on π-electron changes as well as with recent ab initio STO-3G calculations[49] that α values in monosubstituted derivatives of benzene depend 70% on σ-electron and 21% on π-electron charge variation.

Another interesting application of the simple Δ-type approach is the use of the geometries of molecules of p-disubstituted derivatives of benzene with strong cooperative π-electron effects. The dependence of $\Sigma\Delta = \Delta(A) + \Delta(D) = 2b - a - c$ on parameters describing the ability of substituents to participate in through-conjugation [ie, σ^+ (for donors, D) and σ^- (for acceptors, A)] was studied for 16 well-solved structures (ie, with esd for bond lengths ≤ 0.5 pm). The regression reads[46]:

$$\Sigma\Delta = 1.3(\pm 1.3) - (1.6 \pm 1.2)\sigma^- + (3.5 \pm 0.2)\sigma^+ \tag{6-11}$$

with the correlation coefficient $r = 0.971$. Closer inspection into the esd for ρ^+ and ρ^- reveals that the $\rho^-\sigma^-$ term is of negligible significance, whereas $\rho^+\sigma^+$ is large and burdened with low uncertainty. Thus it may be concluded that $\rho^+\sigma^+$ is the only important contribution in the description of $\Sigma\Delta$ variation. This conclusion is additionally supported by comparison of the correlation coefficient for a single plot of $\Sigma\Delta$ versus σ^+, $r = 0.970$. Figure 6-8 shows this dependence. From this finding, we make the following direct conclusion: of two substituents in pD − Ph − A, in which D is releasing π electrons and A is accepting them, changes in the geometry of the ring described by $\Sigma\Delta$ values are determined by the electron-releasing substituent. The much less important effect coming from the side of electron-accepting substituents may be explained, at least in part, as follows: A substituents, if they do not cooperate with p − D substituents, cause an increase of Δ values (Figure 6-3), whereas the isolated D substituents make Δ values more negative by 2 to 3 pm. Thus, since $\Delta(A)$ and $\Delta(D)$ in $\Sigma\Delta$ are the sum of σ- and π-electron effects of substituents on a, b, and c bond lengths, and σ contributions to $\Delta(A)$ are positive while those to $\Delta(D)$ are negative, the $(\Delta^{\pi+\sigma})$ value must be much higher for D substituents than for A substituents.

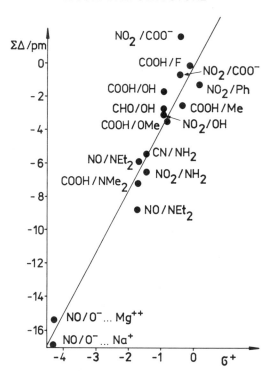

Figure 6-8. Plot of $\Sigma\Delta = 2b - a - c$ versus σ^+ for p-XPhY-disubstituted derivatives of benzene assigned XY. Compare Reference 46.

5. CONCLUDING REMARKS

The geometry of π-electron systems in molecules is an important source of information. Simple geometric parameters Δ, $\Sigma\Delta$, or $\delta\Delta$, as presented in this chapter, may serve as very useful descriptors of the changes in π-electron structure. In turn, they may be used to predict of physicochemical properties as well as to interpret them. The contributions of canonical structure may be estimated directly from the geometry by using the HOSE model. Furthermore they may serve both to predict and to interpret the physicochemical properties of π-electron compounds or fragments of compounds. Careful selection of structural data is recommended for any empirical models based on the geometry of molecules. Sometimes the regularities found for precise data become less convincing or even vanish when less reliable data are added, as exemplified by Figure 6-9, which plots[50] the percentage of Q calculated from a, b, and c bond lengths in the ring versus R_{CN} values: the C—NO_2 bond lengths in a series of p-substituted derivatives of nitroben-

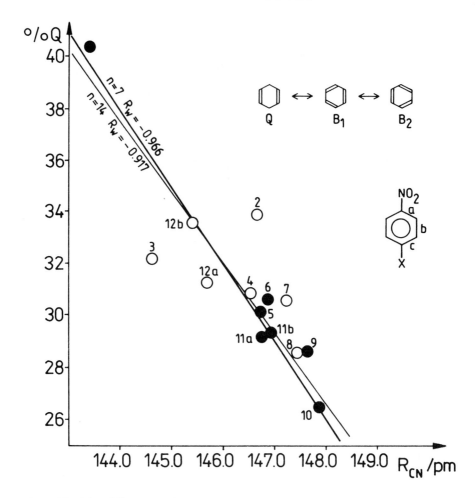

Figure 6-9. Plot of % Q versus C—NO$_2$ bond lengths for *p*-substituted derivatives of nitrobenzene. Only inner ring bond lengths were used in calculating % Q. Data points assigned as follows: 1 = *p*-nitroaniline, 2 = *p*-nitroacetophenone, 3 = *p*-nitrophenol, 4 = *p*-nitrobiphenyl, 5 = *p*-nitrobenzohydroxynol chloride, 6 = *p*-nitrobenzamide, 7 = *p*-nitrobenzaldoxime, 8 = di-(*p*-benzonitro)butadiyne, 9 = *p*-nitrobenzoic acid, 10 = *p*-dinitrobenzene, 11 = *p*-(*O*-acetoxy)-nitrobenzene, 12 = *N*,*N*-dimethyl-*N*′-*p*-nitrophenylformamidine.

zene. Evidently one can expect that the greater the contribution of the quinoid structure "inside the ring," the shorter the C—NO$_2$ bond length R_{CN}. Indeed, for seven highly precise structural data having esd for bond lengths of 0.3 pm or more, the correlation coefficient, R_w, is -0.966, whereas if less precise data are included (with esd ≤ 0.5 pm) then for 14 data points $R_w = -0.917$. In both cases weighted correlation coefficients were calculated with weights = (esd)$^{-2}$. Even more drastic are the results for a plot of Δ versus α for monosubstituted derivatives of benzene.[51] For all 180 structures studied with esd of 0.5 pm or more but without any other restriction, $R = 0.55$. When the data with closer than van der Waals contacts were excluded together with those for which C_2 symmetry of the ring was broken by more than 1.0 pm [ie, $|a - a'|$ or $|b - b'|$ or $|c - c'| \geq 1.0$ pm] for bond lengths and 0.6° [ie, $|\beta - \beta'|$ or $|\gamma - \gamma'| \geq 0.6°$] for angles, the correlation coefficient becomes much better, 0.84 for 27 data points. The same relation for noncrystalline data (ie, electron diffraction, microwave, and NMR in the nematic phase) was found to be most reliable with $R = 0.91$ for 21 data points. Evidently, these results may serve as a serious warning against a propensity to use data of unknown quality or burdened with strong intermolecular interaction in the crystal lattice. Evidently, the substituent effect on ring geometry in monosubstituent benzene derivatives is not strong enough to be easily detected from data influenced by other effects, chiefly packing forces. However, the π-electron cooperative effects are strong enough to be studied quite safely by use of the methods presented in this chapter.

REFERENCES

1. Pauling, L. "The Nature of the Chemical Bond," 3rd ed. Cornell University Press: Ithaca, NY, 1960.
2. Coulson, C. A. "Valence." Oxford University Press: London, 1961.
3. Hiberty, P. C.; Ohanessian, G. *J. Am. Chem. Soc.* **1984**, *106*, 6963.
4. cf. Johnson, C. D. "The Hammett Equation." Cambridge University Press: Cambridge, 1973.
5. Geske, D. H.; Ragle, J. L.; Bambenek, M. A.; Balch, A. L. *J. Am. Chem. Soc.* **1964**, *86*, 987.
6. Hargittai, I.; Hargittai, M. The Importance of Small Structural Differences. In "Molecular Structure and Energetics," Vol. 2; *Physical Measurements*; Liebman, J. F.; and Greenberg, A., Eds.; VCH Publishers: Deerfield Beach, FL, 1987.
7. eg, Coppens, P. *Acta Crystallogr.* **1971**, *B27*, 1931.
8. Cruickshank, D. W. J. *Acta Crystallogr.* **1956**, *9*, 754, 757.
9. For short review, see Dunitz, J. D. "X-Ray Analysis and Structure of Organic Molecules." Cornell University Press: Ithaca, NY, 1979.
10. Cox, E. G.; Cruickshank, D. W. J.; Smith, J. A. S. *Nature* **1955**, *175*, 766.
11. Trueblood, K. N. Program THMB-6. Department of Chemistry, University of California, Los Angeles, 1984.
12. Domenicano, A.; Murray-Rust, P.; Vaciago, A. *Acta Crystallogr.* **1983**, *B39*, 457.
13. Dieterich, D. A.; Paul, I. C.; Curtin, D. Y. *J. Am. Chem. Soc.* **1974**, *96*, 6372.
14. Leiserowitz, L. *Acta Crystallogr.* **1976**, *B32*, 775.
15. Grabowski, S. J.; Krygowski, T. M. *Proceedings of the Fourth Conference on Electric and Related Properties of Organic Solids*, Książ, Poland, Feb. 6–7, 1984; *Mater Sci.* **1984**, *10*, 85. Grabowski, S. J.; Krygowski, T. M. *J. Mol. Struct.* in press.
16. Jeffrey, G. A.; Sax, M. *Acta Crystallogr.* **1963**, *16*, 430.
17. Elvidge, J. A.; Jackman, L. M. *J. Chem. Soc.* **1961**, *8*, 859.
18. Sondheimer, F. *Pure Applied Chem.* **1964**, *7*, 363.
19. Julg, A.; François, P. *Theor. Chim. Acta* **1967**, *7*, 249.
20. Kruszewski, J.; Krygowski, T. M. *Tetrahedron Lett.* **1972**, 3839.
21. Krygowski, T. M.; Więckowski, T. *Croat. Chim. Acta* **1981**, *54*, 193.

22. Więckowski, T.; Krygowski, T. M. *Can. J. Chem.* **1981**, *59*, 1622.
23. Krygowski, T. M.; Anulewicz, R.; Kruszewski, J. *Acta Crystallogr.* **1983**, *B39*, 732.
24. See, eg, Morrison, R. T.; Boyd, R. N. "Organic Chemistry," 5th ed. Allyn & Bacon: Boston, 1987.
25. Kveseth, K.; Seip, R.; Kohl, D. A. *Acta Chem. Scand.* **1980**, *A34*, 31.
26. Breslow, R. "Mechanism of Organic Reactions." Wiley: New York, 1964.
27. Dewar, M. J. S. "The Molecular Orbital Theory of Organic Chemistry." McGraw-Hill: New York, 1969.
28. Hess, B. A.; Schaad, L. J. *J. Am. Chem. Soc.* **1971**, *93*, 305.
29. Nes, G. J. H. van; Vos, A. *Acta Crystallogr.* **1979**, *B35*, 2593.
30. Kuchitsu, K. *J. Chem. Phys.* **1968**, *49*, 4456.
31. Woźniak, K.; Krygowski, T. M. Unpublished, 1986.
32. Karolak-Wojciechowska, J. *Phosphorus Sulphur,* **1985**, *25*, 229.
33. Norden, T. D.; Staley, S. W.; Taylor, W. H.; Harmony, M. D. *J. Am. Chem. Soc.* **1986**, *108*, 7912.
34. Randić, M. *Tetrahedron* **1977**, *33*, 1905.
35. Karolak-Wojciechowska, J. Rozprawa habilitacyjna, P. Ł.; Łódź, Poland, 1986.
36. Krygowski, T. M.; Anulewicz, R. *Mater Sci.* **1984**, *10*, 145.
37. Flandrois, S.; Chasseau, D. *Acta Crystallogr.* **1977**, *B33*, 2744.
38. Long, R. E.; Sparks, R. A.; Trueblood, K. N. *Acta Crystallogr.* **1965**, *18*, 932.
39. Coppens, P.; Guru-Row, T. N. *Ann. N.Y. Acad. Sci.* **1978**, *313*, 244.
40. Konno, M.; Saito, Y. *Acta Crystallogr.* **1974**, *B30*, 1294.
41. Walsh, A. D. *Discuss. Faraday Soc.* **1947**, *2*, 18.
42. Bent, H. A. *Chem. Rev.* **1961**, *61*, 275.
43. Krygowski, T. M. *J. Chem. Res.* **1984**, 238.
44. Domenicano, S.; Vaciago, A.; Coulson, C. A. *Acta Crystallogr.* **1975**, *B31*, 221.
45. Topsom, R. *Acc. Chem. Res.* **1983**, *16*, 292.
46. Krygowski, T. M. *J. Chem. Res.* **1987**.
47. Krygowski, T. M.; Maurin, J. Unpublished, 1986.
48. Anulewicz, R.; Häfelinger, G.; Krygowski, T. M.; Regelman, C.; Retter, Q. *Z. Naturforsch.* Submitted.
49. Krygowski, T. M.; Häfelinger, G.; Schule, J. *Z. Naturforsch.* **1986**, *41b*, 895.
50. Maurin, J.; Krygowski, T. M. *J. Mol. Struct.* in press.
51. Turowska-Tyrk, I.; Krygowski, T. M. Unpublished; in progress.
52. Hoefnagel, A. J.; Wepster, B. M. *J. Am. Chem. Soc.* **1973**, *95*, 5357.

CHAPTER 7

Molecular Aspects in Energetic Materials

Sury Iyer and Norman Slagg
Energetics and Warheads Division, AED, U.S. Army
Armament Research, Development, and Engineering Center
Picatinny Arsenal, New Jersey

CONTENTS

1. Introduction 255
   Glossary 258
2. Sensitivity of Energetic Materials 258
3. Detonations 273
4. Summary 285
References 286

1. INTRODUCTION

Energetic materials are metastable molecules possessing high reactivity. They are usually formed, in a thermodynamic sense, with a positive heat of formation, which is one reason for the high reactivity.[1a] They decompose, producing smaller molecules and large amounts of energy (1–3 kcal/g). In some cases strain energy locked in rings and cages contributes to the energy release. The ease with which decomposition can be started is commonly described by the parameter called "sensitivity."[1] Different modes of initiation of decomposition reactions are described by different sensitivities (heat, shock, friction, etc.).[1b]

Energetic materials are commonly divided into three classes: explosives, propellants, and pyrotechnics. Explosives function by undergoing a very rapid decomposition, with the generation of products at high pressures and temperatures behind a shock front.[1c] Propellants and pyrotechnics function by undergoing a deflagration (a burninglike process), producing products at high temperatures behind a subsonic reaction front. Typical reaction times for explosives and propellants are 10^{-6} and 10^{-2} s, respectively.[1a] Figure 7-1 presents the major steps in the function of energetic materials. It is seen that the application determines the desired chemical and physical state of the reaction products.

In the solid state the properties of isolated molecules can be altered by adjacent atoms and overlapping electrical fields. Furthermore, macroscopic properties such as particle size, the presence of cracks in billets, and the degree of confinement can dramatically alter the sensitivity and the rate of energy release.[1b]

The bulk of the literature on the sensitivity and reactivity of energetic molecules deals with macroscopic properties. It is only in the past 15 years

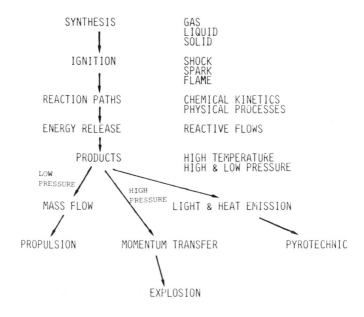

Figure 7-1. Steps in the energetic conversion of materials.

that serious attempts have been made to gain insights into the pertinent molecular processes.[2] These recent efforts are due to development of laser diagnostics and related electronic time-resolving techniques, as well as increased computer capability to handle sophisticated quantum mechanical, molecular dynamical, and complex kinetic programs.

Both sensitivity and the rate of energy release are indices of reactivity. Ignition (the first few endothermic processes) starts the initiation, which leads to exothermic reactions and a rise in temperature. Generally the weak (sensitive) bonds are the first to break, forming radicals and/or ions. Of course electronically and/or vibrationally excited states may be playing a role. As the reaction proceeds and spreads, a front will form and, depending on the rate of energy release, a deflagration or detonation will ultimately develop.

No attempt is made here to present a detailed analysis of our current understanding or to provide an extensive literature survey. Rather, our purpose is to highlight the current trends, emerging techniques, and major issues with regard to the molecular aspects of the behavior of energetic materials.

This chapter uses a number of terms that require definition for those unfamiliar with this subject area. They are presented below.[1a,b]

Energetic material: a substance capable of undergoing very rapid exothermic reactions with the generation of gaseous products at high temperatures.

Explosion: a very rapid outward projection of matter. The term is often applied to events that follow the rupture of steam boilers as well as to the evolutions of high-pressure gaseous products from exothermic reactions.

Thermal explosion: the generation of high pressures due to very rapid exothermic reactions when the heat generation exceeds losses due to heat conduction.

Deflagration: a reaction discontinuity (can be a flame) whose front propagates with a velocity less than the sound speed of the unreacted material.

Shock wave: a pressure discontinuity that propagates with a velocity greater than the speed of sound of the unreacted material. Shock fronts are not necessarily followed by chemical reactions. The strength of the shock wave and the properties of the material determine the likelihood of permanent chemical changes.

Detonation: a reaction front characterized by a pressure discontinuity that moves at a characteristic supersonic velocity relative to unreacted material. Behind the front the reactant is converted to products at very high temperatures (2500–4000°C) and at high pressures (200–400 kbar). The products are usually light gases, such as methane, water, carbon monoxide, carbon dioxide, and nitrogen.

Glossary

3-Amino-TNT	3-Amino-2,4,6-trinitrotoluene
DATB	2,4-Diamino-1,3,5-trinitrobenzene
HMX	1,3,5,7-Tetranitro-1,3,5,7-tetrazacyclooctane
MATB	Same as picramide (2,4,6-trinitroaniline)
PETN	Pentaerythritol tetranitrate
Picramide	2,4,6-trinitroaniline
RDX	1,3,5-Trinitro-1,3,5-triazacyclohexane
TATB	2,4,6-Triamino-1,3,5-trinitrobenzene
TENA	Tetranitroaniline
TET	Tetryl (2,4,6,N-tetranitro-N-methylaniline)
TNB	1,3,5-Trinitrobenzene
TNT	2,4,6-Trinitrotoluene
Trinitroso-RDX	1,3,5-Trinitroso-1,3,5-triazacyclohexane

2. SENSITIVITY OF ENERGETIC MATERIALS

A. Different Kinds of Sensitivity

In the case of energetic materials, sensitivity is the ease with which the substances can be made to undergo explosive decomposition (or detonation) by application of stimuli. Depending on the nature of the stimulus, different kinds of sensitivity can be defined. Thus thermal sensitivity refers to the combination of temperature and rate of heating that results in ignition. When material is subjected to a falling hammer, one talks of the impact sensitivity. The intensity of a shock pulse that can initiate and cause explosion is a measure of shock sensitivity.[1c] Energetic materials can also be initiated by friction and electrostatic charge, which are referred to as friction and electrostatic sensitivity, respectively.

Sensitivities of an energetic material are often related to their thermal characteristics. Furthermore, in the early stages of growth to detonation or explosion, chemical processes occur under mild conditions, particularly with respect to pressure. Hamann[3] pointed out that in the $1-10^{4-6}$ bar regime, normal chemistry occurs. Ignition and initiation processes related to sensitivity occur in the $1-10^5$ bar regime.

Among explosives, there are primary and secondary materials. Primary explosives are more sensitive than secondary explosives. Usually a primary is needed to set off a secondary explosive. Examples of primary explosives are lead azide and mercury fulminate. Secondary explosives include the common ones like TNT and nitramines (eg, RDX and HMX); see Glossary, above, for names.

B. Sensitivity Determination

a. Thermal Sensitivity

Usually thermal sensitivity is appraised by differential thermal analysis (DTA) or by differential scanning calorimetry (DSC). In these tests, the material is subjected to a programmed heat profile (which the detector senses), and the temperature of the sample increases uniformly with time. When the temperature at which the sample is unstable is reached, it will cause exothermic decomposition, and the corresponding heat change can be recorded as an exothermic peak. The lower the temperature at which the material shows exothermic decomposition, the greater its thermal sensitivity.

Attempts are now being made to input thermal pulses of differing intensities (as from a pulsed IR laser, eg, CO_2 laser) in the interest of determining the threshold intensity at which a material will initiate. These attempts are in their infancy, and much progress is still to be made.

Table 7-1 shows the DSC thermal data for various explosives, which are arranged in decreasing order of thermal sensitivity.[4]

b. Impact Sensitivity

The most important single test for energetic materials, which has traditionally been made and is still widely used, is the impact sensitivity test. This test[5] involves subjecting the energetic material to a falling hammer (2.5 or 2 kg weight) from different heights (up to 240 cm) and observing whether initiation occurs. The apparatus used at Picatinny Arsenal (PA) is shown in Figure 7-2. Instead of the die cup assembly shown in the figure, however, 25 mg samples are placed on sandpaper and tested. The height of fall, reported in centimeters, that produces explosions in 50% of the samples tested is

TABLE 7-1. Onset of Exothermic Decomposition of Explosives as Determined by DSC

Explosive	Onset temperature (°C)
Pentaerythritol tetranitrate	155
2,4,6-Trinitro-*N*-methyl-*N*-nitroaniline (Tetryl)	170
RDX	200
Nitroguanidine	200
2,4,6-Trinitrotoluene	250
HMX	260
2,4-Diamino-1,3,5-trinitrobenzene	295
2,4,6-Triamino-1,3,5-trinitrobenzene	350

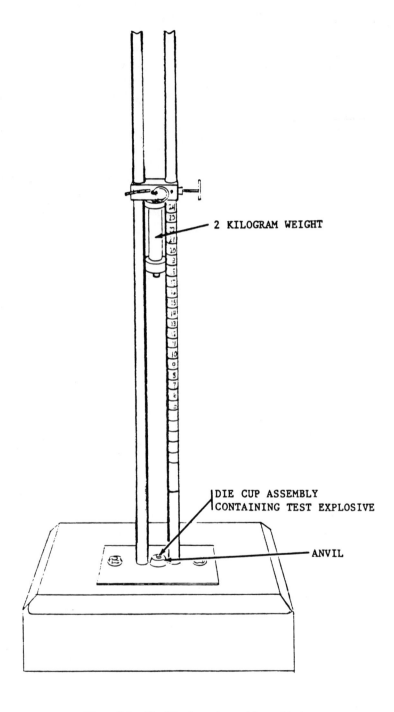

Figure 7-2. The Picatinny Arsenal impact tester.

called the 50% test height and is termed the sensitivity index. At least 25 samples are tested to ensure good statistical distribution. The procedure used to determine the 50% point is the Bruceton up-and-down method. In this method, the sample is tested initially at some arbitrary height in the suspected region. If it goes, the height is decreased by a single unit (usually 1 cm). On the other hand, if it does not initiate, the test height is increased by one unit and the procedure is continued. The ERL (Explosives Research Laboratory, Bruceton, Pennsylvania) impact tester[5] used by the Navy is similar in principle to that described above.

The difficulty one encounters in this test is in determining when an individual test is a "go." The first observation of a black mark (ie, charring) on the sandpaper is used as the indicator in PA unit. In the ERL apparatus, the test setup is hooked to a sound detector, which picks up the detonation noise above the background. Irrespective of the mode of detection of detonation, the overall test itself is not very precise. Nevertheless, it can rank different explosives in the order of their sensitivity.

Since the PA and ERL test units are arbitrarily set to different calibrations (50% test height for TNT = 60 cm in the PA unit and 160 cm in the ERL unit), the absolute values from these test systems should not be compared. However, the relative sensitivity scale should be similar. The 50% impact sensitivity values for various classes of organic explosives, determined in the ERL test unit, are shown in Tables 7-2 and 7-3. These values show interesting structure–sensitivity correlations, which are discussed later in this chapter.

TABLE 7-2. Structure–Sensitivity Relations in Nitroaliphatic Explosives

No.	Compound name	Molecular formula	CB_{100}	$H_{50\%}(cm)$*
	Nitramines			
1.	Methylenedinitramine (MEDINA)	$C_4H_4N_4O_4$	+1.47	13
2.	Ethylenedinitramine (EDNA)	$C_2H_6N_6O_6$	−1.33	34
3.	RDX	$C_3H_6N_6O_4$	0	24
4.	N-Methyl EDNA	$C_3H_8N_4O_4$	−3.65	114
5.	HMX	$C_4H_8N_8O_8$	0	26
6.	3-Nitraza-1,5-pentanedinitramine	$C_4H_{10}N_6O_6$	−2.52	39
7.	3,6-Dinitraza-1,8-octanedinitramine	$C_6H_{14}N_8O_8$	−3.07	53
8.	Tetryl	$C_7H_5N_5O_8$	−1.04	32
	Nitramine plus gem-dinitro			
9.	3,3-Dinitro-1,5-pentanedinitramine	$C_5H_{10}N_6O_8$	−1.42	35
10.	N,3,3,5,5-Pentanitropiperidine	$C_5H_6N_6O_{10}$	+1.29	14
11.	Bis(2,2-dinitropropyl)nitramine	$C_6H_{10}N_6O_{10}$	−0.61	29
12.	Bis(2,2-dinitrobutyl)nitramine	$C_3H_{14}N_6O_{10}$	−2.82	80
13.	2,2,4,7,9,9-Hexanitro-4,7-diazadecane	$C_8H_{14}N_8O_{12}$	−1.45	72
14.	2,2,4,7,7,10,12,12-Octanitro-4,7-diazatridecane	$C_{11}H_{18}N_{10}O_{16}$	−1.47	44
15.	2,2,5,7,7,9,12,12-Octanitro-4,7-diazatridecane	$C_{11}H_{12}N_{10}O_{16}$	−1.47	37
	Nitramine plus trinitromethyl			
16.	Methyl 2,2,2-trinitroethyl nitramine	$C_3H_5N_5O_8$	+2.09	9

TABLE 7-2. Cont.

No.	Compound name	Molecular formula	CB_{100}	$H_{50\%}(cm)^*$
17.	Bis(2,2,2-trinitroethyl)nitramine	$C_4H_4N_8O_{14}$	+4.12	5
18.	2,2,2-Trinitroethyl 3,3,3-trinitropropyl nitramine	$C_5H_6N_8O_{14}$	+2.98	6
19.	N,N'-bis(2,2,2-Trinitroethyl)MEDINA	$C_5H_6N_{10}O_{14}$	+3.46	5
20.	N-Methyl-N'-trinitroethylEDNA	$C_5H_9N_7O_{10}$	+0.30	11
21.	2,2,2-Trinitroethyl N-(2,2,2-trinitroethyl)nitraminoacetate	$C_6H_6N_8O_{16}$	+2.69	9
23.	2,2,2-Trinitroethyl 4-nitrazavalerate	$C_6H_9N_5O_{10}$	-0.97	35
24.	2,2,2-Trinitroethyl 3,3-dinitrobutyl nitramine	$C_6H_9N_7O_{12}$	+0.81	20
25.	Bis(2,2,2-trinitroethyl) 3-nitrazaglutarate	$C_8H_6N_8O_{10}$	+1.58	14
26.	1,1,1,3,6,9,11,11,11-Nonanitro-3,6,9-triazaundecane	$C_8H_{12}N_{12}O_{18}$	+1.42	12
27.	1,1,1,3,6,6,9,11,11,11-Decanitro-3,9-diazaundecane	$C_9H_{12}N_{12}O_{20}$	+1.64	10
28.	1,1,1,1,4,6,6,8,11,11,11-Decanitro-4,8-diazaundecane	$C_9H_{12}N_{12}O_{20}$	+1.64	11
29.	Bis(2,2,2-trinitroethyl) 4-nitraza-1,7-heptanedioate	$C_{10}H_{12}N_8O_{18}$	0	29
30.	1,1,1,3,6,9,12,14,14,14-Decanitro-3,6,9,12-tetrazatetradecane	$C_{10}H_{16}N_{14}O_{20}$	+0.61	19
31.	Bis(2,2,2-trinitroethyl 3,6-dinitraza-1,8-octanedioate	$C_{10}H_{12}N_{10}O_{20}$	+0.67	29
32.	1,1,1,3,6,6,8,10,10,13,15,15,15-Tridecanitro-3,8,13-triazapentadecane	$C_{12}H_{16}N_{16}O_{26}$	+1.50	23
	Nitramine plus nitrate ester			
33.	N-(2,2,2-Trinitroethyl)nitraminoethyl nitrate	$C_4H_6N_6O_{11}$	+3.19	7
34.	3-[N-2,2,2-Trinitroethylnitramino]propyl nitrate	$C_5H_8N_6O_{11}$	+1.83	12
35.	3,5,5-Trinitro-3-azahexyl nitrate	$C_5H_9N_5O_9$	-0.35	21
	Nitramides			
36.	N-Nitro-N-(3,3,3-trinitropropyl)2,2,2-trinitroethyl carbamate	$C_6H_6N_8O_{16}$	+2.70	9
37.	2,2,2-Trinitroethyl 2,5-dinitrazahexanoate	$C_6H_9N_7O_{12}$	+0.27	15
38.	2,2,2-Trinitroethyl 2,5,5-trinitro-2-azahexanoate	$C_7H_9N_7O_{14}$	+0.72	22
39.	2,2,2-Trinitroethyl 2,4,6,6-tetranitro-2,4-diazaheptanoate	$C_7H_9N_9O_{16}$	+1.47	18
40.	N,N'-Dinitro-N,N'-bis(3,3,3-trinitropropyl)oxamide	$C_8H_8N_{10}O_{18}$	+2.26	9
41.	2,2,6,9,9-Pentanitro-4-oxa-5-oxo-6-azadecane	$C_8H_{12}N_6O_{12}$	-1.56	47
42.	1,1,1,5,7,10,14,14,14-Nonanitro-3,12-dioxa-4,11-dioxo-5,7,10-triazatetradecane	$C_9H_{10}N_{12}O_{22}$	+1.88	11
43.	N,N'-Dinitro-N,N'-bis(3,3-dinitrobutyl)oxamide	$C_{10}H_{14}N_8O_{14}$	-1.28	37
44.	1,1,1,5,8,11,14,18,18,18-Decanitro-3,16-dioxa-4,15-dioxo-5,8,11,14-tetrazaoctadecane	$C_{12}H_{16}N_{14}O_{24}$	+0.54	19
45.	N,N'-Dinitro-N,N'-bis(3-nitrazabutyl)oxamide	$C_8H_{14}N_8O_{10}$	-2.61	90
	Trinitromethyl compounds			
46.	2,2,2-Trinitroethyl carbamate	$C_3H_4N_4O_8$	+1.79	18
47.	Methyl 2,2,2-trinitroethyl carbonate	$C_4H_5N_3O_9$	+1.25	28
48.	1,1,1,3-Tetranitrobutane	$C_4H_6N_4C_8$	+0.84	33
49.	Methylenebis(2,2,2-trinitroacetamide)	$C_5H_4N_8O_{14}$	+3.50	9
50.	1,1,1,3,5,5,5-Heptanitropentane	$C_5H_5N_7O_{14}$	+3.36	8
51.	Bis(2,2,2-trinitroethyl) carbonate	$C_5H_4N_6O_{15}$	+3.60	16

MOLECULAR ASPECTS

TABLE 7-2. Cont.

No.	Compound name	Molecular formula	CB_{100}	$H_{50\%}(cm)$*
52.	N,N'-Bis(2,2,2-trinitroethyl)urea	$C_5H_6N_8O_{13}$	+2.60	17
53.	5,5,5-Trinitropentanone-2	$C_5H_7N_3O_7$	−1.36	125
54.	2,2,2-Trinitroethyl 4,4,4-trinitrobutyrate	$C_6H_6N_6O_{14}$	+2.07	18
55.	Ethyl 2,2,2-trinitroethyl carbonate	$C_5H_7N_3O_9$	−0.39	81
56.	Tris(2,2,2-trinitroethyl) orthoformate	$C_7H_7N_8O_{21}$	+3.80	7
57.	1,1,1,7,7,7-Hexanitroheptanone-2	$C_7H_8N_6O_{13}$	+1.04	34
58.	Nitroisobutyl 4,4,4-trinitrobutyrate	$C_8H_{12}N_4O_{10}$	−3.09	279
59.	4,4,4-Trinitrobutyric anhydride	$C_8H_8N_6O_{15}$	+0.93	30
60.	N,N'-Bis(3,3,3-trinitropropyl)oxamide	$C_8H_{10}N_8O_{14}$	+0.45	45
61.	Bis(2,2,2-trinitroethyl) succinate	$C_8H_8N_6O_{16}$	+0.90	30
62.	Tetrakis(2,2,2-trinitroethyl) orthocarbonate	$C_9H_8N_{12}O_{28}$	+3.83	7
63.	Methylenebis(4,4,4-trinitrobutyramide)	$C_9H_{12}N_8O_{14}$	−0.44	113
64.	Ethylene bis(4,4,4-trinitrobutyrate)	$C_{10}H_{10}N_6O_{16}$	−0.43	120
	Trinitromethyl plus gem-dinitro			
65.	N-(2,2,2-Trinitroethyl)-3,3,5,5-tetranitropiperidine	$C_7H_8N_8O_{14}$	+1.40	18
66.	2,2,2-Trinitroethyl 4,4-dinitrovalerate	$C_7H_9N_3O_{12}$	−0.28	70
67.	2,2-Dinitropropyl 4,4,4-trinitrobutyramide	$C_7H_{10}N_6O_{11}$	−0.57	72
68.	2,2-Dinitrobutyl 4,4,4-trinitrobutyrate	$C_8H_{11}N_5O_{12}$	−1.33	101
69.	2,2,2-Trinitroethyl, 4,4-dinitrohexanoate	$C_8H_{11}N_5O_{12}$	−1.33	138
70.	N,N-Bis(2,2-dinitropropyl) 4,4,4-trinitrobutyramide	$C_{10}H_{14}N_8O_{15}$	−0.82	72
71.	2,2-Dinitropropane-1,3-diol bis(4,4,4-trinitrobutyrate)	$C_{11}H_{12}N_8O_{20}$	+0.35	50
72.	Bis(2,2,2-trinitroethyl) 4,4-dinitroheptanedioate	$C_{11}H_{12}N_8O_{20}$	+0.35	68
73.	Bis(2,2,2-trinitroethyl) 4,4,6,6,8,8-hexanitroundecanedioate	$C_{15}H_{16}N_{12}O_{28}$	+0.74	32
	gem-Dinitro compounds			
74.	3,3,4,4-Tetranitrohexane	$C_6H_{10}N_4O_8$	−2.25	80
75.	2,2,4,6,6-Pentanitroheptane	$C_7H_{11}N_5O_{10}$	−1.54	56
76.	2,2,4,4,6,6-Hexanitroheptane	$C_7H_{10}N_6O_{12}$	0	29
77.	Bis(2,2-dinitropropyl) exalate	$C_8H_{10}N_4O_{12}$	−1.70	227
78.	2,2-Dinitropropyl 4,4-dinitrovalerate	$C_8H_{12}N_4O_{10}$	−3.09	>320
79.	3,3,9,9-Tetranitro-1,5,7,11-tetraexaspiro[5,5]undecane	$C_7H_8N_4O_{12}$	+0.59	66

* 25-Shot determinations on Type 12-tools.
Source: Reference 6.

c. Shock Sensitivity

Shock sensitivity is measured in a standard small-scale gap test (SSGT) setup (Figure 7-3).[8] In this method, a standard shock pulse is generated by using a donor charge (RDX) on the upper portion of the setup. This shock pulse can be attenuated to variable extents (see variable gap in Figure 7-3) by using spacers made of Lucite (PMMA). The shock pulse after the required attenuation is incident on the test explosive (acceptor loaded at the lower

TABLE 7-3. Structure–Sensitivity Relations in Nitroaromatic Explosives

No.	Compound	Molecular Formula	Mol.Wt.	OB_{mole}	OB_{100}	$h_{50\%}$ [cm]
Compounds with no alpha C-H linkage						
1.	2,3,4,5,6-pentanitroaniline	$C_6H_2N_6O_{10}$	318	+6	+1.88	15
2.	2,2,2-trinitroethyl-2,4,6-trinitrobenzoate	$C_9H_4N_6O_{14}$	420	+4	+0.95	24
3.	2,4,6-trinitroresorcinol	$C_6H_3N_3O_8$	245	+1	+0.41	43
4.	2,3,4,6-tetranitroaniline	$C_6H_3N_5O_8$	273	+1	+0.37	41
5.	2,2,2-trinitroethyl-3,5-dinitrosalicylate	$C_9H_5N_5O_{13}$	391	+1	+0.26	45
6.	2,2,2-trinitroethyl-3,5-dinitrobenzoate	$C_9H_5N_5O_{12}$	375	−1	−0.28	73
7.	picric acid	$C_6H_3N_3O_7$	229	−1	−0.44	87
8.	2,4,6-trinitro-3-aminophenol	$C_6H_4N_4O_7$	244	−2	−0.81	138
9.	2,2',4,4',6,6'-hexanitrobiphenyl	$C_{12}H_4N_6O_{12}$	424	−4	−0.94	85
10.	2,4,6-trinitrobenzoic acid	$C_7H_3N_3O_8$	257	−3	−1.12	109
11.	2,2-dinitropropyl-2,4,6-trinitrobenzoate	$C_{10}H_7N_5O_{12}$	389	−5	−1.28	214
12.	1,3,5-trinitrobenzene	$C_6H_3N_3O_6$	213	−3	−1.46	100
13.	2,4,6-trinitrobenzonitrile	$C_7H_2N_4O_6$	238	−4	−1.68	140
14.	picramide	$C_6H_4N_4O_6$	228	−4	−1.75	177
15.	4,6-dinitroresorcinol	$C_6H_4N_2O_6$	200	−4	−2.00	>320
16.	2,4-dinitroresorcinol	$C_6H_4N_2O_6$	200	−4	−2.00	296
17.	2,4,6-trinitroanisole	$C_7H_5N_3O_7$	243	−5	−2.06	192
18.	1,3-dimethoxy-2,4,6-trinitrobenzene	$C_8H_7N_3O_8$	273	−7	−2.56	251
19.	3-methoxy-2,4,6-trinitroaniline	$C_7H_6N_4O_7$	258	−6	−2.32	>320
20.	1,3-diamino-2,4,6-trinitrobenzene	$C_6H_5N_5O_6$	243	−5	−2.06	320
21.	2,2',4,4',6,6'-hexanitrodiphenylamine	$C_{12}H_5N_7O_{12}$	439	−5	−1.14	48
22.	2,4,6-trinitrophloroglucinol	$C_6H_3N_3O_9$	261	+3	+1.15	27
23.	3-hydroxy-2,2',4,4',6,6'-hexanitrobiphenyl	$C_{12}H_4N_6O_{13}$	440	−2	−0.45	42
24.	3,3'-dihydroxy-2,2',4,4',6,6'-hexanitrobiphenyl	$C_{12}H_4N_6O_{14}$	456	0	0.00	40
25.	3,3'-diamino-2,2',4,4',6,6'-hexanitrobiphenyl	$C_{12}H_6N_8O_{12}$	454	−6	−1.32	132
Compounds with alpha C-H linkage						
26.	1-(2,2,2-trinitroethyl)-2,4,6-trinitrobenzene	$C_8H_4N_6O_{12}$	376	+4	+1.07	13
27.	1-(3,3,3-trinitropropyl)-2,4,6-trinitrobenzene	$C_9H_6N_6O_{12}$	390	0	0.00	21
28.	1-(2,2,2-trinitroethyl)-2,4-dinitrobenzene	$C_8H_5N_5O_{10}$	331	−1	−0.30	31
29.	2,4,6-trinitrobenzaldehyde	$C_7H_3N_3O_7$	241	−3	−1.24	36
30.	2,2',4,4',6,6'-hexanitrobibenzyl	$C_{14}H_8N_6O_{12}$	452	−12	−2.64	114
31.	TNT	$C_7H_5N_3O_6$	227	−7	−3.08	160
32.	2,4,6-trinitrobenzyl alcohol	$C_7H_5N_3O_7$	243	−5	−2.06	52
33.	2,4,6-trinitrobenzaldoxime	$C_7H_4N_4O_7$	256	−4	−1.56	42
34.	2,4,6-trinitro-m-cresol	$C_7H_5N_3O_7$	243	−5	−2.06	191
35.	3-methyl-2,2',4,4',6,6'-hexanitrobiphenyl	$C_{13}H_6N_6O_{12}$	438	−8	−1.81	53
36.	3,3'-dimethyl-2,2',4,4',6,6'-hexanitrobiphenyl	$C_{14}H_8N_6O_{14}$	452	−12	−2.64	135
37.	3-methyl-2,2',4,4',6-pentanitrobiphenyl	$C_{13}H_7N_5O_{10}$	393	−13	−3.30	143
38.	3,5-dimethyl-2,4,6-trinitrophenol	$C_8H_7N_3O_7$	257	−9	−3.50	77

Source: Reference 7.

portion of the apparatus) and initiates it. The test involves determining the strength of the critical shock pulse that is just necessary to detonate the test explosive. This provides the index of shock sensitivity. As in the case of the impact test, multiple tests are carried out following the statistical Bruceton up-and-down method, and 50% firing stimulus is evaluated. The dimensions of the donor and acceptor (test explosive) charges are shown in Figure 7-3. An RP 87 EBW (exploding bridge wire) igniter is normally used to fire RDX donor charges (Figure 7-3 shows a model MK70 detonator).

Critical shock pressures for initiation are expressed in units of decibangs (DBG)[8b]:

$$DBG = 30 - 10 \log X$$

where X is the Lucite gap thickness in mils (0.001 in.). Decibangs are related to calibrated shock pressures (kbar) and X (mils) as shown in Table 7-4.

MOLECULAR ASPECTS

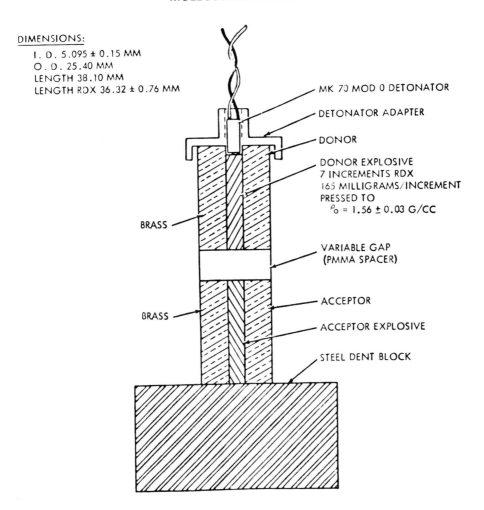

Figure 7-3. The SSGT shock sensitivity tester.

Figure 7-4 shows shock sensitivity data for various explosives as a function of pressed (from powder) density in decreasing order of sensitivity from bottom to top. It is interesting to note that for a particular explosive, the shock sensitivity decreases as the pressed density is increased. This effect is attributed to decrease in porosity of the pressed explosive with increase in density (less porosity provides fewer chances for hot spot creation). Adiabatic compression of air at vacancies resulting in high temperature is a common explanation for the shock initiation process. In nitroaromatic series (TNB, PA, DATB, TATB, and TNT and 3-amino-TNT), shock sensitivity

TABLE 7-4. Relations Between Various Units Used in Small-Scale Gap Tests

Decibangs	Kilobars	Gap (mils)
3	8.8	501
4	12.2	398
5	17	316
6	24	251
7	32	200
8	44	159
9	62	126
10	84	100
	216	0

decreases progressively as —NH_2 substitution of ring hydrogens increases. This observation is further discussed in Subsection D.

d. Friction and Electrostatic Sensitivities

The friction and electrostatic sensitivities of energetic materials to friction and electrostatic charge, like others mentioned above, become very impor-

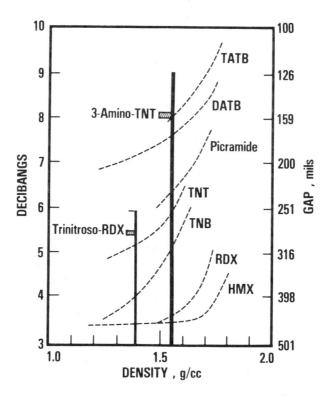

Figure 7-4. Shock sensitivity data for various explosives. (From Reference 9.)

C. Mechanism of Initiation

Traditionally initiation has been thought to be caused by creation of high temperature by the stimuli in the bulk or at localized sites, depending on the homogeneity of the material. When the rate of heat generation exceeds that of heat loss in the region of the hot spots, propagation and finally explosion occur.[10a] Presumably initiation arises from initial chemical reactions at the molecular level. Thus, it becomes important to understand the relationship of molecular properties of the energetic materials to sensitivity.

D. Molecular Properties of Energetic Materials Related to Sensitivity

a. Nitroaromatics

In nitroaromatic series, the molecule starts becoming an explosive when the degree of nitro substitution is 3 per ring. Thus trinitrobenzene and additionally nitrated benzenes are explosives, while dinitro and mononitro substitutions do not make the benzene ring explosive. This is also true for the toluene series. Recent pioneering investigations in cage molecule chemistry (eg, cubanes, bishomocubanes, adamantanes)[10b] also indicate a point of nitro substitution above which the molecules are explosives. What is true for nitro groups is also true for other chemical functions that have the ability to confer explosive property to the molecule (these are called energetic groups or "plosophores," eg, nitrate and azide groups etc). In general, as the power of the explosive molecule is increased via increased introduction of these energetic groups, its sensitivity increases proportionately. The challenge is to keep the loss in explosive power to the minimum while attempting to desensitize a very sensitive explosive.

In nitroaromatic series, significant understanding has been realized in correlating molecular properties to sensitivity. The molecular properties that have been correlated are:

1. The total electron donating ability of all groups in the molecule[11]
2. Bond energies (Figure 7-5) of the C—NO_2 bond (which seems to be the bond that breaks first in the initiation process)[11,12]
3. The electrostatic potential between the C and N of the critical C—NO_2 bond[11,12]

"tant in the case of primary explosives, which are the least stable energetic materials. Methods of determining friction and electrostatic sensitivities have been extensively described.[1b] The reader is referred to that source for further information."

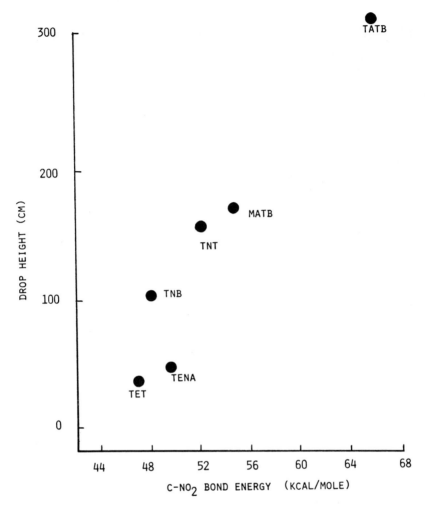

Figure 7-5. The relationship of molecular property (C—NO$_2$ bond energy) to sensitivity in nitroaromatics.

These correlations show important results. In a series of nitroaromatics with equal number of nitro groups, the higher the total electron-donating ability of all other groups, more insensitive the molecule. This suggests that the electron-donating groups have a tightening effect on the C—NO$_2$ bond and that the best way to desensitize nitroaromatics is via increasing the number of electron-donating groups. This is borne out in both the TNB and TNT series (Figure 7-4). As shown in Figure 7-5,[11] C—NO$_2$ bond energy correlations indicate that it is the first bond likely to be broken. The electrostatic potentials between the C and N atoms, a measure of bond energy, are also consistent with the C—NO$_2$ bond energies and observed sensitivity.[12]

Given some understanding of the factors that play roles in determining sensitivity in an explosive nitroaromatic molecule, further attempts to design explosive sensitivity around molecular aspects have also been successful. Thus in TNT,[9,13] by increasing the substitution of the aromatic ring with electron-donating groups (eg, —NH$_2$), the sensitivity has been reduced by factors of 3 or 4. Furthermore, the amino substitutions in TNT bring an added benefit. Because of the increased dipolar attractions (eg, hydrogen bonding) of the —NO$_2$ groups with the —NH$_2$ groups of the neighboring molecule, increases in bulk density are realized in the amino-substituted TNBs and TNTs, which increase the power, while still providing desensitization, as a function of the square of the increase in density.

b. Nitramines

As a class of explosives, nitramines are more sensitive than nitroaromatics. Thus desensitization is even more important when dealing with nitramines. The important military nitramine explosives are RDX and HMX.

RDX has three nitro groups that are responsible for the instability of the molecule. Desensitization of RDX was achieved[9] by converting these nitro groups to nitroso groups without resulting in any significant loss in power. Since the first step in the explosive decomposition of RDX is the breakage of N—NO$_2$ bond, it seems that desensitization has occurred via tightening of the N—N bond, which is consistent with the fact that in nitrosamines, there is actually a partial double bond between the two N atoms. (*Note:* There could be some N—N double-bond character in nitramines, but it would be much less pronounced than in nitrosamines.)

c. Empirical Sensitivity–Structure Correlations

Kamlet and co-workers[6,7] have developed for similar classes of C, H, N, and O organic explosives (ie, nitroaliphatics, nitroaromatics, etc) empirical relationships connecting impact sensitivity with chemical structure, that is, the relative number of C, H, N, and O atoms and thus the "oxygen balance," which is the amount of oxygen available in the molecule for combustion of C and H atoms in the molecule to CO/CO$_2$ and H$_2$O. Figures 7-6 and 7-7 show these correlations for several classes of nitroaliphatic explosives and for polynitroaromatics.

d. Reactive Intermediates Generated on Initiation

Considerable effort has been made to identify reactive intermediates generated in the initiation process. Given the difficulty of identifying directly the transient species in thermally initiated, uncontrolled explosions in powders by methods like transmission optical spectroscopy, attempts are usually made to relate to the actual explosion process the transients generated in explosives via light and electron impact excitation under subinitiation condi-

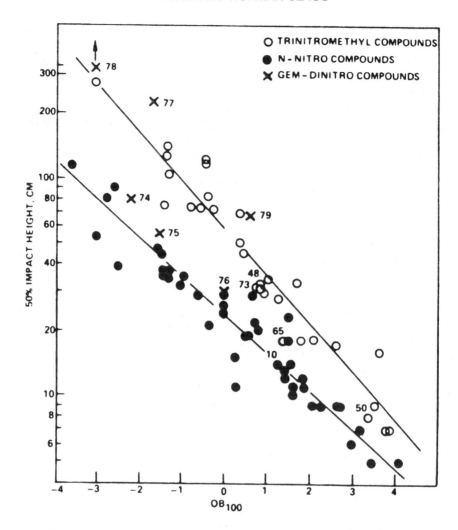

Figure 7-6. Structure–sensitivity relations: nitroaliphatics. (From Reference 6.)

tions. Recent advances in diagnostics are changing this picture, as seen in Section 3, on detonations.

TNT produces under UV light excitation, a species resulting from a transfer of H atom from the —CH_3 group to the ortho-nitro group (Scheme I). This transient exists in protonated form (the aci form) in nonpolar media (millisecond lifetime) and in proton-dissociated form in polar solvents (this is the 2,4,6-trinitrobenzyl anion of TNT, with a lifetime of seconds).[14] Electronic excitation of TNT in the gas phase generates a transient species that has been assigned to the 2,4,6-trinitrobenzyl radical.[15a,b]

MOLECULAR ASPECTS

Highly reactive (more so than those produced from nonexplosives) metastable positive ions of TNT and RDX have been observed in electron impact mass spectrometry.[16] In the case of TNT, the molecular ion of TNT was

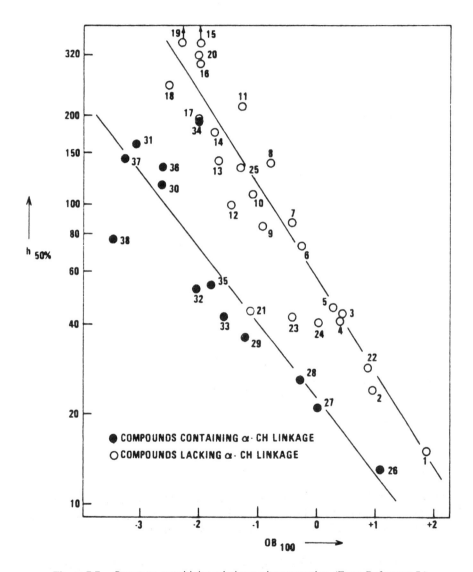

Figure 7-7. Structure–sensitivity relations: nitroaromatics. (From Reference 7.)

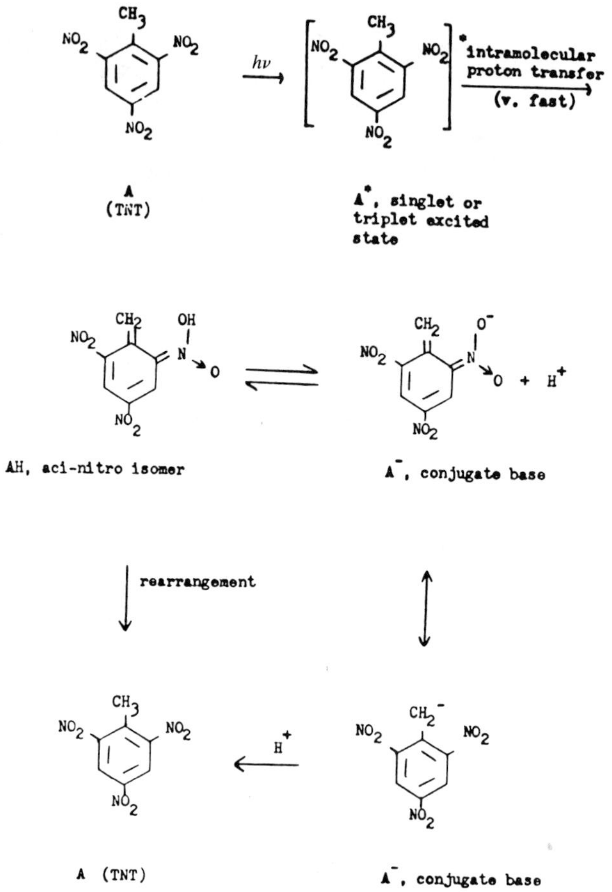

Scheme I

found to undergo a rapid hydrogen rearrangement and loss of a hydroxyl radical to give the ion[16]:

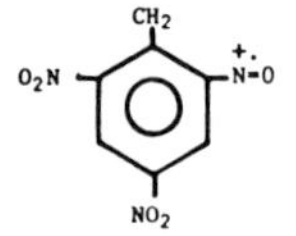

Similarly, in the case of RDX, three principal primary bond-breaking steps have been identified in the breakdown of electron impact generated positive ion of RDX. They are loss of oxygen atom, loss of hydroxyl radical, and the loss of NO_2 group.[16]

Investigations of the thermal decomposition of TNT via electron paramagnetic resonance (EPR) techniques have led to the observation of a radical derived from TNT whose identity is indefinite.[17a] Also, the CH bond breakage of the CH_3 group in 2,4,6-TNT is a very likely rate-determining step in TNT initiation.[17c]

3. DETONATIONS

A. Physical Chemical Overview

It is generally agreed that the energy supporting the shock front of a detonation originates in a region behind it, where chemical changes occur (reactants → products).[18,19] However, we know very little about the sequence of chemical steps leading to the products. Both chemical and physical factors influence the rate and the manner in which reactions occur. This is particularly true in condensed-phase detonation, where consideration must be given to diffusion of reactive species and surface area for packed powders.

In the case of organic explosives, where the intermolecular forces are small, the reactive unit is the molecule.[20] For inorganic explosives, lattice structure and distances between cationic and anionic groups are significant.[10]

For organic explosives the first step is expected to consist of the breaking of a bond, leading to free radicals or ions. The formation of the radicals and/or ions can proceed through electronically excited states, or an isomerized metastable molecule. The radicals and ions can be created in a variety of energetic states. This primary process is endothermic. It is generally accepted that explosions proceed via a chain mechanism. Thus, whatever occurs in the primary step must eventually lead to a chain process.[21,22a–c,23a]

The rapid reactions most often associated with explosions are of the thermal, chain-isothermal, and chain-thermal types.[22a–c,23a,b] In the case of thermal reactions the acceleration occurs through self-heating; for chain-isothermal reactions the branching processes exceed the rate of removal of chain carriers. When the exothermic nature of the processes influences the onset of explosion, so that both the heat of reaction and the coefficients of branching and termination reactions appear in the condition for explosion, we have a chain-thermal type.

In 1935 Semenov pointed out that degenerative chain mechanisms may be operating in both gaseous and solid explosives.[23a,b] From studies of the thermal decomposition and explosion of mercury fulminate, barium azide, and nitroglycerine, Semenov concluded: "The intermediate products formed during the first stages of the explosive reaction activate the neighboring molecules at the cost of the energy liberated during the decomposition, thus promoting the development of the chain," (Reference 23a, p. 438).

B. Detonation Model

In detonation, the energy release manifests itself in the form of high pressures and shock waves. The energy change between initial reference state of an explosive molecule and the equilibrium state of the detonation products is called the heat of detonation. The mechanical energy contained in the heat of detonation is used to move matter. It can be defined as the *PV* work as a result of an adiabatic expansion of the products from the end of the reaction zone to a final expansion state. The pressure P_{CJ} is an indicator of the explosive power output (ie, energy release rate). Kamlet and Jacobs[24] derived a semiempirical expression for P_{CJ} and D as follows:

$$P_{CJ} = K\rho_0^2 \phi \quad (7\text{-}1)$$

$$D = A\phi^{1/2}(1 + B\rho_0) \quad (7\text{-}2)$$

where

P_{CJ} = Chapman–Jouguet pressure
D = detonation velocity
$\phi = NM^{1/2}Q^{1/2}$
ρ_0 = initial density of the explosive [g/cm³ (mg/m³)]
$K = 15.58$
$A = 1.01$
$B = 1.30$
N = moles of gaseous detonation products per gram of explosive
M = average molecular weight of detonation product gas (g gas/mol gas)
Q = chemical energy of the detonation reaction (cal/g)

Values of N, M, and Q can be estimated from the H_2O–CO_2 decomposition assumption:

$$C_aH_bN_cO_d \rightarrow \frac{c}{2} N_2 + \frac{b}{2} H_2O + \left(\frac{d}{2} - \frac{b}{4}\right) CO_2 + \left(a - \frac{d}{2} + \frac{b}{4}\right) C$$

Detonation rates indicated by the values of detonation velocities (this is the velocity of the steady state propagation of detonation wave in a column of usually confined and packed explosive) are indices of explosive reactivity.

It can thus be seen that from both reactivity and energy release points of view, the important properties for high explosives are the packing density, the detonation velocity, and the detonation pressure (ie, the CJ pressure). Detonation velocity varies as the first power of density, while the CJ pressure is a function of ρ_0^2. Table 7-5 shows measured detonation velocities and CJ pressures of common explosives. The most widely accepted model for detonations is the Chapman–Jouguet model, a one-dimensional model based

TABLE 7-5. Measured Detonation Velocities and CJ Pressures of Common Explosives

Explosive	Pressed or cast density (g/cm³)	Detonation velocity (mm/μs)	CJ pressure (kbar)
TNB	1.64	7.27	219
TNT	1.64	6.95	190
Tetryl	1.61	7.58	226
PETN	1.57	7.79	240
RDX	1.80	8.59	341
HMX	1.89	9.11	390

on the assumption that there exists a region behind the reaction front where the flow velocity is equal to the local velocity of sound, and the products are in thermodynamic equilibrium. The reactants are assumed to be instantaneously converted to products. Detailed discussion of detonation models can be found in several good texts and articles.[20,21,28-31]

When the conservation equations for mass momentum and energy are combined, the Rankine–Hugoniot equation (commonly called the Hugoniot equation) is obtained:

$$e_1 - e_0 = \tfrac{1}{2}(p_1 + p_0)(V_0 - V_1) \qquad (7\text{-}3)$$

where e, p, and V refer to the specific energy, pressure, and volume, respectively, and the subscripts 0 and 1 refer to the initial and final states. The Hugoniot gives the relation between thermodynamic quantities that determine p_1 as a function of V_1 for a specified initial pressure p_0, initial volume V_0, and C_0 the velocity of sound in a given explosive. The Hugoniot relation is also commonly written in the form[20]:

$$h_1 - h_0 = \tfrac{1}{2}(p_1 - p_0)(V_1 + V_0) \qquad (7\text{-}4)$$

when h is the specific enthalpy. The mechanical conservation conditions can be combined to yield the equation of a straight line:

$$-\rho_0 D^2 = (p_1 - p_0)(V_1 - V_0) \qquad (7\text{-}5)$$

where ρ_0 is the density and D is the shock or detonation velocity.

The line called the Rayleigh line is the chord connecting the final conditions behind the discontinuity to the initial conditions. Since the equation is negative on the left-hand side, it is apparent that the pressure and volume must change in opposite directions. Therefore, the Hugoniot curve is seen to have two branches as illustrated in Figure 7-8. According to the C–J theory, the Rayleigh line tangent to the Hugoniot portion where p_1 exceeds p_0 for the product determines the observed detonation velocity.

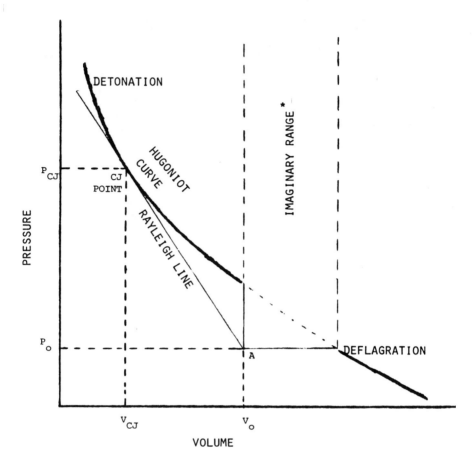

Figure 7-8. The Hugoniot curve and the Raleigh line. Asterisk indicates imaginary range which has no physical reality.

Points above this CJ point yield strong detonations and those below yield weak detonations. For the branch at which p_1 is less than p_0, the deflagrations that occur are called weak or strong depending whether the slope of the Rayleigh line is greater or smaller than the tangent.

The heat detonation for the products that yield the appropriate Hugoniot is given by the following expression:

$$q = \sum_i n_i(\Delta H_f)_i - \sum_j m_j(\Delta H_f)_j \tag{7-6}$$

where n_i is the number of moles of i species in the products and m_j is the number of moles of unreacted explosive, and ΔH_f is the heat of formation in the standard state. It should be noted that the explosive can consist of one or more ingredients.

As the products expand and the pressure drops, shifts in equilibrium can result in the change of products. There exists a temperature below which the reactions are so slow that further drops in temperature and pressure do not change the composition. Table 7-6 presents the heats of detonation of common explosives.

Researchers in the 1940–1950 decade thought that the conversion to products was not instantaneous and proposed a model of detonation that contained a reaction zone, but still was essentially in agreement with the Chapman–Jouguet model. This new version became known as the ZND for its basis in the work of Zeldovich, von Neumann, and Doering.[26-28] Calculations to resolve the condensed-phase reaction zones have been carried out by Mader and Craig, Fickett and Wood, and Erpenbeck.[29-31] Mader has presented a detailed analysis of these studies in his text. Essentially, an Arrhenius rate law, the Navier–Stokes model of fluid dynamics, and suitable equations of states are used in numerical computer programs. Typical values for the reaction zone length are around 10^{-5} cm. The search for suitable equations of state for condensed-phase detonations ($\approx$ 4000 K and $\approx$ 300 kbar) has involved a considerable effort over the past 40 years, since no hydrostatic pressure–velocity–temperature data exist for this regime. Numerous assumptions are required. Extensive calculations carried out by Mader[32,33] and Cowan and Fickett[34] involved the assumptions that the products were gases (CO, CO_2, H_2O, H_2, NH_3, CH_4, N_2, NO, etc) and solid carbon in the form of graphite. It is not clear whether there is sufficient time for graphite to form. There exists the possibility of the formation of the C_n species. Besides, at high pressures it is not known whether there is a gaseous state or a liquid state or whether strong interactions lead to a solidlike state. The common procedure is to calibrate the computation by using a posteriori equation of state (EOS) to give the observed result. Normally calculations of the detonation velocities of ideal CHNO explosives (essentially instantaneous conversion to products) yield values within 5% of the experimental values. A useful discussion of the various aspects of EOSs is given by Alster and co-workers.[35]

Recent results have demonstrated the failure of the one-dimensional, (1-

TABLE 7-6. Heats of Detonation of Common Explosives

Explosive	Heat of detonation (kcal/g)
RDX	1.51
PETN	1.49
HMX	1.48
Tetryl	1.25
Nitromethane	1.23
TNT	1.09

D), steady state CJ detonation model. The slow build up of CJ pressure and the scaling failure are possibly explained by an unstable periodic wave.[25] The existence of such waves in gas phase detonations was demonstrated more than 20 years ago.[19]

One of the fundamental assumptions of the 1-D detonation theory is that the products are in thermodynamic equilibrium at the end of the reaction zone. If an unstable periodic wave exists as in accordance with gas phase detonations, there will be a temperature–time profile associated with an elementary explosive volume.[19,36-38,41] Only an average temperature is likely to be observed under such conditions.

C. Initiation of Detonations

Initiation can occur via light, heat, or shock. However, each mode will have different requirements depending on whether it is measured in terms of energy, density, or power. This difference is expected to result from the different efficiencies of conversion of the energy form into reactive species. In other words, energy or power must be put into the explosive in a manner capable of generating some minimum concentration of certain reactive species. The primary modes of decomposition of the parent molecule are expected to involve the formation of free radicals (neutral or ionic), an isomerization step followed by free radical formation, the formation of an energetic parent molecule followed by decomposition, or ionization.

In the past 15 years there has been a dramatic upsurge in the number of papers offering a molecular view of shock phenomena. This upsurge is clearly evident in the papers presented at the last four Symposiums of Detonation starting in 1970. The fifth symposium, held in 1970, contained two papers that could be related to molecular processes; the sixth, held in 1976, had two, the seventh, held in 1981, had 6; the latest symposium, held in 1985, had two sessions related to molecular processes, for a total of eight papers.[39-42] One session was devoted to molecular dynamics and the other to detonation spectroscopy. Advances in electronics, laser technology, spectroscopy, and computers were mainly responsible for these new efforts.

Shock initiation phenomena are often discussed in terms of the homogeneity or nonhomogeneity of the energetic material. In the case of nonhomogeneous materials, a strong shock wave begins to accelerate at the point of entry, gaining strength until a steady velocity is obtained, which is referred to as a CJ detonation. This process is characterized by a smooth transition (Figure 7-9a). In the case of a homogeneous material, the entering shock decays slightly until the end of the induction period, at which time energy from chemical reactions begin to support the shock front. The shock wave accelerates, achieving momentary velocities above that of the steady state value. The acceleration is due to the rapid release of energy, an explosion, behind the front. This mechanism is characterized by abrupt changes in velocities (Figure 7-9b). Solid explosive formulations are nonhomogeneous. Thus when the shock is sufficiently strong, chemical energy release starts immediately, accelerating the shock wave until a stable detonation is obtained. As pointed out earlier, in the 10^{4-6} bar regime, chemical processes are normal. The detonation regime extends from 100 to 1000 kbar.

To obtain kinetic information, thermal explosion theory has been applied to shock processes in homogeneous explosives. This theory is due to the work of Frank-Kamenetskii and Semenov (see References 22a, 23b, and 27). Briefly, it states there will be an explosion whenever the heat generated by chemical reaction exceeds heat losses by conduction. The basic equation is:

$$\rho c \frac{\partial T}{\partial t} = \lambda \nabla^2 T + \rho Q \frac{(d\varepsilon)}{(dt)} \qquad (7\text{-}7)$$

self-heating | heat loss by conduction | production of heat by chemical reaction

where

T = solid temperature (K)
c = specific heat (cal/g K)
ρ = density (g/cm^3)
λ = thermal conductivity, which is assumed to be temperature independent (cal/cm-K-s)
Q = heat of reaction (cal/g)
ε = fraction of explosive reacted (dimensionless)

Generally, good agreement is found between theory and experiment for purely thermal conditions. In the case of shock initiation, data on the Hugoniot, an equation of state, and thermal kinetic data for the explosive are required. With assumptions on the equations of state, induction times were calculated for nitromethane, liquid TNT, and a single crystal of PETN (Table 7-7). These results demonstrate the consistency of the thermal explosion theory but do not prove it.

A related study pointed out the importance of experimental conditions. It was observed that when the gases are not permitted to escape, the times to explosion obtained differed from those found in the more usual condition that permits venting of product gases. The heat flow model with a single Arrhenius term does not adequately describe the results.[45]

These studies employed activation energies and rate constants obtained from ambient pressure and isothermal studies, and they touched only slightly the area of molecular processes. The advances in diagnostics mentioned earlier began yielding molecular insights.

The kinetic isotope effect has been shown by Bulusu and colleagues[46] to be a powerful technique for unraveling complex solid state processes. Data

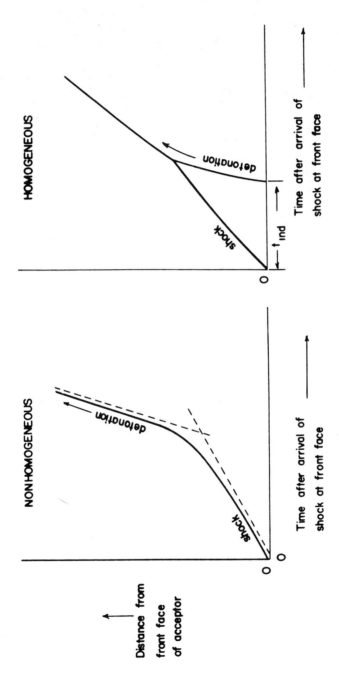

Figure 7-9. (a) Nonhomogeneous and (b) homogeneous detonations.

MOLECULAR ASPECTS

TABLE 7-7. Theoretical and Experimental Induction Times

Material and initial shock pressure	Induction time (μs)		
	Calculated[a]	Calculated[b]	Observed
Nitromethane (86 kbar)	2.31	1.34	~1.0
Liquid TNT (125 kbar)	0.5	0.68	0.7
Single-crystal PETN (110 kbar)		0.34	0.3

[a] Enig and Petrone.[44]
[b] Mader.[43]

were obtained for RDX and HMX showing that the rate-controlling step, a hydrogen transfer to NO_2 of an adjacent molecule, may be the same for thermal decomposition and shock initiation.[46]

Furthermore, in the Fifth Symposium on Detonation, Berke and associates[47] presented a paper on the shock initiation times for some liquid explosives. They suggested that methyl nitrate has a smaller reaction time than nitromethane because the —C—NO_2 bond is stronger than the C—O—NO_2 bond. In addition, in the case of difluoraminoalkanes, it was proposed that vicinal fluorine compounds can undergo a low activation energy, four-center process with the elimination of HF.

As pointed out earlier, new experimental techniques and computer programs have been used to gain insights into chemical reaction. Yang and Menichelli used pulses from a switched laser to directly initiate detonations in PETN and RDX.[49] Approximately 37 J/cm^2 was required.[49] No attempt was made in this study to relate the frequency of radiation to absorption characteristics. Capellos and co-workers, employing RDX, were able to gain evidence from absorption that a specific bond may lead to a detonation and/or explosion.[50] Van Hulle and Tarver explored a time-resolved, two-band IR technique to measure emission from shocked propellants and explosives.[51] The rates of emission were related to chemical reactions and compared well with computer simulations. Evidence was obtained for a time delay between the incident shock and the onset of detonation for nitromethane that is consistent with the thermal explosion model discussed earlier. Raman scattering was also shown to be of potential value in measuring temperatures behind shock waves.[52] Duval employed reflection and absorption spectroscopy to measure changes in the spectral characteristics of liquid CS_2 when shocked.[53] The observed internal absorption was not evident in high-pressure and higher temperature state studies, suggesting differences between chemical processes at dynamic high pressures and processes occurring under conventional conditions.

Recently EPR and X-ray photoelectron studies have been used to shed light on chemical processes in the solid state. EPR studies have revealed radicals in molten TNT at 200°C. One of the radicals appears to be the parent TNT less one NO group (methyl dinitrophenoxyl radical).[54]

XPS (X-ray photoelectron spectroscopic) studies of shocked RDX revealed that N—NO_2 bonds were broken.[55] In the case of TNT, increased bonding occurs between hydrogens of CH or CH_3 groups and oxygen atoms on adjacent NO_2 group in the parent TNT molecule.[55] Some distortion of the lattice is also suggested. EPR results demonstrated the formation in shocked TNT of paramagnetic species that could not be positively identified. Similar measurements on samples that were thermally initiated, pointed to clear differences from shocked samples. Another EPR study on copper tetrammine nitrate showed signals from shocked samples indicating the formation of new complex ions.[56] Other examples of the application of new experimental techniques were presented at the 1983 American Physical Society Conference on Shock Waves in Condensed Matter.[57]

In the Sixth Symposium on Detonation there is a very valuable discussion on the decomposition of nitromethane. One of the major issues was whether the aci form of nitromethane $\begin{matrix} \diagdown \\ \diagup \end{matrix} C = N - OH$, plays a major role in the detonation regime. It is suggested that nitromethane is converted to the aci form, which then can react exothermically with normal nitromethane. Others suggested the aci form results from metal wall catalysis, that observations require the rejection of single-step process, and that a bimolecular process gives the aci form more appropriately. It is also pointed out that a bimolecular process would have a negative volume of activation, which is consistent with a more rapid reaction at high pressures. A recent paper[40b] by Englke, Earl, and Rohling presents microscopic evidence that aci ion formation is the rate-determining step in the detonation of liquid nitromethane. Evidence from ^{13}C NMR spectroscopy is presented that the only new chemical species generated by bases that sensitize nitromethane is the aci ion.

Theoretical studies by Mader clearly demonstrate the differences between homogeneous and inhomogeneous (pressed or plastic bonded) explosives.[32,33,48] As pointed out previously, the behavior of homogeneous explosives can be shown to be consistent with thermal explosion theory; however, the action of inhomogeneous explosives is not. The variance is believed to be due to the formation of hot spots and local thermal discontinuities, thus making it difficult to define a temperature. Mader showed that by employing run to detonation experimental data, heterogeneous explosives could be modeled.

The theoretical papers now appearing in increased numbers can be arranged into two groups: quantum mechanical and molecular dynamical. With regard to molecular dynamical studies, two groups have been active in the last few years. One is the group of Trevino and Tsai of the U.S. Army and the National Bureau of Standards,[58-60] respectively, and the other consists of Walker and Karo of Lawrence Livermore National Laboratory and Hardy of the University of Nebraska, respectively.[61-63] Trevino and Tsai have concentrated on a system consisting of 256 particles arranged in pairs as diatomic molecules in a three-dimensional cube. With regard to energy transport they found that intra- and intermolecular energy contributions among the degrees of freedom were highly inefficient and that no thermal equilibrium existed during dissociation. They were also able to show that their equilibrium model contains kinetics that is consistent with thermodynamic considerations. Besides, at constant pressure, the additive relation between the Arrhenius energy of reaction, ΔE, the potential energy change upon reaction, and the work done due to the volume change upon reaction, $P\Delta V$, is satisfied. Recently they were able to show that both thermally and shock-initiated reactions combine to form a propagating reaction front whose velocity depends on the rate of energy release. The compression of a reaction front approximating a Chapman–Jouguet wave was observed.

In early papers the Karo–Hardy–Walker group described the manner in which energy in a shock front is dissipated by interactions with voids, crystal defects, and atoms at surfaces.[61-63] In later work they described computer codes capable of handling arbitrary arrangements of atoms and following the energy flux in any specific region of a lattice. They also were able to show that shock fronts can transmit large amounts of translational energy to the internal modes. The coupling is extremely rapid, occurring within 10^{-10} second. The resultant internal energy is one to two orders of magnitude larger than the shocked lattice. Large polyatomic molecules may thus be shocked significantly above their dissociation energy.

Dremin and co-workers pointed out that typical shock front in a condensed material is 10 to 15 Å wide or 3×10^{-13} second in time, and that there is no translational equilibrium.[64] They offer an explanation for the two maxima observed from measurements of pressure versus time behind shock waves in terms of partial chemical reactions during shock compression. As the pressure initially increases, heat is evolved from fast chemistry and subsequently often some decay, and at this point the slow processes begin to play a role. Pastine, Kamlet, and Jacobs, using a statistical thermodynamic approach, developed equations to describe the volume and pressure dependence of kinetic processes.[65] They considered both monomolecular and bimolecular processes. For monomolecular (bond-breaking) processes, only a slight volume–pressure dependence is indicated. In the case of a bimolecular process, pressure suppresses the reaction until the critical volume is approached, whereupon the reaction accelerates. The investigators suggest that this mechanism is more likely where free rotation is hindered, thus indicating that it is not probable for liquids.

Walker and Wasley have advanced a qualitative generalized model for shock nitration calling for the formation of transients very early behind the shock front.[66] Earlier, they had developed a very useful macroscopic interior for shock initiation based quantity $P^2\tau$, where P is the shock pressure and τ is the shock duration, which also lends itself to interpretation in molecular terms.[67]

In regard to quantum mechanical calculations, studies have been performed to determine bond strengths, electron distributions, possible isomerizations, distortions, and activation energies. In 1979 Alster and others used semiempirical techniques (MINDO/3 and UMINDO/3) and activated complex theory to highlight possible shock reaction paths for alkyl nitro and nitrate compounds.[68] They pointed out the possibility of a bimolecular reaction path as discussed in the Sixth Symposium on Detonation. Subsequent more detailed computations by Dewar and co-workers supported the possibility of bimolecular reaction pathways.[69] Bardo employed the CNDO/S-CI method to investigate the role of electronic states in detonation reactions.[70] It was observed that high pressure reduced activation energies. Monomolecular, bimolecular, and termolecular reaction paths were discussed. It was pointed out that at high pressures and temperatures, excited vibrational modes could lead to intermolecular bonds.

As indicated above, certain fundamental questions are essentially unresolved. Two of the more often discussed questions are as follows:

Can chemical kinetics in the high-pressure/high-temperature detonation regime be related to chemical processes obtained from lower temperature data at ambient pressures?

To what extent can processes in inhomogeneous explosives, where temperature discontinuities apply, be related to processes at the molecular level?

D. Detonation Measurements

Essentially no measurements that have been taken of the reaction zone of a propagating detonation in the condensed phase bear on molecular processes. The overwhelming experimental and theoretical modeling has been on initiation processes as discussed above or on the nature of the products and their states. Two types of measurement that have been conducted are the measurement of temperature of the products via time-resolved emission spectroscopy and calorimetric measurements that yield the heat of detonation to the products.

Burton, Hawkins, and Hooper, studying transparent liquid explosives, measured several narrow emission bands across the visible region simultaneously using photomultiplier detectors.[71] Graphite was used to calibrate the system. The radiation was found to have blackbody nature. Good agreement with calculations of Chapman–Jouguet temperatures was obtained, as seen in Table 7-8. Thus, global support for the CJ model was obtained. In the publication cited, references from earlier and similar efforts are discussed and listed. Typical experimental detonation temperatures are given in Table 7-8.

In a similar study, Kato, Bauer, Brochet, and Bouriannes employed a four-color pyrometer to determine the detonation temperatures for nitromethane–tetranitromethane liquid mixtures (NM-TNM) and hydrocarbon–O_2–N_2 gaseous mixtures.[74] For the NM-TNM, the detonation front behaved

TABLE 7-8. Experimental Detonation Temperatures

Material	Temperature (°K)	Reference
Nitroglycerine	4023	Gibson et al.[72]
	3470	Mader[32,33]
	4250	Burton, Hawkins, and Hooper[71]
Nitromethane	3800	Gibson et al.[72]
	3380	Mader[32,33]
	3300	Urtiew[73]

like a blackbody. Good agreement between these detonation temperatures and computed equilibrium values was obtained. In a subsequent paper, Kato, Mori, and Sakai determined the detonation temperatures of TNT, Tetryl, PETN, RDX, and HMX.[75] In addition to measuring the temperature of the front, measurements of the products were taken after reflection. Generally good agreement between experimental and computer temperatures was obtained. Results with different equations of states are presented in the reference.

He, Han, and Kang employed a two-color pyrometer with optical fibers to measure detonations in NM and the solid explosives TNT, Tetryl, HMX, and PETN. They investigated the effects of additives and gaseous environment.[76] They raised questions about the validity of the measurements because of erroneous calibration, variation with different EOSs, and impedence mismatches.

Ornellas spearheaded an effort to use calorimetric techniques to determine heats of detonation and products.[77,78] It was observed that the composition freezes in the 1500–1800K range, and thus with appropriate equations of state, Ornellas was able to calculate the product distribution at the CJ point and obtain agreement between theory and experiment. The agreement thus obtained does uniquely define the Chapman–Jouguet model. The product distribution would be expected on the basis of thermal equilibrium of the explosive atoms at high temperatures and pressures, but they do not define the end of the reaction zone.

4. SUMMARY

The combination of advances in numerical simulations, quantum mechanical calculations, and laser diagnostics has increased our understanding of the molecular basis of the behavior of energetic materials. On the other hand, the advances in experimental techniques have not been fully exploited and offer the possibility of even greater understanding. Tying together the macroscopic and microscopic observations is an exciting area that can lead to even better and safer control of energetic materials.

REFERENCES

1. (a) "Encyclopedia of Explosives and Related Items," Vols. 1–10. ARDEC: Dover, NJ. (Available from National Technical Information Service, Springfield, VA 22151.) (b) Fair, H. D.; and Walker, R. F., Eds.; "Energetic Materials," Vols. 1 and 2. Plenum Press: New York, 1977. (c) Fickett, W.; Davis, W. C. "Detonation." University of California Press: Berkeley, 1979. (d) Meyer, R. "Explosives." VCH Publishers: Deerfield Beach, FL, 1977.
2. Capellos, C.; Walker, R. F., Eds. "Fast Reactions in Energetic Systems." Reidel: Dordrecht, 1981.
3. Hamann, J. N. "Advances in High Pressure Research," Vol. 1. Academic Press: New York, 1966.
4. Gibbs, T. R.; Popolato, A., Eds. "LASL Explosive Property Data." University of California Press: Berkeley, 1980.
5. Walker, G. R., Ed. "Manual of Sensitiveness Tests." Canadian Armament Research and Development Establishment: Valcartier, Quebec, 1966.
6. Kamlet, M. J. Sixth International Symposium on Detonation, Naval Surface Weapons Center, Silver Spring, MD, 1976. (Available from Government Printing Office, Washington, DC 20402.)
7. Kamlet, M. J.; Adolph, H. J. *Propellants Explos.* **1979**, *4*, 30.
8. (a) Cachia, G. P.; Whitebread, E. C. *Proc. Roy. Soc.* **1958**, *246A*, 268–93. (b) Prince, D.; Liddiard, T. P., Jr. Naval Ordnance Laboratory Technical Report, **1966**, 66–87. White Oak, MD. Available from National Technical Information Service, Alexandria, VA 22304.
9. Iyer, S. *Propellants Explos. Pyrotechnics* **1982**, *7*, 37–39.
10. (a) This is the basis of hot spot theory of initiation. See Bowden, F. P.; Yoffe, A. D. "Fast Reactions in Solids." Butterworths: London, 1958. (b) Sollott, G. P.; Alster, J.; Gilbert, E. E.; Sandus, O.; Slagg, N. *J. Energ. Mater.* **1986**, *4*, 5.
11. Owens, F. J. *J. Mol. Struct. (THEOCHEM)* **1985**, *121*, 213.
12. Owens, F. J.; Jayasuriya, K.; Abrahmsen, L.; Politzer, P. *Chem. Phys. Lett.* **1985**, *116*, 434.
13. Iyer, S. *J. Energ. Mater.* **1984**, *2*, 151.
14. Iyer, S.; Capellos, C. *Int. J. Chem. Kinet.* **1974**, *6*, 89.
15. Capellos, C.; Iyer, S. (a) In *Proceedings of the International Conference on Combustion and Detonation*, Fraunhofer Institute for Propellants and Explosives (ICT), Karlsruhe, 1979, p. 611. (b) In Reference 2, pp. 401–418.
16. Capellos, C. In Reference 2, pp. 33–45.
17. (a) Owens, F. J. *Mol. Cryst. Liq. Cryst.* **1983**, *101*, 235. (b) Owens, F. J. In Eighth International Symposium on Detonation, Los Alamos National Laboratory: Albuquerque, NM, 1985. (Available from Government Printing Office, Washington DC, 20402.) (c) Bulusu, S.; Autera, J. R. *J. Energ. Mater.* **1983**, *1*, 133. (d) Shackelford, S. A.; Beckmann, J. W.; Wilkes, J. S. *J. Org. Chem.* **1977**, *42*, 4201.
18. U.S. Army Material Command. "Engineering Design Handbook." AMCP 706-180, HQ: Washington DC, 1972.
19. Strehlow, R. A. "Fundamentals of Combustion." International Textbook Co.: Scranton, PA, 1968.
20. Bawn, C. E. H. "Decomposition of Organic Solids." In Chemistry of the Solid State series, Garner, W. E., Ed.; Butterworths: London, 1955.
21. Sokolik, A. S. Self-ignition, flame, and detonation in gases. *Izv. Akad. Nauk SSSR* **1960**. Translated from Russian, OTS 63-11179. U.S. Department Commerce: Washington, DC.
22. (a) Ben-Ami, R.; Lucquin, M. In "Oxidation and Combustion Reviews," Vol. I; Tipper, C. F. H., Ed.; Elsevier: New York, 1965. (b) Ben-Ami, R.; Lucquin, M. In "Oxidation and Combustion Reviews," Vol. II; Tipper, C. F. H., Ed.; Elsevier: New York, 1965. (c) Frank-Kamenetskii, D. A. "Diffusion and Heat Transfer in Chemical Kinetics." Plenum Press: New York, 1969.

23. (a) Semenov, N. N. "Chemical Kinetics and Chain Reactions." Oxford University Press: New York, 1935. (b) Semenov, N. N. "Some Problems in Chemical Kinetics and Reactivity," Vols. I and II. Princeton University Press, Princeton, NJ, 1959.
24. Kamlet, M. J.; Jacobs, S. J. *J. Chem. Phys.* **1968,** *48,* 23.
25. Grusehka, H. D.; Weeken, F. "Gasdynamic Theory of Detonation." Gordon & Breach: New York, 1971.
26. Zeldovitch, B.; Kompaneets, A. S. "Theory of Detonation." Academic Press: New York, 1960.
27. von Neumann, J. U.S. Office of Scientific Research and Development, Report 541, 1942.
28. Doering, W. *Ann. Phys.* **1943,** *43,* 421.
29. Mader, C. L.; Craig, B. G. Nonsteady-State Detonations in One-Dimensional, Diverging and Converging Geometries. Los Alamos Scientific Laboratory Report LA-5865. Los Alamos, NM, 1975.
30. Fickett, W.; Wood, W. Numerical calculations of one-dimensional, unstable pulsation detonations. *Phys. Fluids* **1966,** *9,* 903.
31. Erpenbeck, J. *Phys. Fluids,* **1964,** *7,* 684; **1962,** *5,* 602.
32. Mader, C. L. Detonation Performance Calculations Using the Kistiakowsky–Wilson Equation of State. Los Alamos Scientific Laboratory Report LA-2613, Los Alamos, NM, 1961.
33. Mader, C. L. Detonation Properties of Condensed Explosives Computed Using the Becker–Kistiakowsky–Wilson Equation of State. Los Alamos Scientific Laboratory Report LA-2900, Los Alamos, NM, 1963.
34. Cowan, R. D.; Fickett, W. *J. Chem. Phys.* **1956,** *24,* 932.
35. Alster, J.; Downs, D. S.; Gora, T.; Iqbal, Z.; Fox, P. G.; Mark, P. In Reference 1b, Vol 1, p. 449.
36. Edwards, D. H.; Jones, A. T. *Phys. D Appl. Phys.* **1978,** *11,* 155.
37. Chiu, K. W.; Lee, J. H. *Combust Flame* **1976,** *26,* 353.
38. Edwards, D. H. In Twelfth Symposium (International) on Combustion, Combustion Institute, Pittsburgh, PA, p. 819.
39. Fifth International Symposium on Detonation, 1970. Naval Surface Weapons Center, White Oak, Silver Spring, MD.
40. (a) Sixth International Symposium on Detonation, 1976. Naval Surface Weapons Center, White Oak, Silver Spring, MD. (b) Engelke, R.; Earl, W. L.; Rohling, C. M. *J. Chem. Phys.* **1986,** *84,* 142.
41. Seventh International Symposium on Detonation, 1981. Naval Surface Weapons Center, White Oak, Silver Spring, MD.
42. Eighth International Symposium on Detonation, 1985. Los Alamos National Laboratory, Los Alamos, NM.
43. Mader, C. L. *Phys. Fluids* **1963,** *6,* 375.
44. Enig, J. W.; Petrone, F. J. *Phys. Fluids* **1966,** *9,* 398.
45. Catalano, E.; McGuire, R.; Lee, E.; Wrenn, E.; Ornellas, D.; Walton, J. In Reference 40a, p. 214.
46. Bulusu, S.; Weinstein, D.; Autera, J.; Velicky, R. W. *J. Phys. Chem.* **1986,** *90,* 4121.
47. Berke, J. G.; Shaw, R.; Tegg, D.; Seely, L. B. In Reference 39, p. 168.
48. Mader, C. L. "Numerical Modeling of Detonations." University of California Press: Berkeley, 1979.
49. Yang, L. C.; Menichelli, V. J. In Reference 40a, p. 612.
50. Capellos, C.; Lee, S.; Bulusu, S.; Gamms, L. A. In "Advances in Chemical Reaction Dynamic;" Rentzepis, P. M.; and Capellos, C., Eds.; Reidel: Dordrecht, 1986, p. 395.
51. von Hulle, W. G.; Tarver, C. M. In Reference 41, p. 993.
52. Bisard, F.; Tombini, C.; Menil, A. In Reference 41, p. 1010.
53. Duval, G. E. Shock-Induced Chemical Reactions in Condensed Matters. ONR Contract N 00014-77c-0232, August 1982. Office of Naval Research, 800 N. Quincy St., Arlington, VA, 22217.
54. Guichy, R. M.; Davis, L. P. *Thermochim. Acta* **1979,** *32,* 1.

55. Owens, F. J.; Sharma, J. *J. Appl. Phys.* **1980,** *51,* 1494.
56. Owens, F. J. *J. Chem. Phys.* **1982,** *77,* 5549.
57. American Physical Society Conference on Shock Waves in Condensed Matter. Sandia National Laboratories, Santa Fe, NM, July 18–21, 1983.
58. Trevino, S. T.; Tsai, D. H. In Reference 42.
59. Trevino, S. T.; Tsai, D. H. *J. Chem. Phys.* **1984,** *81,* 248, 5636.
60. Tsai, D. H.; Trevino, S. F. *J. Chem. Phys.* **1983,** *79,* 1684.
61. Karo, J. M.; Hardy, J. R. In Reference 2, p. 611.
62. Walker, F. G.; Karo, A. M.; Hardy, J. R. In Reference 41, p. 777.
63. Karo, A. M.; Walker, F. E.; DeBoni, T. M.; Hardy, J. R. "Gas Dynamics of Detonations and Explosion," Vol. 94 in IAA Progress in Astronautics & Aeronautics; AIAA: New York, 1984, p. 405.
64. Dremin, A. N.; Klinenko, V. Yu.; Michaelyuk, K. M.; Trodinov, V. S. In Reference 41.
65. Pastine, D. J.; Kamlet, M. J.; Jacobs, S. J. In Reference 40a, p. 305.
66. Walker, F. E.; Wasley, R. J. *Propellants Explos.* **1976,** *1,* 73.
67. Walker, F. E.; Wasley, R. J. *Explosivestoffe* **1969,** *17,* 9.
68. Alster, J.; Slagg, N.; Dewar, M. J. S.; Ritchie, J. P.; Wells, C. C. In *Proceedings of the International Conference on Combustion and Detonation,* 1979 [see Reference 15a], p. 695.
69. Dewar, M. J. S.; Ritchie, J. P.; Alster, J. *J. Org. Chem.* **1985,** *50,* 1031.
70. Bardo, R. D. In Reference 41, p. 93.
71. Burton, J. T. A.; Hawkins, S. J.; Hooper, G. In Reference 41, p. 759.
72. Gibson, F. C.; Bowser, M. L.; Summers, C. R.; Scott, F. H.; Mason, C. M. *J. Appl. Phys.* **1958,** *29,* 628.
73. Urtiew, P. A. *Acta Astronaut.* **1976,** *3,* 555.
74. Kato, Y.; Bauer, P.; Brochet, C.; Bouriannes, R. In Reference 41, p. 768.
75. Kato, Y.; Mori, N.; Sakai, H. In Reference 42.
76. He, X.; Han, C.; Kang, S. In Reference 42.
77. Ornellas, D. *J. Phys. Chem.* **1968,** *72,* 2390.
78. Ornellas, D.; Carpenter, J. H.; Gunn, S. *Rev. Sci. Instrum.* **1968,** *37,* 907.

CHAPTER 8

Ultraviolet Photoelectron Spectroscopy and Matrix Isolation: A Combined Approach to the Study of Reactive Species

Reinhard Schulz‡ and Armin Schweig*
University of Marburg, Marburg, West Germany

CONTENTS

1. Methods... 289
2. Molecules 295
3. Outlook 354
   Acknowledgments 356
   References... 356

1. METHODS

A. Variable Temperature Photoelectron Spectroscopy

Physical methods have steadily gained importance in chemistry. They already dominate in many areas that, in former times, were covered by classical chemical procedures. This is especially true for the many spectroscopic methods for structure determination and instrumental analysis.

‡ Present address: Industrial Chemicals Division, E. Merck, Darmstadt, West Germany

* To whom correspondence should be sent.

Thus UV photoelectron spectroscopy (PES),[1] introduced in about 1970, essentially contributed not only to a deeper understanding of molecular and electronic structures but also to the elucidation of new reaction pathways and reactive intermediates.

The investigations of highly reactive organic π systems (intermediates) presented in this chapter pertain to hydrocarbons as well as oxygen, sulfur, selenium, silicon, and germanium compounds. Some of these compounds have become cornerstones of modern organic chemistry, such as cyclobutadiene, *o*-benzyne, *o*-xylylene, small-sized strained heterocyclic ring systems, and systems with silicon–carbon double bonds. It is really fascinating that all these structures can be generated in the gas phase by simple thermal decomposition of appropriate precursor molecules, frequently much more selectively than by other methods.[2] Moreover, for problems involving molecular properties of free molecules, the ideal medium (ie, the gas phase) is available. This is particularly so when the thermal reactions are conducted at sufficiently low pressure (eg, <0.1 mbar) and short contact (reaction) time (eg, <0.1 s) in a gas flow system. Under these conditions even kinetically very unstable compounds can be studied for many hours.

Therefore the installation of a suitable thermal flow reactor in the gas inlet system of a UV photoelectron spectrometer was suggested, as an approach to the study of reactive (transient or intermediate) species[3] immediately after their production, using the method of flash vacuum pyrolysis (FVP).[2] In addition, this so-called variable temperature photoelectron spectroscopy (VTPES) provides the opportunity for investigating equilibria (distributions) of vibrating molecules, rotamers, tautomers, or other isomers, as well as for measuring rate constants.[4] This new technique supplements other methods for monitoring reactive species including microwave,[5] infrared (IR),[6] Raman, and electronic (UV-VIS) absorption and emission spectroscopy, as well as mass spectrometry (MS)[2a,7] or, in some cases, electron paramagnetic resonance (EPR, ESR) and laser magnetic resonance (LMR)[8] spectroscopy. Among these methods, IR and ultraviolet–visible (UV-VIS) spectroscopy are particularly widely used in conjunction with the low-temperature matrix isolation technique (MIT) (see Subsection C). Since there is much experience using MIT, this technique lends itself to comparison purposes and thus independent confirmation of VTPES results.

We cannot present here a comprehensive overview of the extensive investigations of VTPES and MIT of unstable species.[3,9] Instead we concentrate on a few systems whose choice is largely determined by the scientific interests in the authors' group.

B. Interpretation of UV Photoelectron Spectra

Many readers will be familiar with the process of photoionization, which forms the basis of photoelectron spectroscopy. Light sources most frequently used in UV photoelectron spectroscopy to ionize molecular species in the gas phase are the HeI (21.2 eV) or HeII (41.8 eV) gas discharge lamps. The electrons emitted from the sample gas perpendicular to the incident radiation are separated according to their kinetic energy in the energy analyzer and then counted. Accordingly, a photoelectron (PE) spectrum is a plot of the photoelectron count rate (intensity) versus the ionization energy (IE), which is the difference between the energy of the ionizing radiation (eg, ≈ 21 eV for HeI radiation) and the measured kinetic energy of a respective portion of photoelectrons separated in the analyzer. Each band of a photoelectron spectrum can be assigned to the transition (as usual) from the electronic ground state of the neutral molecule to an ion state. Thus the first band corresponds to the transition to the electronic ground state of the ion, the next bands to transitions to electronically excited ion states. The vertical ionization energy (VIE) is defined as the energy difference between the molecule in its ground state and the ion in a particular electronic state, but with the nuclei in the same positions as they had in the neutral molecule. In practice, the vertical ionization energy is equated to the ionization energy with maximum count rate (intensity) of a band.

Using Koopmans' theorem[10] the VIEs of a molecule are most easily calculated as the negative orbital energies of the occupied molecular orbitals (MOs) obtained from a Hartree–Fock (HF), approximate HF (AHF), or semiempirical valence electron calculation for the molecule. Within this model an ion state is simply and fully described as the doublet state corresponding to an electronic configuration with all MOs doubly occupied except one, which is only singly occupied, and all virtual orbitals (VOs) unoccupied (such as configurations B and C of Figure 8-1).

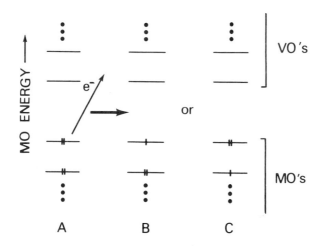

Figure 8-1. Schematic representation of the electronic configurations of the molecular ground state configuration A and two Koopmans ion configurations B and C. Ejection of one electron denoted by e⁻-labeled arrow; transition from A to B or C shown by bold arrow.

Provided Koopmans' theorem is valid, inductive or conjugating effects of substituents, or effects from building up molecules from molecular subunits, on ionization energies can correctly be traced back to such effects on molecular orbitals. These effects can be pictured in so-called orbital correlation (or interaction) diagrams or quantified using well-known MO variants of the variation method or perturbation theory [such as the LCMO (linear combination of molecular orbitals),[11,12] PMO (perturbational MO),[11-13] or FO (frontier orbital)[11-13] methods].

In general the assumptions made by Koopmans are not fulfilled. The orbitals of the ions are different from the orbitals of the molecule and electron correlation in the molecular ground state, and ion states cannot be neglected. When these effects are taken into account, agreement between calculated and measured VIEs is generally improved. Moreover, electron correlation accounts for the so-called non-Koopmans ionizations [seen in the PE spectra as shakeup bands (or satellites)]. These are "extra ionizations" (or, to put it differently, there are more ionizations than occupied orbitals) not explicable in the Koopmans picture. According to Figure 8-1, a so-called Koopmans ionization is a one-electron process. Figure 8-2 exhibits two electronic configurations B' and C' that give rise to non-Koopmans doublet configurations (B' to just one and C' to two ones). Clearly, ionizations to such final states are two-electron processes: one electron is ejected and a second one promoted from a MO to a VO: thus the doublet state based on B' is the so-called HOMO (highest occupied MO) LUMO [lowest unoccupied MO (or VO)] or, to be more precise, $HOMO^0 LUMO^1$ (indicating the number of electrons

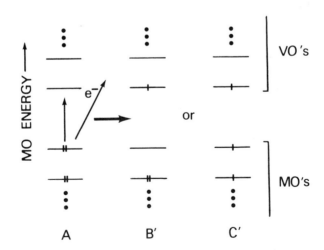

Figure 8-2. Schematic representation of the electronic configurations of the molecular ground state configuration A and two non-Koopmans configurations B' and C'. Ejection of one electron denoted by e^--labeled arrow; vertical arrow denotes excitation of a second one, and transition from A to B' or C' shown by bold arrow.

occupying these MOs by superscripts) non-Koopmans configuration. Such two-electron processes are forbidden in PES, but they can get intensity due to mixing with appropriate Koopmans configurations. Figure 8-2 pictures non-Koopmans ionizations that owe their existence to final state (ie, ion state) electron correlation. Much less is known (and experimentally not yet confirmed) about non-Koopmans ionizations that may arise due to electron correlation in the molecular ground state (or initial state) as a consequence of mixing doubly excited singlet configurations into the closed-shell ground state configuration. For more details about the latter type of non-Koopmans ionization, the reader is referred to References 14a and 15.

Although we cannot give a comprehensive overview of the various methods in use for the calculation of reliable VIEs (and thus a quantitative interpretation of the localization of PE bands),[14] one point must be stressed. Regarding the large organic molecules and reactive systems presented in this chapter, semiempirical valence electron methods including electron correlation for the calculations of VIEs are indispensable. To this end, the PERTCI (perturbational configuration interaction) method,[16a] a large-CI approach including all singly and doubly excited configurations, was combined with semiempirical valence electron methods, namely the LNDO/S (local neglect of differential overlap, for spectroscopy),[16b] MNDO (modified neglect of diatomic overlap),[16c] and CNDO/S (complete neglect of differential overlap, for spectroscopy)[16d] approaches, to the LNDO/S PERTCI, MNDO PERTCI, and CNDO/S PERTCI procedures. The methods have proved very helpful both for the prediction of Koopmans[17] and non-Koopmans[18] ionizations, and thus also for the identification of novel reactive species. For other semiempirical approaches, the reader is referred to Reference 19. The theoretical ion state energies obtained are usually presented in the form of a stick diagram (an ionization spectrum) along with the corresponding PE spectrum. A state is further specified by giving the symmetry symbol of the irreducible representation to which the wave function of this state belongs. Occasionally, measured VIEs and/or calculated VIEs of related compounds are grouped together to so-called ionization energy (or ion state) correlation diagrams to compare experimental and theoretical results, to find support for the assignment of bands to particular ion states, or to illustrate relations between the ionization processes of related systems or of a system and related subunit systems. Examples will be found in subsequent sections.

Finally two notes concerning PE band intensities must be added. As mentioned above, a non-Koopmans ionization gets intensity from a Koopmans ionization. As a consequence, the intensity of the original Koopmans ionization (arbitrarily set equal to 1) is reduced. The resulting intensities of both sorts of ionization are given as I_{rel} in the respective stick diagrams (ionization spectra) of this chapter.

If second-row or higher atoms are present in a molecule, considerable help in assignment or identification problems can come from the comparison of observed and calculated (or qualitatively expected) changes in relative integrated count rates (band intensities) when going from HeI to HeII excitation. For details the reader is referred to References 1 and 20.

C. Low-Temperature Matrix Isolation Technique

The method introduced by Pimentel[6] has rapidly developed due to improved cryogenic techniques.[21] The method is based on isolating reactive molecules in an inert (transparent) solid matrix (eg, in an argon matrix at ≈10 K). The reactive species can be generated in various ways. For the transient species described below, two techniques have been used. First, a variable temperature thermal flow reactor (same in principle as that used with VTPES) is installed as near as possible to the condensation region of the cryostat. The product molecules of the thermal reaction are very rapidly cooled and isolated one from another by an inert medium (eg, solid argon). In the solid matrix thermal bimolecular reactions (except for atoms or very small molecules participating in a reaction) or even unimolecular reactions of the reactive species (unless the activation energy is less than ≈3 kcal/mol) are practically excluded. Because of these properties, MIT in conjunction with a VT reactor is a technique (subsequently referred to as FVP [Flash Vacuum Pyrolysis] MIT) truly complementary to VTPES, with the advantage that additional information on novel systems can easily be gained by recording IR and UV-VIS (or if wanted in particular cases, Raman, ESR, or NMR) spectra, thus improving the chance of identifying new species. Hence a combined VTPES-MIT approach using identical thermal reactors is a powerful route for producing and characterizing new reactive species. Moreover, MIT offers a unique advantage insofar as reactive species can also be generated photochemically by preparing the low-temperature matrix with suitable precursor molecules and irradiating the matrix with light at an appropriate wavelength.

The interpretation of UV-VIS and IR absorption spectra has a much longer tradition than the interpretation of PE spectra. Nevertheless, the possibilities of calculating reliable spectra of larger organic molecules (the calculations are then necessarily of the semiempirical valence electron type) are very limited. A number of packages presently available for obtaining theoretical UV-VIS spectra are CNDO/S (CI),[22a-c] CNDO/S SECI [singly excited CI],[22d] LNDO/S PERTCI,[16b] and HAM/3 (hydrogenic atoms in molecules version 3) CI.[22e] Other combinations such as MNDO PERTCI[22f] or CNDO/S PERTCI,[22g] tested in the authors' group, are improper. The most serious drawback, from a practical point of view, is the limited availability of parametrized atoms: as a rule these are the atoms of the first row and, additionally, Cl [CNDO/S (CI)] and Cl, P, and S [CNDO/S SECI]. The quality of the semiempirical MNDO and AM1 (Austin model 1)[23] methods for calculating IR spectra of organic molecules is not well established at present (no reliability tests available). Therefore, for the interpretation of UV-VIS and IR absorption spectra, comparisons with appropriate reference compounds are even more needed than for the interpretation of PE spectra.

2. MOLECULES

A. Cyclobutadiene

As the parent compound of carbocyclic 4n π-electron systems, cyclobutadiene (**1**) is considered to be one of the most fascinating systems in organic chemistry. The results of numerous experimental and theoretical studies on the properties of cyclobutadiene and stable substituted derivatives are summarized in Reference 24. Questions of ground state structure and multiplicity have always been of central interest.

Whereas the IR spectrum of **1** in an argon matrix at 10 K was interpreted at first in terms of a square geometry,[25] the photoelectron spectra of the derivatives **2**[26a] and **5**[26b] indicated a rectangular structure. Moreover, X-ray analyses of **3**[27a] and **5**[27b] give rectangular structures. Room temperature X-ray data of **4** were interpreted in terms of a square structure,[28a] whereas the PE spectrum[28b] of this compound is virtually identical to those of **5** and **6** (in the essential part of the spectra), implying that **4** should also possess a rectangular geometry.[28c] Low-temperature X-ray data of **4** were subsequently found to be consistent with a rectangular structure.[28d] Thorough calculations[29a] as well as a reinterpretation of the matrix IR spectrum[29b] of unsubstituted cyclobutadiene (**1**) indicate a rectangular structure. A study of the PE spectrum of unsubstituted cyclobutadiene is expected to provide a further fundamental contribution to the aforementioned structural problem.

The generation of free cyclobutadiene in the gas phase was not possible for a long time because promising precursors were not stable enough, intermediates were formed that did not react as wanted, or unexpected competitive pathways were followed. For instance, the pyrolysis of the anhydride **7** (using the VTPES technique) quantitatively yielded cyclopentadienone (**8**)[30] (see Subsection C of Section 2), which cannot be decarbonylated thermally (Scheme I). In light of this, all other attempts to generate **1** from precursors

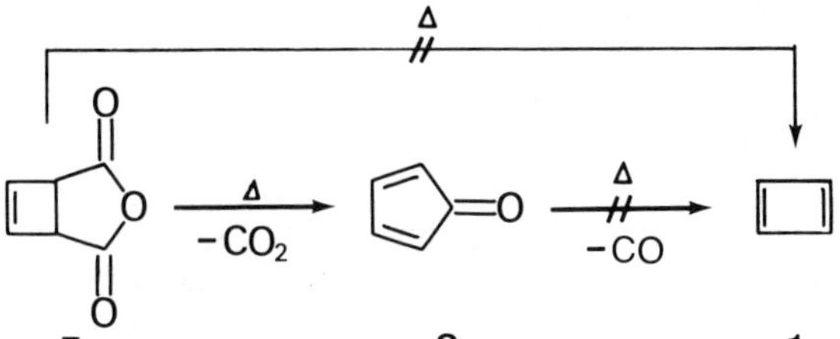

Scheme I

that lead to **8** were unsuccessful. Thermal cleavage of photo-α-pyrone (**9**) was unsuccessful, too,[31] despite the fact that **9** was previously described as a source of free **1**.[32] A thorough VTPES study yielded only the trivial product α-pyrone (**10**)[31] (Scheme II). Only the tricyclic α-pyrone isomer **11** proved to

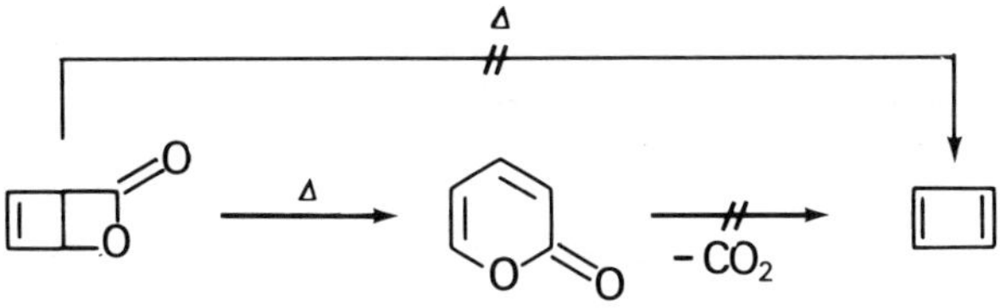

Scheme II

be the appropriate precursor for free (gaseous) **1**.[15,33] The VTPES experiment[15] yields **10**, **1**, CO_2, benzene, acetylene, and supposedly small amounts of **12** (dimeric cyclobutadiene, *anti*-tricyclo[4.2.0.0^{2,5}]octa-3,7-diene). The same products can be inferred from IR spectra using FVP MIT (Scheme III).

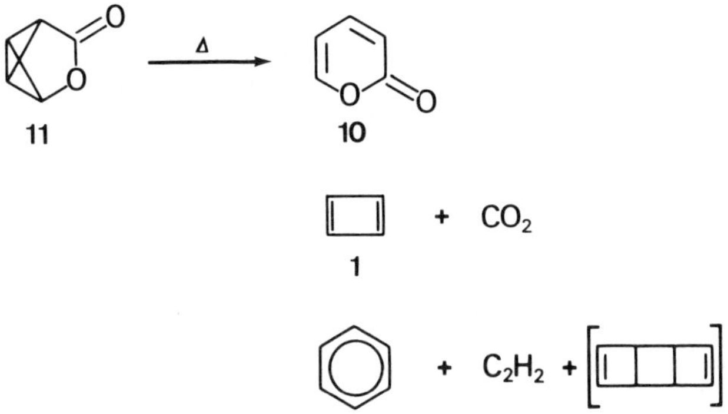

Scheme III

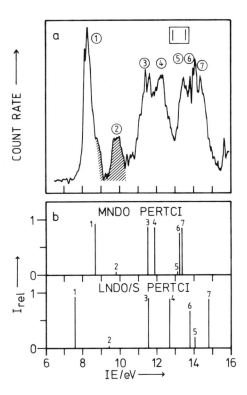

Figure 8-3. (a) HeI photoelectron spectrum and (b) ionization spectrum of cyclobutadiene (**1**) obtained using the MNDO PERTCI and LNDO/S PERTCI methods. The numbering of ionizations is based on the sequence of MNDO ionizations obtained. The ionizations are assigned as follows: (1) $^2B_{2g}(\pi)$, (2) $^2B_{3g}(\pi)$, (3) $^2B_{2u}(\sigma)$, (4) $^2B_{1u}(\pi)$, (5) $^2B_{3u}(\sigma)$, (6) $^2B_{3u}(\sigma)$, and (7) $^2A_g(\sigma)$. (Reproduced with permission from Reference 15.)

Figure 8-3a exhibits the PE spectrum of virtually pure cyclobutadiene at 580°C (after subtracting the PE spectra of by-products) and Figure 8-3b the ionization spectra (calculated spectra) using the MNDO PERTCI and LNDO/S PERTCI methods. Note the shoulder (VIE at ≈8.9 eV) on the high-energy side of the first peak and the second peak in the spectrum (VIE at ≈9.9 eV); both could be due to **12**.

It is clear that experiment and theory (based on the MNDO-optimized rectangular structure) are in reasonable agreement. Moreover the VIE (8.24 eV) of the first peak (band) is in excellent agreement with previous experimental estimates of 8.29[34] and 8.3 eV.[28b] An estimate of the VIE of the $^2B_{1u}(\pi)$ ionization amounts to 11.95 eV,[34] in good agreement with the measured value of 12.2 eV.

What can be learned from the PE spectrum of cyclobutadiene and from the fact that the spectrum could be recorded under the conditions indicated above? First, the conditions applied demonstrate that free **1** is stable up to temperatures of several hundred degrees! Second, the ionization spectra of Figure 8-3b exhibit two ionizations of very low intensity, namely ionization 2 [$^2B_{3g}(\pi)$] with I_{rel} = 2–5% and ionization 5 [$^2B_{3u}(\sigma)$] with I_{rel} = 4–20%, of the non-Koopmans type. The $^2B_{3u}(\sigma)$ non-Koopmans ionization is due to CI between ion configurations [a Koopmans configuration due to removal of an electron from the $b_{3u}(\sigma)$ orbital and two non-Koopmans doublet configurations based on the electron configuration NHOMO^1 (next HOMO), HOMO^1, LUMO^1 (cf. C' of Figure 8-2)]. Of particular interest, however, is the $^2B_{3g}(\pi)$ non-Koopmans ionization. It owes its existence to configuration interaction in the molecular ground state (ie, mixing in doubly excited configurations), in the present case the doubly excited configuration due to the electron configuration HOMO^0 LUMO^2, leading to a non-Koopmans ion state based on the electron configuration HOMO^0 LUMO^1 (cf. B' of Figure 8-2). Shakeup ionizations of this type are usually of very low intensity. Cyclobutadiene is an interesting particular case, however, since, for hypothetical square singlet cyclobutadiene the $^2B_{2g}(\pi)$ and $^2B_{3g}(\pi)$ (shakeup state) ion states will degenerate into the $^2E_g(\pi)$ ion states, with both states of the degenerate pair contributing the relative intensity of 0.5 to the full (relative) intensity of 1 of the corresponding ionization (for more about hypothetical square cyclobutadiene, see below). Presently, the experimental spectrum (Figure 8-3) provides little opportunity to localize any shakeup structure (with regard to non-Koopmans ionization 2 because of the interfering by-product ionization). From the shape of the first photoelectron band of cyclobutadiene, a simple argument for the rectangular geometry can be derived. If it is supposed that cyclobutadiene possesses a square singlet electronic ground state, the first ion state would be degenerate (^2E_g, see above) and thus would undergo a (first-order) Jahn–Teller (JT) distortion in this state with the result that—instead of a sharp band with a relative intensity of 1 as observed—a double band, with each part having a relative intensity of, say, 0.5 or at least a broad band with markedly reduced height should occur. Nowadays it is known from matrix IR spectroscopy, in accord with ab initio calculations (see above) that square singlet cyclobutadiene undergoes a pseudo- (or second-order) JT distortion to a rectangular geometry and is thus a hypothetical system. Nevertheless, the photoelectron spectroscopic argument is as legitimate as that based on the previous IR work and constitutes a simple independent confirmation. Besides o-xylylene and some of its derivatives (see next section), cyclobutadiene is one of the still rare examples of a compound whose first excited ion state is a shakeup state. This state is only about 1.1 eV (MNDO PERTCI) to 1.8 eV (LNDO/S PERTCI) above the ground state, which is less than for all other cases known, certainly a consequence of the small HOMO–LUMO energy gap. In accord with this is the chemical behavior of **2**, which may react as a diene, dienophile, electrophile, or nucleophile, as well as a diradical,[35a–c] and also the relatively low-lying first electronically excited singlet state [$^1B_{1g}(\pi\pi^*)$] of **1** at 2.4–4.1 eV (500–300 nm)[16b,24b] or the lowest triplet state of **6** at about 1.2 eV (adiabatic transition at $\approx$0.5 eV).[35d]

Frequently, an unstable compound is stabilized by benzoannelation. Nevertheless, benzocyclobutadiene (13) can be detected only in the gas phase or at low temperatures dissolved in a solid solvent. Free 13 can be generated by dehalogenation of cis-1,2-diiodobenzocyclobutene (14) over zinc powder[36a] or more elegantly by dehalogenation of α,α,α',α'-tetrabromo-o-xylene (15) over sublimed magnesium[36b] in a flow system (Scheme IV), and then monitored by PE spectroscopy[36b] or isolated in an argon matrix and investigated using IR and UV-VIS spectroscopy.[36a]

Scheme IV

According to HAM/3 CI[36b] and LNDO/S PERTCI,[18a] the PE bands of 13 are due to Koopmans ionizations. A shakeup ionization [$^2A_2(\pi)$, ≈2 eV above the ion ground state][18a] of very low intensity ($I_{rel} < 0.01$) is predicted, which eventually can be seen in the UV-VIS spectrum of the radical cation of 13.

MNDO calculations for 13 as well as for 16 revealed a rather pronounced single/double bond localization, in the sense of a bismethylenecyclobutene structure (17).[37] Thus 13 and 16 are not true cyclobutadienes. In particular it was shown for 16 that a true cyclobutadiene subunit in this system corresponds to a doubly excited configuration that lies 150 kcal/mol above the ground state configuration.[37]

According to MNDO calculations both the HOMO and LUMO of 16 are sizably localized at the four-ring double bond. Therefore both electrophilic and nucleophilic reagents should primarily react at this double bond; experimental results for dimethyl-substituted (at the double bond) 16 have confirmed this prediction.[38]

B. Quinomethanes and Related Compounds

Quinomethanes (xylylenes) and their heteroanalogues play an important role in synthetic and theoretical chemistry, as well as in spectroscopic investigations.[2g,39]

Whereas *p*-xylylene (**18**) is rather stable in the gas phase at reduced pressure but relatively high temperatures, *o*-xylylene and hetero-*o*-xylylenes (**19**) undergo rapid secondary reactions, resulting in a loss of the *o*-quinoid structure. Nonetheless, quite a few of these species could be thermally generated and monitored, starting from appropriate precursor systems.

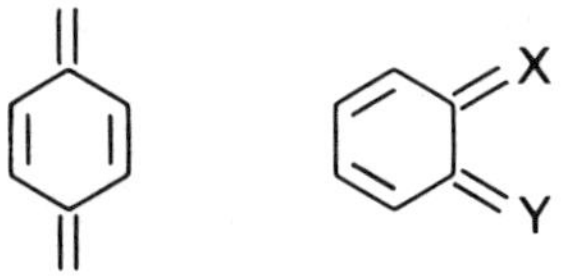

X, Y = CH$_2$, NH, O, S, etc.

18   **19**

a. *p*-Xylylene

The simplest route to free *p*-xylylene (**18**) is the gas phase pyrolysis of [2.2]paracyclophanes (**20**).[40] This reaction has found technical application for producing polyparaxylylene passivation layers for electronic devices (Scheme V).[41] An original interpretation of a shoulder at 9.8 eV in the PE

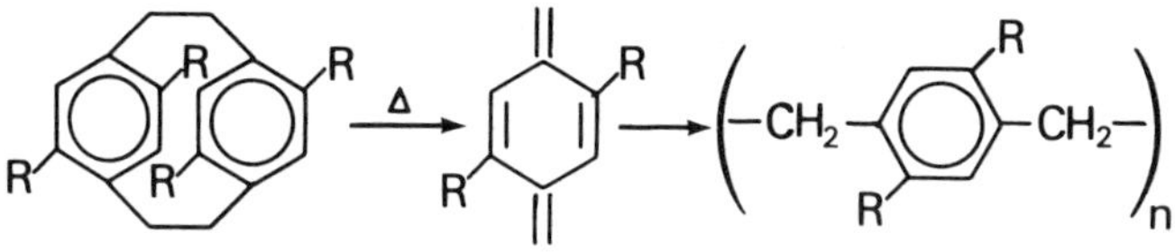

20   **18**

R = H, CH$_3$, Cl

Scheme V

spectrum of *p*-xylylene (**18**) assigned it the HOMO LUMO shakeup state from the very beginning of the photoelectron spectroscopic study of this compound.[40c] This interpretation was later supported by a photoelectron spectroscopic study of the 2,5-dimethyl derivative (**21**)[40d] and by many theoretical studies between 1983 and 1986.[14b,18a,19c–f,42]

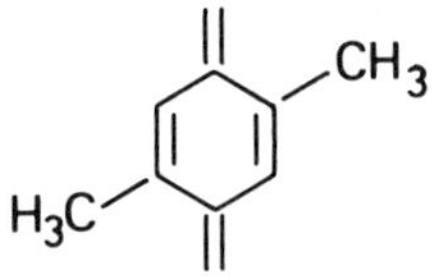

21

b. *o*-Xylylene and Isoindene Derivatives

Being the parent compound of all *o*-quinoid hydrocarbons, *o*-xylylene (**22**) has aroused great interest for now more than 40 years.[39,43] For a long time, only indirect evidence for **22** [based on the observation of secondary products, as benzocyclobutene (**23**) or the dimers **24** and **25**] could be found, as a result of the high reactivity of this species.[44] Attempts to thermally generate

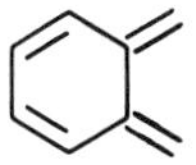

22 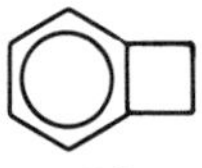 **23**

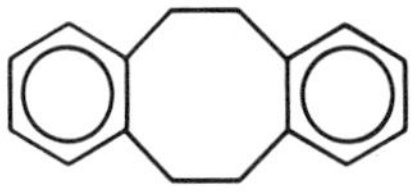

24 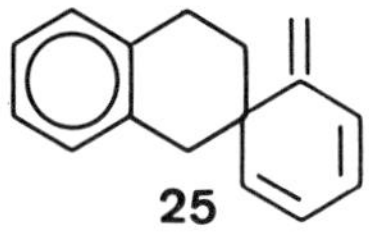 **25**

22 are normally unsuccessful due to the pronounced tendency of this compound for rapid ring closure to **23**. It is just the latter possibility that is nowadays broadly used to synthesize various benzocyclobutene derivatives[21,44,45] (Scheme VI). Whereas the UV-VIS spectrum of **22** could early be

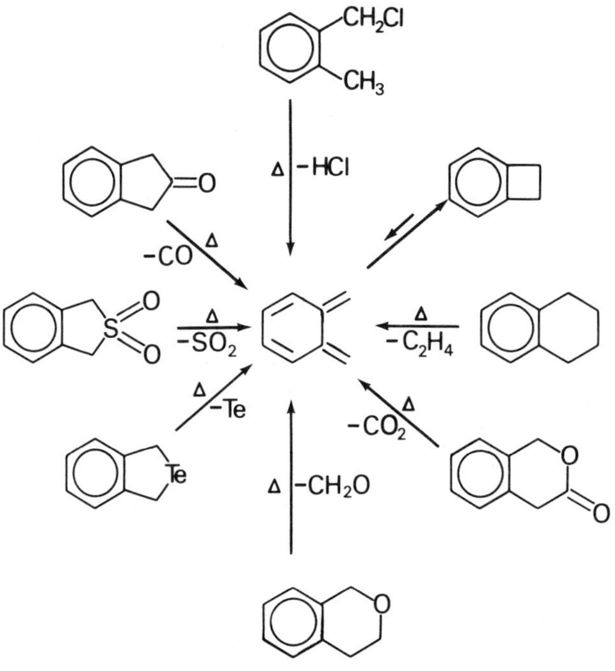

Scheme VI

recorded after photochemical generation of this product in an organic glass at low temperature,[46] the continuous generation in a flow system making use of the reaction of excited alkali metal atoms with α,α'-dibromo-o-xylene (**26**) was reported in 1977.[47] In this way the IR and Raman spectra of **22** isolated in an argon matrix could be recorded. In addition, two bicyclic compounds, namely **27**[48] and **28**,[45] are available, and these could be particularly suitable precursors for producing **22** under mild (thermal) conditions in high yields (Scheme VII). Actually, the VTPES experiment using **28** as the precursor

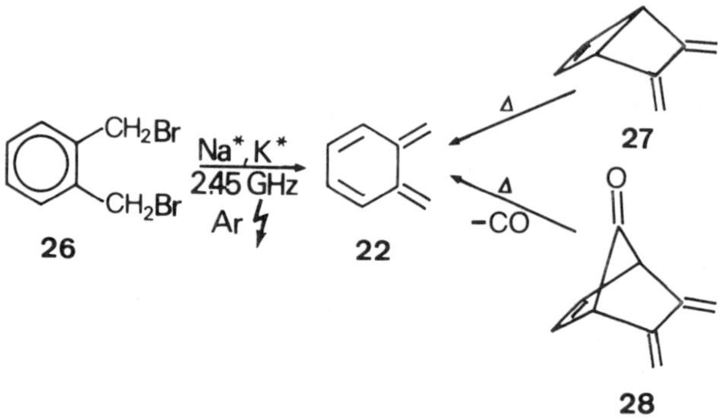

Scheme VII

system was successful (at 225°C and 0.1 mbar).[18c] Figure 8-4a shows the PE spectrum of **22** and Figure 8-4b the calculated ionization spectra using the MNDO PERTCI, CNDO/S PERTCI, and LNDO/S PERTCI approaches. According to the calculations, the first Koopmans ionization (ie, 1) is followed by a group of closely spaced ionizations (ie, 2, 3, and 4). The second is the HOMO LUMO shakeup ionization, which borrows intensity from the Koopmans ionization 4. Numbers 3, 5, and 6 are practically pure Koopmans ionizations. The points of interest here were the three ionizations 2, 3, and 4. Inspection of the corresponding region in the PE spectrum of **22** (ie, in the ionization region of 9–11 eV) revealed the obviously low intensity of band 4 and the unusually shaped low-energy flank of band 3, indicating that, probably, an additional ionization (ie, band 2) was hidden under a common envelope. The spectrum between 9 and 12 eV was therefore Gaussian-fitted over this range (Figure 8-5). The result of the fit, along with the good agreement with the calculations provided strong evidence for low-energy (1.9 eV above the ion ground state), non-Koopmans ionization of **22**. The assignment is further supported by recent UV-VIS spectra of **23** irradiated with X-rays in an argon matrix.[18e]

In the present context, the PE spectrum of the linearly conjugated 8π electron system all-*trans*-octatetraene (**29**) is relevant, too. Thorough inves-

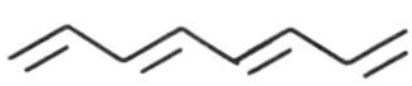

tigations have revealed that the first non-Koopmans ion state of **29** (of the

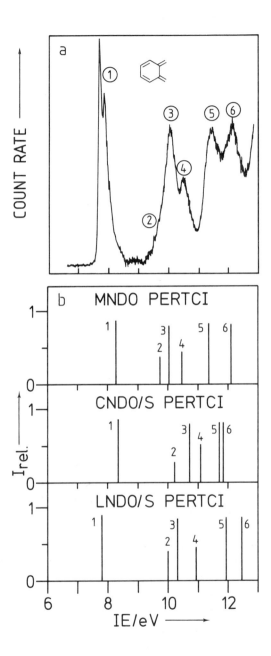

Figure 8-4. (a) HeI photoelectron spectrum and (b) ionization spectrum of o-xylylene (**22**) obtained using the MNDO PERTCI, CNDO/S PERTCI, and LNDO/S PERTCI approaches. The ionizations are assigned as follows: (1) $^2A_2(\pi)$, (2) $^2B_1(\pi)$, (3) $^2A_2(\pi)$, (4) $^2B_1(\pi)$, (5) $^2A_1(\sigma)$, and (6) $^2B_2(\sigma)$. The numbers denote the ionizations in the order of increasing energy. (Reproduced with permission from Reference 18c.)

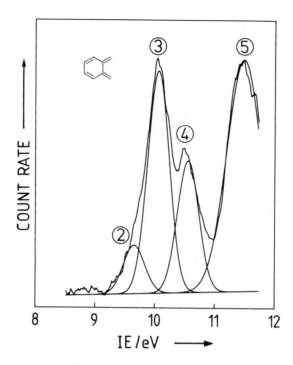

Figure 8-5. Curve fitting of the PE spectrum of *o*-xylylene (**22**) in the ionization region of 9–12 eV using four symmetrical Gaussian functions. For further details, consult Kreile et al.[18c] (Reproduced with permission from Reference 18c.)

HOMO LUMO shakeup type) lies about 3.0 eV above the ion ground state.[18b,49] This value is distinctly higher than the corresponding value (1.9 eV in each case) for **18** and **22**. This is in accord with the conception of a relatively low-lying LUMO in xylylenes. On the other hand, the ion ground state energies of **18**, **22**, and **29** are practically the same, which means, in the Koopmans picture, that the HOMO energy of **18** and **22** exhibits no remarkable deviation from the one of **29**.

Until the present time, only indirect evidence could be obtained for the transient species isoindene (**30**).[50] Isoindene possesses approximately the same π system as *o*-xylylene but undergoes rapid rearrangement to more stable indene (**31**) (Scheme VIII). The corresponding rearrangement of 2,2-dimethylisoindene (**32**) is slow enough to permit the spectroscopic character-

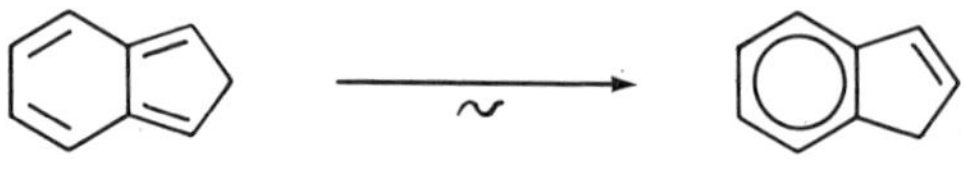

Scheme VIII

32

ization of this compound, even at room temperature.[51] The PE spectrum of **32** as well as the UV-VIS spectrum of the corresponding radical cation were recorded.[52] The PE spectra of **22** and **32** are very similar in the π ionization region, apart from the fact that the spectrum of **32** is shifted to lower ionization energies (by ≈ 0.5 eV), as a result of the alkyl substitution.

The LNDO/S PERTCI ionization spectrum of **32** was calculated.[18a] This spectrum is practially identical to the ionization spectrum of **22** (apart from the fact that the shift of 0.5 eV to lower ionization energies for **32** is not reproduced, as a result of an underestimation of alkyl group effects by semiempirical valence electron methods). The close correspondence between the theoretical spectra of **32** and **22** (in the essential parts) suggested a reinterpretation of both PE and cation UV-VIS experimental spectra of **32**.[18a] With regard to the PE spectrum, this essentially means that the second PE peak consists of three bands: a band at about 9.8 eV due to the $^2A_2(\pi)$ Koopmans ionization, flanked by two bands of low intensity at about 9.4 and 10.2 eV corresponding to the two $^2B_1(\pi)$ non-Koopmans ionizations (cf, eg, Figures 8-4 and 8-5, showing the analogous situation for **22**). The UV-VIS cation spectrum (consisting of two band systems) is then perfectly explicable by assigning the $1^2A_2(\pi) \rightarrow 1^2B_1(\pi)$ transition to the first band system and the $1^2A_2(\pi) \rightarrow 2^2A_2(\pi)$ and $1^2A_2(\pi) \rightarrow 2^2B_1(\pi)$ transitions to the second band system. Thus there is a perfect reconciliation between the PE spectrum of **32** and the UV-VIS spectrum of its cation, and the UV-VIS cation results give a clear independent proof of the mixed Koopmans/non-Koopmans structure of the second PE band system of **32**.

Isobenzofulvene derivatives (**33**) cannot rearrange to indenes. Therefore enhanced thermal stability is expected. Nonetheless, the characterization of

33   X = CH_2, C = CH_2

33 was unsuccessful for a rather long time.[53] Only one stable compound (**34**) of this type was known.[54] Pyrolysis of **35** at about 600°C and 0.1 mbar yielded

34

a blue product of the formula $C_{10}H_8$ that was stable at 77 K and gave the endo dimer **37** of isobenzofulvene (**36**) when heated to $-100°C$. When raising the temperature to 700°C during pyrolysis, naphthalene was also obtained[55] (Scheme IX).

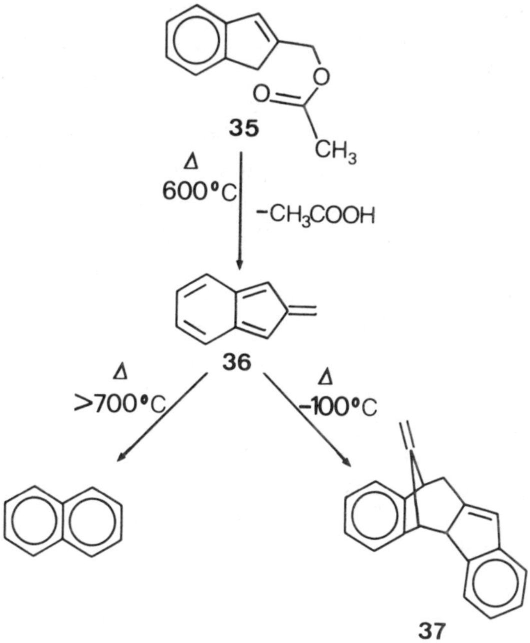

Scheme IX

Direct detection and characterization of **36** was based on the excellent agreement of the calculated and measured VIEs (applying the VTPES technique; for the relevant PE spectra, see Figure 8-6) of the first three photoelectron bands of **36** (Table 8-1). The first VIE of **22**, **36**, and **38** (see below) are practically the same [because the corresponding ion state is an $A_2(\pi)$

TABLE 8-1. CNDO/S PERTCI, MNDO PERTCI, LNDO/S PERTCI, and Observed Ion State Energies (eV) for Isobenzofulvene (**36**)

Ion state[a]	CNDO/S PERTCI	MNDO PERTCI	LNDO/S PERTCI	Observed
$1^2A_2(\pi)$	8.05	7.99	7.48	7.32
$2^2B_1(\pi)$	9.60	9.20	9.06	9.10
$3^2A_2(\pi)$	10.50	9.82	10.13	9.90

[a] Koopmans ion states numbered in order of increasing energy.

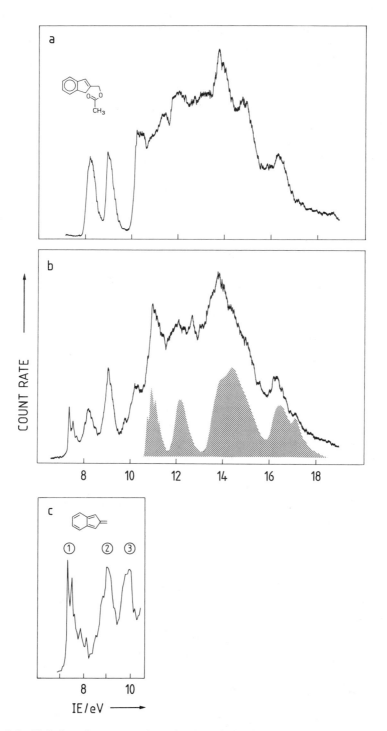

Figure 8-6. HeI photoelectron spectrum of (*a*) 2-1*H*-indenylmethyl acetate (**35**) at 25°C, (*b*) a mixture of **35** and its pyrolysis products at 590°C (the spectrum of acetic acid is shown hatched), and (*c*) isobenzofulvene (**36**). (Reproduced with permission from Reference 55.)

state]. Note that for benzofulvene the first ionization should occur at energies higher than 8 eV, as concluded from comparisons with reference compounds (eg, styrene).

The PE spectrum of **36** can be fully understood in terms of Koopmans' theorem. LNDO/S PERTCI calculations predict a non-Koopmans ionization at 9.6 eV ($^2B_1$) with I_{rel} = 0.04 which, however, plays no significant role because of the low intensity.[18a]

The calculated electronic excitation (absorption) spectrum of **36** using the LNDO/S PERTCI and CNDO/S SECI approaches is similar to that of the azulene isomer. Note, however, that the excitation energies are heavily dependent on the geometry used. Minor reductions of bond alternation cause pronounced shifts of the $^1A_1(\pi) \rightarrow {}^1B_2(\pi\pi^*)$ absorption (first band in the spectrum) to longer wavelengths. Interestingly, the ionization energies are hardly influenced by these changes.

The FVP of naphtho[b]cyclopropene (**39**) provided a first hint to the formation of isobenzofulvenallene (**38**) (Scheme X). A blue product was

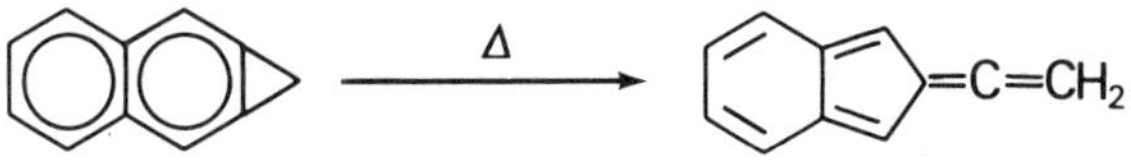

Scheme X

formed, which after condensation at 77 K or in an argon matrix exhibited an intense absorption at 1935 or 1952 cm^{-1}, respectively.[56] Further strong absorptions occur at 1483, 910, 837, and 799 cm^{-1} (in the argon matrix). The technique of VTPES yielded at 500°C and about 0.1 mbar (after subtraction of the PE spectrum of small amounts of unreacted **39**) the PE spectrum of Figure 8-7a. As for isobenzofulvene (**36**), the excellent agreement between the measured PE spectrum and the calculated ionization spectrum (Figure 8-7b) provides a clear structural proof of **38**. Shakeup ionizations are negligible for the interpretation of the PE spectrum; the lowest one occurs at 9.8 eV with I_{rel} of only 0.02 (therefore not shown in Figure 8-7b).[18a] The interpretation of the spectrum is further supported by the ion state correlation diagram (identical to the orbital interaction diagram if use of Koopmans's theorem is made) of appropriate subunit ion states (MOs). For completeness, the diagram is extended to isobenzofulvene (**36**) in Figure 8-8.

c. Hetero-o-Xylylenes and Related Compounds

o-Quinoid compounds of type **19** tend to participate either in thermal rearrangements leading to fulvene systems or in rapid thermal valence isomerisations to benzenoid systems **40** (depending on the type of heteroatoms) (Scheme XI; see also Scheme XX below). Most of the systems of types **19** or

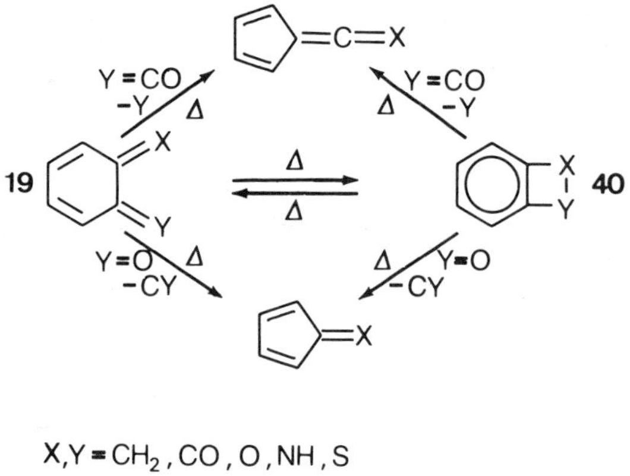

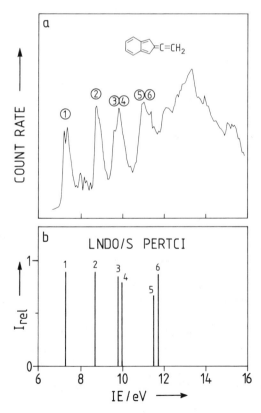

Scheme XI

Figure 8-7. (a) HeI photoelectron spectrum and (b) ionization spectrum of isobenzofulvenallene (2-vinylidene-2H-indene) (**38**) using the LNDO/S PERTCI method. The ionizations are assigned as follows: (1) $^2A_2(\pi)$, (2) $^2B_1(\pi)$, (3) $^2B_2(\sigma)$, (4) $^2A_2(\pi)$, (5) $^2B_1(\pi)$, and (6) $^2A_1(\sigma)$. (Reproduced with permission from Reference 56.)

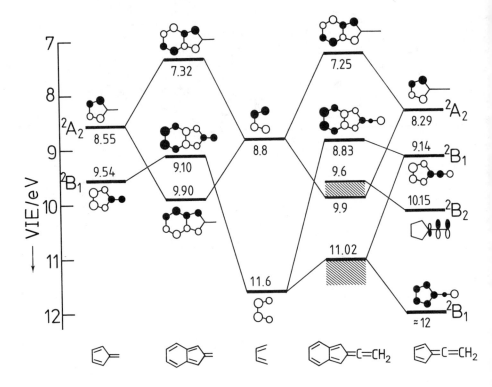

Figure 8-8. Ion state correlation diagram of fulvene (**93**), isobenzofulvene (**36**), butadiene, isobenzofulvenallene (**38**), and fulvenemethylene (**95**). The orbitals shown are those from which ionization occurs (in the frame of Koopmans' theorem).

40 are reactive intermediates that have not been studied much so far. Suitable FVP precursors are summarized in Scheme XII. For these routes, the amount of by-products is generally low (apart from small gaseous molecules such as H_2O, CO, CO_2, SO, SO_2, C_2H_4). Thus, the decomposition reactions depicted in Scheme XII are not only advantageous for synthetical purposes but also for spectroscopic investigations.

For example, applying the methods of Scheme XII, o-benzoquinonemethide (**41**) was thoroughly characterized at low temperature using MIT and IR spectroscopy[57] as well as applying the VTPES technique.[58] No hint of isomeric benzoxete (**42**) was found. At temperatures above $-50°C$ **41** rapidly

trimerizes. On the other hand, the parent oxete rapidly rearranges at room temperature to thermally more stable acrolein.[59]

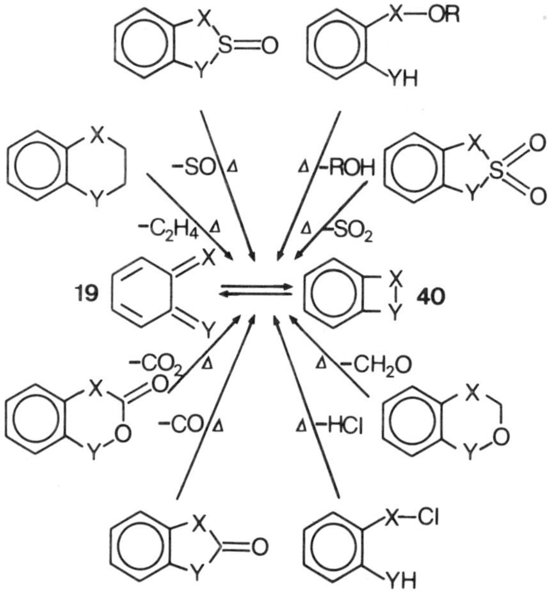

Scheme XII

Replacement of one methylene group (in *o*-xylylene) by an oxygen atom leads to a pronounced perturbation of the electronic structure. As the result of the inductive effect of the carbonyl group and its diminished conjugative ability with respect to the remaining π system, an increase in the VIEs of **41** was found; π ion states are observed at 8.8, 10.6, and 12 eV and the lone pair ion state at 9.3 eV. Neither the PE spectrum nor the calculated ionization spectra gave any evidence for low-lying shakeup states.

Another related *o*-quinoid intermediate is the oxoketene **43** (Scheme XIII). Long ago, **43** was assumed to be an intermediate in the pyrolysis of

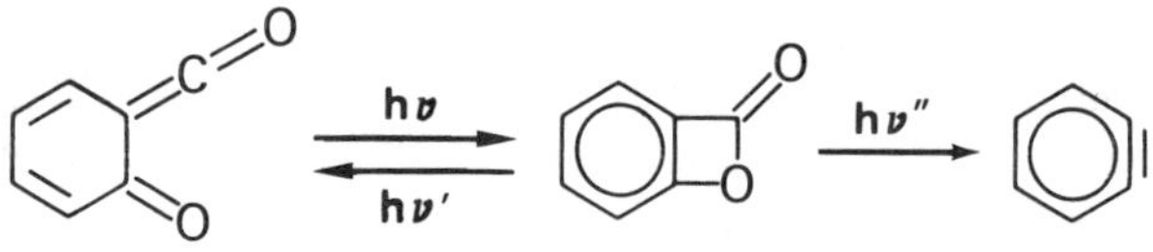

Scheme XIII

methyl salicylate. However, trapping experiments or MS analysis did not succeed in proving the existence of **43**. [Note that a ketene band found in the IR spectrum of the pyrolysis products of salicylic chloride does not permit one to distinguish **43** from its decomposition product 6-fulvenone (**44**); compare Subsection C].[60] On the other hand, low-temperature photochemical generation (in an argon matrix) and IR and UV-VIS spectroscopic characterization of **43** were possible.[61] Moreover, in this study a photochemical equilibrium between **43** and the isomeric lactone **45** was detected; **45** could be converted to *o*-benzyne (**46**) (Scheme XIII).

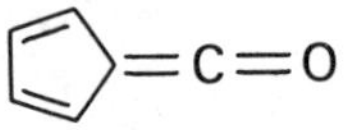

$$\text{44}$$

After all, it could be shown with the VTPES technique that free **43** is indeed the primary pyrolysis product of compounds **47** and **48**, at about 600°C and 0.1 mbar (Scheme XIV).[62] Figure 8-9 shows the PE spectrum of

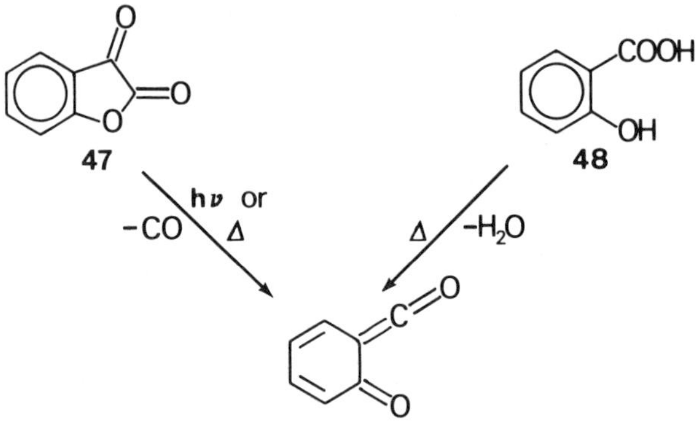

Scheme XIV

the pyrolysis products of the reaction **48**→**43** (with **43** as the principal product). The band pattern as well as the measured VIEs would not be explicable on the basis of a lactone structure (**45**). The measured VIEs at 8.43 [band 1, $^2A''(\pi)$], 9.38 [band 2, $^2A'(n)$], 10.31 [band 3, $^2A''(\pi)$], and 11.56 [band 4, $^2A''(\pi)$] eV are in good agreement with the corresponding calculated data using the semiempirical PERTCI methods. In addition, the IR spectra of the pyrolysis products from the reactions **47**→**43** and **48**→**43** in argon matrix confirm that the formation of **43** is quantitative in both reactions.[63]

The analogous thio compound **49** has not been detected until now. In fact, it was postulated as the thermolysis product of **50** and **51** because of the dimer **52** isolated in these reactions[64] (Scheme XV). However, the latter is formed from the thiolactone (**53**), at temperatures above 230 K (Scheme XV). Photochemically generated **53** (at 77 K) was identified from its carbonyl absorption at 1803 cm^{-1} in the IR spectrum.[65] The PE spectrum recorded (applying the VTPES technique to **50** and **51**; the spectrum of the pyrolysis products from the reaction **51**→**53** is shown in Figure 8-10) can be unequivocally assigned to **53**, based on comparisons of measured with calculated

UV-PES/MATRIX ISOLATION APPROACH   313

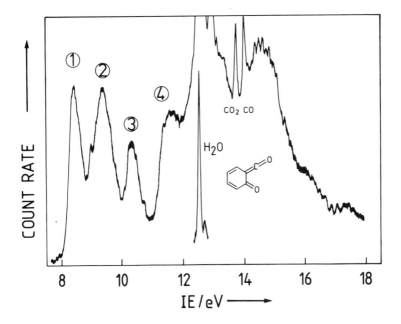

Figure 8-9. HeI photoelectron spectrum of the pyrolysis products (essentially the oxoketene **43**) from salicylic acid (**48**).

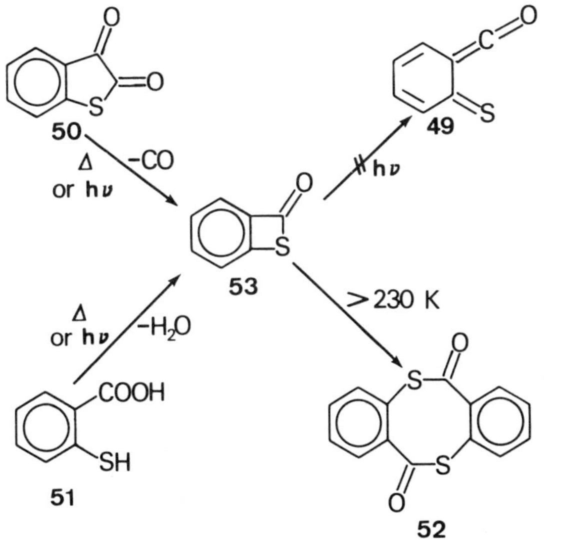

Scheme XV

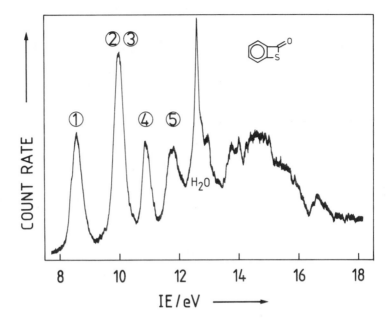

Figure 8-10. HeI photoelectron spectrum of the pyrolysis products (essentially the thiolactone **53**) from thiosalicylic acid (**51**).

semiempirical PERTCI VIEs, of measured VIEs with corresponding data from the reference compound **54**, and of the HeI and HeII PE spectra of **53**.[21,62,63] The IR spectra in an argon matrix are also available (carbonyl absorption at 1828 cm^{-1}).[63] Neither in the VTPES pyrolysis nor after the irradiation of matrix-isolated **53**, could any hints to the formation of **49** and/or **46** be obtained. Instead, when raising the pressure in the pyrolysis region, thioxanthone (**55**) is observed during the gas phase pyrolysis of **50**.[66] The latter reaction is analogous to the formation of xanthone in the gas phase pyrolysis of **47** at higher pressures.[67]

Following the routes of Schemes XIV and XV, the analogous naphthoannelated compounds **56** and **57** are accessible, too. In addition, lactone **58** was detected.[68] Obviously, additional annelation favors the lactone over the ketene structure.

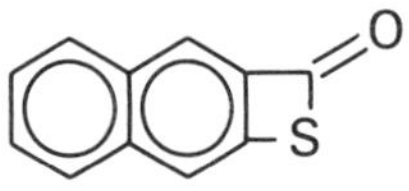

56 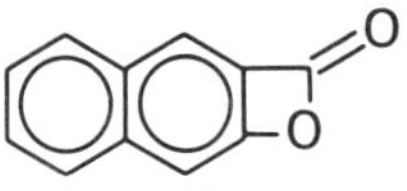 **58**

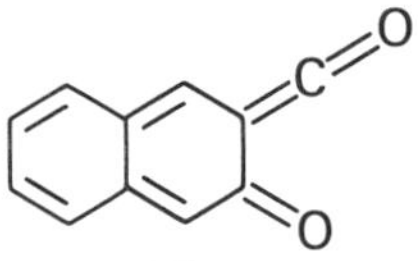

57

Thio-*o*-quinonemethide (**59**) is thermodynamically less stable than its valence isomer, benzo[*b*]thiete (**60**). Therefore **59** is rapidly converted to **60** and has never been observed directly before (see also Subsection C).[69] However, when argon matrix isolated benzo[*b*]thiete (**60**) is irradiated at 10 to 20 K with UV light at 280 nm, the matrix gets dark red, and the IR bands of **60** [1450 (s), 1045 (s), 733 (vs), and 709 (s) cm^{-1}] lose intensity. The dark red photoproduct [with IR absorptions at 1410 (w), 1363 (w), 1210 (w), 1163 (w), 1147 (w), 995 (w), 933 (w), and 756 (s) cm^{-1}] is reversibly reconverted to **60** on irradiating at wavelengths exceeding 300 nm within seconds[70] (Scheme XVI). Upon irradiation or heating in solution, the intermediate **59** can be

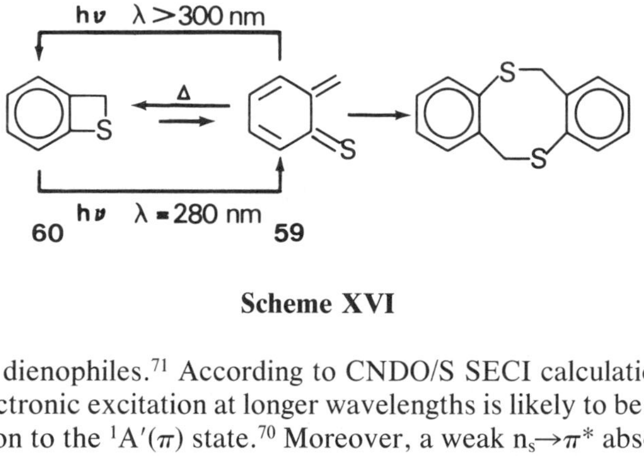

Scheme XVI

trapped by dienophiles.[71] According to CNDO/S SECI calculations, the intense π-electronic excitation at longer wavelengths is likely to be ascribed to the transition to the $^1A'(\pi)$ state.[70] Moreover, a weak $n_s \rightarrow \pi^*$ absorption can be expected to occur at about 500 to 600 nm (as in other thiones).

A very similar photochemical transformation was observed for the system thio-*o*-benzoquinone (**61**)⇌benzoxathiete (**62**).[70] Upon UV irradiation of argon matrix isolated **63** (Scheme XVII) at 280 nm, **63** decomposes to CO, yellow-green **61**, and small amounts of colorless **62**. Further irradiation at wavelengths exceeding 370 nm yields quantitatively **62**, which at λ = 280 nm is predominantly reconverted to **61** (photostationary equilibrium). Photolysis at 254 nm yields the biscumulene **64** (Scheme XVII).[70] The products were identified from their characteristic IR absorptions (in cm^{-1}): **61**: $\nu_{C=O}$ 1652

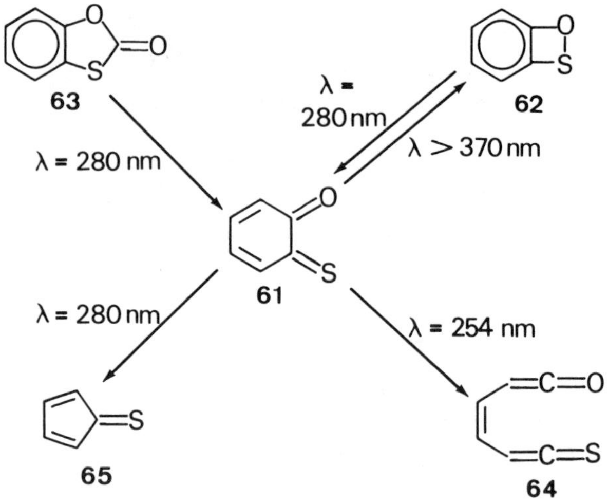

Scheme XVII

(s), **62**: $\nu_{C=O}$ 1200 (s), $\nu_{CH\ \delta_{oop}}$ 732 (s), and **64**: $\nu_{C=C=O}$ 2130 (s), $\nu_{C=C=S}$ 1760, 1750 (s). It is interesting to note that the photolysis of the 5-methyl derivative of **63**, in an alcohol glass, produced only small amounts of the corresponding oxathiete. At present it cannot be answered whether this result is due to an effect of medium or of temperature. After prolonged irradiation of **61** or **63** at 280 nm, the additional IR absorptions of cyclopentadienethione (**65**, see Subsection C) are found. It is formed by extrusion of CO from **61**. The best way, however, to synthesize **65** is FVP (see Subsection C). CNDO/S SECI calculations predict, for the lowest energy π transition, an absorption at 300 nm for **62** and at 350 nm for **61**. The colors observed as well as the measured UV-VIS spectra of the alkyl-substituted derivatives[72] essentially agree with the predictions.

Contrary to the thio-*o*-benzoquinonemethide (**59**) case, thio-*o*-benzoquinone (**61**) can also be generated thermally because it is markedly more stable than benzoxathiete (**62**) (standard enthalpies of formation for **61** and **62** using the MNDO method: 21 and 40 kcal/mol, respectively. The IR spectra of the pyrolysis products of **63**, **66**, and **67** exhibit the absorption bands

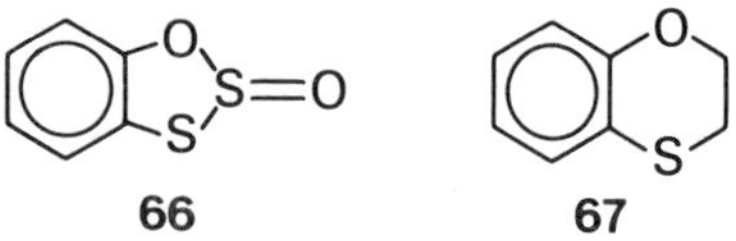

typical of thio-*o*-benzoquinone. Subsequent thermal decarbonylation leads to cyclopentadienethione (**65**, see below). The PE spectrum of **61** (recorded using the VTPES method) exhibits ionizations at 8.85 eV [$^2A''(\pi)$], 9.45 eV [$^2A''(\pi)$ and $^2A'(n_s)$], and at energies higher than 11.6 eV [$^2A''(\pi)$].

Thermal or photochemical generation of dithio-*o*-benzoquinone (**68**), and thus the spectroscopic characterization of this compound, in analogy to the routes shown in Scheme XVII, was not possible. According to MNDO calculations, benzodithiete (**69**) is more stable than dithio-*o*-benzoquinone (**68**) by about 25 kcal/mol. In harmony with this, the thermal decomposition of suitable precursors leads exclusively to **69** (see Subsection E). In addition, benzodithiete (**69**) isolated in an argon matrix was photostable; neither ring opening nor extrusion of S_2 occurred (Scheme XVIII).

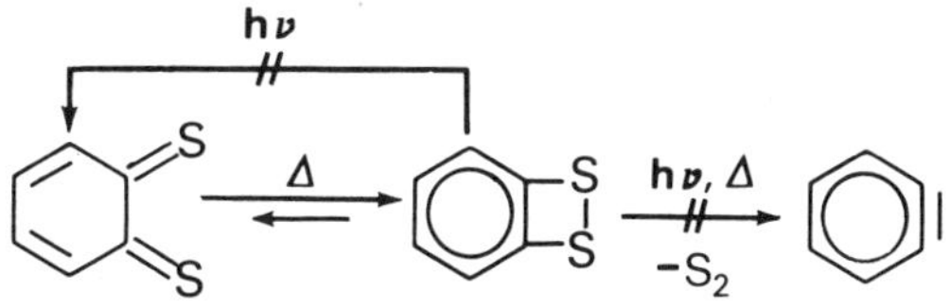

Scheme XVIII

The chemical behavior of some related systems containing nitrogen has been known for some time. Thus pyrolysis of **70** yields both iminoketene **71** and lactam **72** (identified after condensation of the products at low temperatures by IR spectroscopy: bands at 2080 and 1710 cm^{-1} and 1825 and 1710 cm^{-1}, respectively).[73] Secondary thermal decarbonylation finally yields the cyclopentadiene derivative **73**.[73] Obviously, the stability difference between

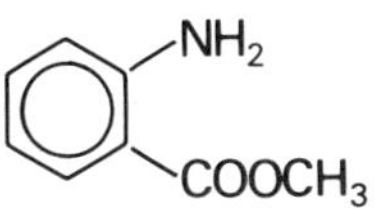

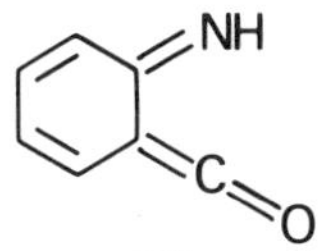

the open (ie, *o*-quinoid) and closed (ie, benzenoid) structures is, for the benzazete case, just in between the benzoxete and benzothiete cases. The results of EHMO (extended Hückel MO)[74] calculations give this tendency right; however, the stability of the *o*-quinoid systems is drastically overrated.[75] MNDO calculations are much better, insofar as the "stability differences" (ie, the difference in the standard enthalpies of formation or the standard reaction enthalpies of the reaction benzenoid system →*o*-quinoid system) are obtained with the correct sign; however, the absolute values are

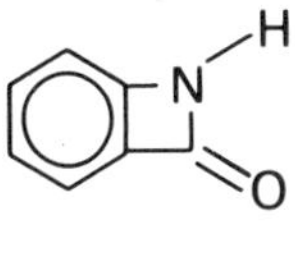

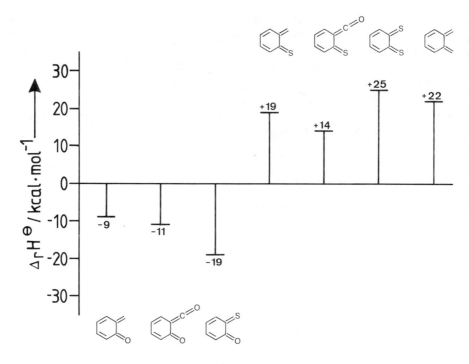

Figure 8-11. MNDO standard reactions enthalpies ($\Delta_r H^\ominus$) of the reaction closed (benzenoid) form → open (o-quinoid) form (ie, differences in "stability") for o-xylylene → benzocyclobutene and the various related systems discussed in the text (**41**, **57**, **61**, **59**, **49**, **68**, and **22**).

too high[76] (see Subsection D). Figure 8-11 shows the MNDO "stability difference" for the various systems discussed above. There are further particularly interesting o-quinoid systems (eg, **74**, **75**, and **76**), which cannot be

discussed here in detail. Only one actual example of the structural type **76**, namely 2,3-dimethylene-2,3-dihydrofuran (**76a**), is presented briefly. Three thermal pathways to **76a** are available at present: extrusion of ethylene from 4,5,6,7-tetrahydrobenzofuran (**77**),[77] of benzoic acid from 2-methyl-3-furylmethylbenzoate (**78**),[78] and most elegantly and for spectroscopic purposes most appropriately, of HCl from 2-methyl-3-chloromethylfuran (**79**) (Scheme XIX).[76] Little was known about **76a** (indeed, only that it exists).[76,78]

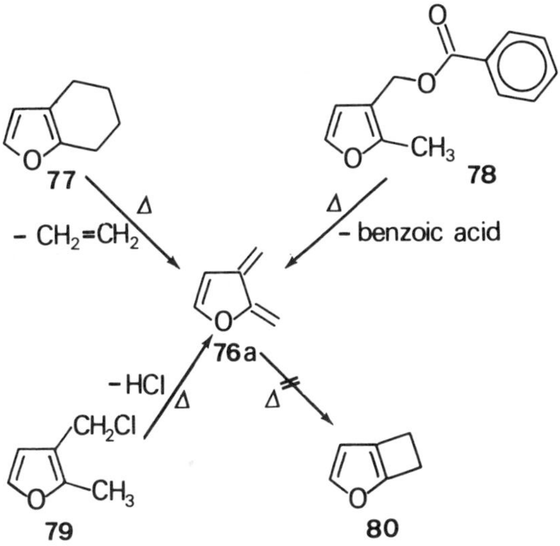

Scheme XIX

Meanwhile, the PE spectrum (applying the VTPES technique) and the IR and UV-VIS spectra in an argon matrix (making use of FVP MIT) are available.[76] The PE spectrum is shown in Figure 8-12 together with the LNDO/S PERTCI ionization spectrum. Ionizations 3 and 4 are essentially a mixture of a Koopmans and the HOMO LUMO non-Koopmans ionization. Both the shape and the reduced overall intensity of the second band system (consisting of bands 2 and 3) are in accordance with the theoretical prediction. The UV-VIS spectrum exhibits a vibrationally well-resolved first absorption band at 318 nm, which is very similar to the first absorption band of o-xylylene (**22**).[47] The UV-VIS spectrum is well reproduced by LNDO/S SECI, LNDO/S PERTCI, and HAM/3 electronic excitation energies and oscillator strengths (Subsection D). The chemically most relevant result, however, is: there is no chance to generate furanocyclobutene (**80**) from **76a** thermally.[76] Compound **76a** is much more stable than **80**. This is fundamentally different from the analogous o-xylylene⇌benzocyclobutene case (**22**⇌**23**, see Subsection D).

C. Fulvenelike Systems

It has already been mentioned (see Scheme XI, above) that compounds of types **19** and **40**, with X or Y a carbonyl group or an oxygen atom, undergo

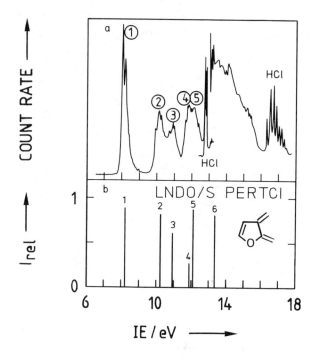

Figure 8-12. (a) HeI photoelectron spectrum and (b) LNDO/S PERTCI ionization spectrum of 2,3-dimethylene-2,3-dihydrofuran (**76a**) The ionizations are assigned as follows: (1) $^2A''(\pi)$, (2) $^2A''(\pi)$, (3) $^2A''(\pi)$, (4) $^2A''(\pi)$, (5) $^2A'(\sigma)$, and (6) $^2A'(\sigma)$. (Reproduced with permission from Reference 76.)

facile ring contraction to fulvenelike compounds **81** or **82** (for a comprehensive collection of the various reaction pathways, see Scheme XX).

In addition, as Scheme XX indicates, thermal and photolytic fragmentations of compounds of type **83** make further reactive intermediates easily

$X = CH_2, O, S, Se$

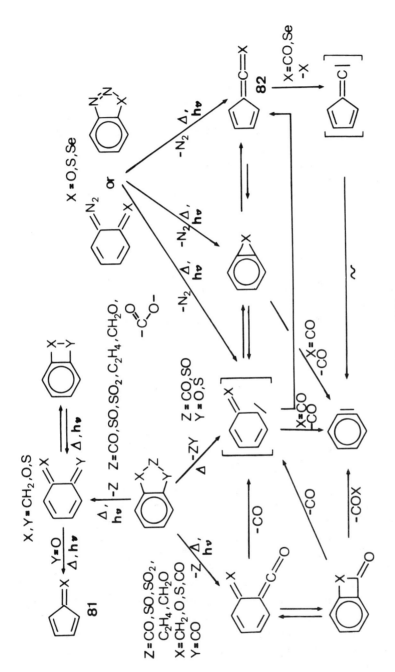

Scheme XX

accessible, such as the systems just discussed in Subsection B, plus o-benzyne, benzothiirene, and benzoselenirene, which will be discussed in the Subsection D.

a. Cyclopentadienone

Cyclopentadienone (**8**) is a highly reactive intermediate that can be studied only in the gas phase at low pressures or isolated in a solid matrix at very low temperatures. Suitable reactions for the thermal generation of **8** (by FVP) are collected in Scheme XXI, making use of the precursors: o-quinone (**84**),[79] 3-cyclobutene-1,2-dicarboxylic acid anhydride (**85**),[30] benzodioxane (**86**),[80] o-phenylene carbonate (**87**),[81] o-phenylene sulfite (**88**),[79] o-phenylene sulfate (**89**),[63] bicyclobutane-3,4-dicarboxylic acid anhydride (**90**),[82] and the acetate **91**.[83] Detection of the dimer of cyclopentadienone (**8**) has been reported.[79–81] Free cyclopentadienone (**8**) was directly seen (spectroscopically detected) by microwave spectroscopy using the reaction **84**→**8**[84] and by photoelectron spectroscopy (by use of the VTPES technique) making use of the decomposition reactions **84**→**8**,[30,85] **85**→**8**,[30] **88**→**8**,[30,85] **86**→**8**,[70] and **89**→**8**.[63] Cyclopentadienone (**8**) isolated in a matrix (using VT MIT) was investigated by IR[63,86,87] and UV-VIS[86] spectroscopy making use of the various reactions of Scheme XXI and, additionally, of the decomposition of **92**.[87]

X = O, CH$_2$

92

The photoelectron spectrum of **8** exhibits bands at 9.4 eV [$^2A_2(\pi)$], 9.9 eV [$^2B_2(n_0)$], and 11.8 eV [$^2B_1(\pi)$].[30,85] In the frame of Koopmans' theorem, the first VIE is in accord with a pure inductive effect exerted by the carbonyl group on the adjacent butadiene π orbital (ie, 0.8 eV if the HOMO of cyclopentadiene or fulvene is taken as reference system). This conclusion is based on previous work carried out in our laboratory, where it was shown that the inductive effect of a carbonyl group on the energy of a π orbital, adjacent to the group, amounts to 0.7 to 0.8 eV (as derived from VIE measurements in the frame of Koopmans' theorem).[88] The second (n_0) VIE has a magnitude comparable to that of other ketones. The third VIE is comparably high, in harmony with the low-lying $b_1(\pi)$ MO of, say, formaldehyde. These findings, as well as the measured dipole moment of 3.13 debyes[84] (not very different from the one of cyclopentanone: 3.30 D) vote for only poor orbital interactions between the π orbitals of the butadiene and carbonyl group subunits, hence for poor energetic and structural effects due to such interactions (for a more quantitative estimation, see below).

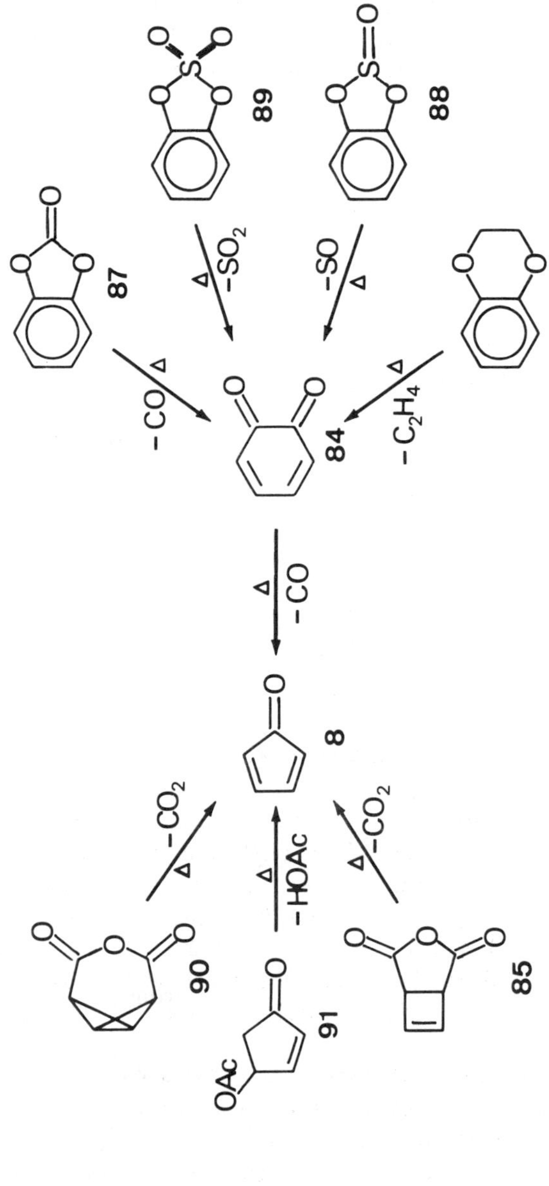

Scheme XXI

b. Cyclopentadienethione

Some years ago, hints to the existence of alkyl-substituted cyclopentadienethiones, based on UV-VIS absorptions in a solid organic glass at 77 K, were communicated.[72] Shortly afterward, free unsubstituted cyclopentadienethione (**65**) was detected using the VTPES method and thus characterized for the first time.[89] The thermal reactions used for this purpose (Scheme XXII) are similar to corresponding reactions of Scheme XXI. In addition,

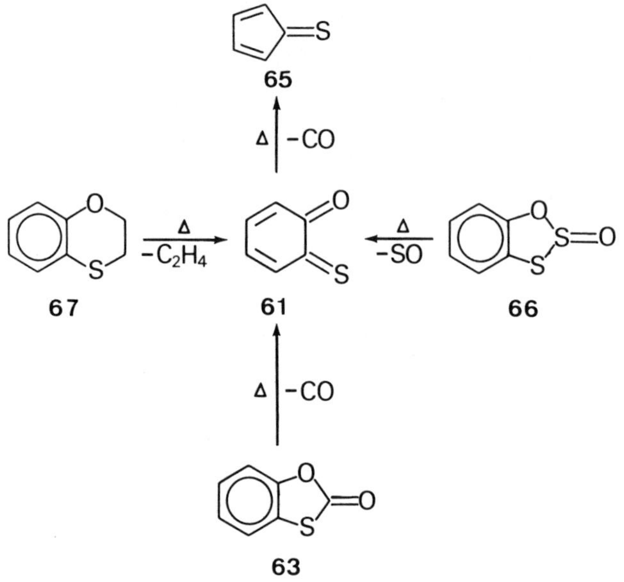

Scheme XXII

cyclopentadienethione (**65**) was isolated in an argon matrix and investigated by IR spectroscopy. The PE spectra as well as the IR spectra of the decomposition products of the three different precursor compounds of Scheme XXII are identical (except for the trivial by-products SO and its secondary products SO_2 and S_2O, CO, or C_2H_4, respectively) and exclude thiobenzoquinone (**61**) as the major final product, which is strongly indicative of cyclopentadienethione as the common product. For the PE spectra, see Figure 8-13.

The first photoelectron band of **65** occurs at 8.87 eV [$^2B_2(n_S)$]. Both for its location and shape, it is typical of an n_S (thiocarbonyl) ionization (and this assignment is fully backed by the observed relative intensity changes when going from HeI to HeII excitation, see Figure 8-13b). The second band [at 9.18 eV, $^2A_2(\pi)$] occurs—at only slightly lower (by 0.2 eV) energy, due to the slightly lower inductive effect of the thiocarbonyl group (compared to the carbonyl group).[88c] Finally, the observed position of the third photoelectron

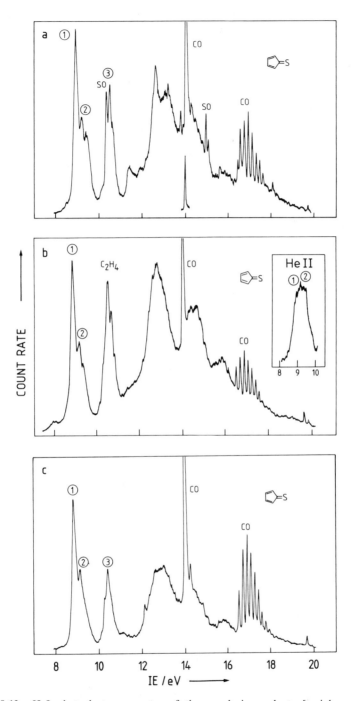

Figure 8-13. HeI photoelectron spectra of the pyrolysis products [mainly cyclopentadienethione (**65**)] of (*a*) 1,2,3-benzoxadithiol-2-oxide (**66**), (*b*) 1,4-benzoxathiene (**67**), and (*c*) 1,3-benzoxathiol-2-one (**63**). The insert in (*b*) is the HeII spectrum of the pyrolysis products of **67** shown in the range of the first two ionization bands. (Reproduced with permission from Reference 89.)

band [at 10.35 eV, $^2B_1(\pi)$] is in excellent agreement with the position predicted from the empirical relation[63,90]:

$$\text{VIE } [^2B_1(\pi), \text{\small⬠}=X] = (3.12 \pm 0.08) \text{ eV} + (0.61 \pm 0.01)$$
$$\cdot \text{VIE } [^2B_1(\pi), CH_2=X]$$

(for the meaning of X as well as a graphical representation of the relation, see Figure 8-14). MNDO PERTCI and CNDO/S PERTCI calculations have confirmed the empirical (however, from chemical point, most relevant) interpretation of the PE spectrum of **65**.[89]

Figure 8-15 shows the IR spectrum of the reaction products (of the thermal decomposition reaction **66→65** using FVP MIT) isolated in an argon matrix at 10 K. The spectrum exhibits bands of CO (2138 cm^{-1}, not shown in Figure 8-15), SO (1138 cm^{-1}), and its secondary products SO_2 (1352 cm^{-1}) and S_2O (1157 cm^{-1}), and, moreover, absorptions at 3140 (w), 3098 (w), 1358 (s), 1250

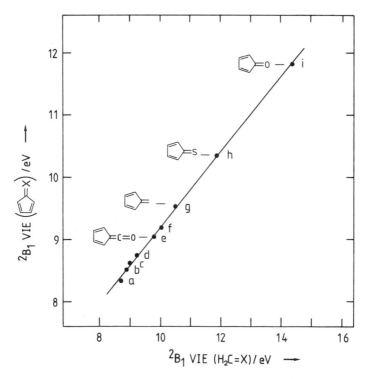

Figure 8-14. $^2B_1(\pi)$ vertical ionization energies (VIE) of fulvenelike molecules as a function of the $^2B_1(\pi)$ vertical ionization energy of the respective $H_2C=X$ systems {fitted line: VIE $[^2B_1(\pi),$ ⬠$=X] = (3.12 \pm 0.08)$ eV $+ (0.61 \pm 0.01) \cdot$ VIE $[^2B_1(\pi), CH_2=X]$}: a = 6-fulveneselone (**97**), b = 6-fulvenethione (**96**), c = diazocyclopentadiene (**94**), d = 6,6-dimethylfulvene, e = 6-fulvenone (**44**), f = fulvenemethylene (**95**), g = fulvene (**93**), h = cyclopentadienethione (**65**) and i = cyclopentadienone (**8**). (Reproduced with permission from Reference 90.)

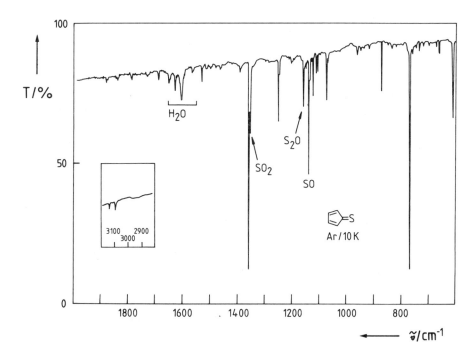

Figure 8-15. IR absorption spectrum of the pyrolysis products [mainly cyclopentadienethione (**65**)] of 1,2,3-benzoxadithiole-2-oxide (**66**) isolated in an argon matrix at 10 K. The absorption band of CO (2138 cm^{-1}), which is a further product of the decomposition of **66**, is located outside the range shown.

(m), 1122 (m), 1074 (m), 875 (m), 770 (s), and 612 (s) cm^{-1}, which are ascribed to cyclopentadienethione (**65**). In the region of the thiocarbonyl absorption (1000–1200 cm^{-1}), there is no intense absorption band. From this point of view and simply because thiocarbonyl IR absorptions strongly vary in intensity and position (see above), the IR spectrum is less appropriate for a first identification of **65** than is the PE spectrum. A Raman spectrum of **65** would be desirable. However, the IR matrix spectrum provides us with valuable additional information: **65** is stable in the matrix up to 100 K, which means that **65** appears to be more stable (less reactive) than cyclopentadienone (**8**), which dimerizes at temperatures as low as 40 K.[63,86] Both the MIT experiments (regarding the stability of **65** and **8**) and PE spectra [regarding the extent of orbital interactions as derived from the $^2B_1(\pi)$ ionization, cf Figure 8-14] suggest that cyclopentadienethione (**65**) occupies some medium position between cyclopentadienone (**8**) and fulvene (**93**). This result can be further corroborated by a simple quantum chemical estimation.[88a,c,91] These systems [and, in addition, 6-fulvenone (**44**) and diazocyclopentadiene (**94**), see below] were subject to two calculations (using the CNDO/S procedure), first a normal one (thereafter referred to as "a calculation without interruption of conjugation") and second a calculation when the $p_\pi p_\pi$ overlap integrals of the C—C bonds connecting the butadiene and C=X subunits are set to zero (thereafter termed "a calculation with interruption of conjugation"). Then the changes in the π charge of both subunits (expressed in terms of the π charge transfer Δq from the exocyclic π systems to the butadiene π system) produced by "switching conjugation on" and the corresponding gain ΔE in total energy are evaluated. These data are listed in Table 8-2, along with the experimental VIE [$^2B_1(\pi)$] of CH_2=X and the MNDO bond lengths r, for the various systems considered.[63] As indicated, the data in the four rows of Table 8-2 closely parallel each other in a sensible way, clearly confirming that the conjugative interaction between the respective subunits increases from cyclopentadienone (**8**) to 6-fulvenone (**44**) or diazocyclopentadiene (**94**), and, in particular, that cyclopentadienethione (**65**) is expected to possess electronic properties somewhat intermediate between **8** and fulvene (**93**).[63]

c. Fulvenemethylene and Heterofulvenemethylenes

Whereas fulvenemethylene (**95**) and diazocyclopentadiene (**94**) are fairly stable compounds, 6-fulvenone (**44**),[92] 6-fulvenethione (**96**),[93] and 6-fulvenese-

TABLE 8-2. $^2B_1(\pi)$ Vertical Ionization Energy (VIE) of CH_2O, CH_2S, C_2H_4, CH_2CO, and CH_2N_2, π Charge Transfer from the Exocyclic π System into the Butadiene π System (Δq) Using the CNDO/S Method in Conjunction with the Method of Conjugation Interruption (see text), Decrease in Energy as a Result of Conjugative π Coupling (ΔE), and MNDO-Optimized Bond Lengths (r) for Cyclopentadienone (**8**), Cyclopentadienethione (**65**), Fulvene (**93**), 6-Fulvenone (**44**), and Diazocyclopentadiene (**94**)

	8 (O)	65 (S)	93 (CH$_2$)	44 (C=O)	94 (N=N)
VIE (CH$_2$=X) (eV)	14.38	11.89	10.51	9.80	9.00
Δq (units of an electron charge)	−0.01	0.02	0.04	0.14	0.14
ΔE (kcal/mol)	23	30	30	39	39
r(pm) o	151.7	149.9	149.1	147.5	146.9
p	135.8	136.2	136.6	137.3	137.7
q	148.9	148.9	147.7	146.7	146.3

lone (**97**)[90] are reactive intermediates, which can directly be observed only in the gas phase at low pressures or isolated in matrices at low temperatures.

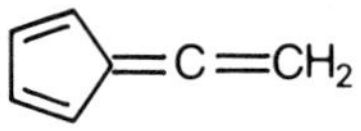

95

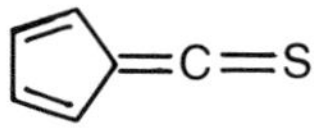

96

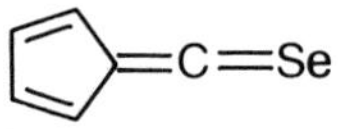

97

Scheme XXIII presents reactions that are particularly useful for the generation and PE and IR spectroscopic characterization of **44** [**43**→**44**,[62,63] **98**→**44**[62,63,94] (this reaction was known before; however, **44** was only indi-

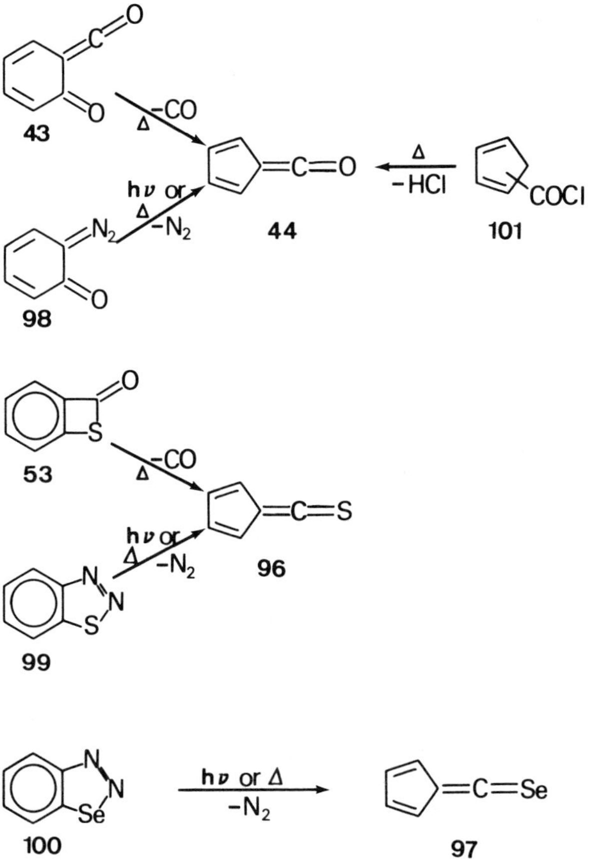

Scheme XXIII

rectly confirmed[95]), **101→44**[96]], **96** [**53→96**,[62,63] **99→96**,[62,63,97] (this reaction was known before; however, **96** was only indirectly confirmed[93])], and **97** (**100→97**[90]).

Isolated at low temperatures, 6-fulvenone (**44**) is colorless, 6-fulvenethione (**96**) red, and 6-fulveneselone (**97**) purple. UV-VIS absorption of **44** occurs at 262 nm (4.73 eV).[92a] According to CNDO/S SECI calculations, this absorption is assigned to the $^1A_1(\pi\pi^*)$ excited state (predicted at 4.22 eV with $f = 1.29$).[63] For **96** a strong absorption was found at 325 nm (3.82 eV), and a very weak one at 545 nm (2.27 eV).[63] The latter is responsible for the red color of **96**. The theoretical data using the CNDO/S SECI method, $^1A_1(\pi\pi^*)\lambda = 353$ nm (3.51 eV) with $f = 0.58$ [$^1A_1(\pi\pi^*)$] and $\lambda = 535$ nm (2.32 eV) with $f = 0$ [$^1A_2(n\pi^*)$, symmetry forbidden], are in fair agreement with the observed data.[63] For **97**, neither experimental nor theoretical UV-VIS data have been determined so far.

6-Fulvenone (**44**) was repeatedly identified by means of its ketene vibration [in N_2 matrix at 10 K (2130 cm^{-1}) and at 77 K (2120 cm^{-1})].[92a,b] Further absorptions have been reported at 1445 (m), 1325 (m), 898 (m), 578 (w), and 520 (w) cm^{-1}.[92c] Four more absorptions (in argon matrix at 10 K after both thermal and photolytic generation of **44**) were found at 1582 (w), 1459 (m), 1406 (m), and 725 (s) cm^{-1}.[63] For 6-fulvenethione (**96**) isolated in an argon matrix, IR absorptions were observed at 1760 (vs), 1737 (s), 1700 (s), 1450 (s), 1370 (m), 1307 (m), 1155 (m), 1077 (m), 1056 (m), 870 (s), and 737 (s) cm^{-1}.[63] IR bands ascribed to benzothiirene, when radiation having a wavelength of 265 nm was used in photolysis experiments,[97] did weakly occur under pyrolytic conditions[98] (see Subsection D). The IR spectrum of **97** isolated in an argon matrix is available, too.[98] The antisymmetric C—C—Se stretching mode occurs at 1710 cm^{-1}; further bands lie at 1685 (w), 1650 (m), 1450 (s), 1360 (w), 1105 (w), 1033 (m), 843 (s), 733 (s), and 590 (w) cm^{-1}.[98] It is interesting to note that the IR spectrum of **97** is quite similar to the one of **96**, and that the spectrum of **97** does not exhibit any of the known bands of *o*-benzyne, showing that the elimination of N_2 from **100** without loss of Se is a clean reaction leading to **97** as the sole product.

Figure 8-16 presents the experimental and MNDO PERTCI ion state correlation diagrams for 6-fulveneselone (**97**), 6-fulvenethione (**96**), 6-fulvenone (**44**), fulvenemethylene (**95**), cyclopentadienethione (**65**), and cyclopentadienone (**8**). The agreement between both sorts of diagram is as good as can be expected on the basis of semiempirical calculations, hence it offers firm and convincing additional confirmation of all the identification work described above.

The unstability of the above-mentioned ketenes increases in the order of **44**, **96**, and **97**. At −120°C, **96** is no longer stable as a monomer. At markedly lower temperatures ($\approx$ −190°C), **97** can react spontaneously (explosively) to give a brown polymer.[90]

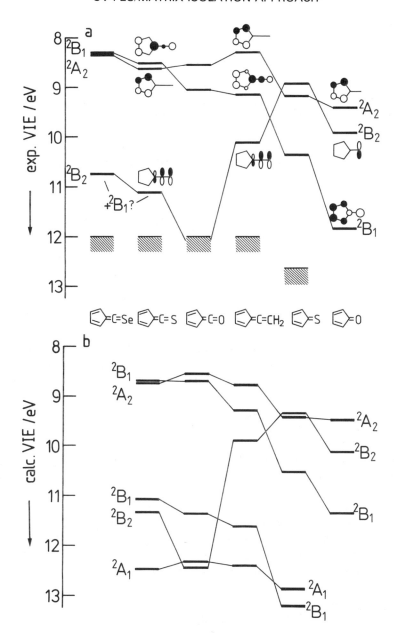

Figure 8-16. Ion state correlation diagrams of (*a*) measured (exp) and (*b*) MNDO PERTCI (calc) vertical ionization energies (VIE) for 6-fulveneselone (**97**), 6-fulvenethione (**96**), 6-fulvenone (**44**), fulvenemethylene (**95**), cyclopentadienethione (**65**), and cyclopentadienone (**8**).

D. Strained Systems

a. Benzothiirene and Benzoselenirene

Nowadays, the formation of thiirenes on photolysis of 1,2,3-thiadiazoles is well documented (among other things, by isotope scrambling).[99] For a long time, however, the question of whether thiirenes can also originate as a result of thermal cleavage of 1,2,3-thiadiazoles, had been unsettled.[100]

In particular the controversy regarding thermolysis of 1,2,3-benzothiadiazole (**99**) to form benzothiirene (**102**) (Scheme XXIV) has been discussed.[100b,101]

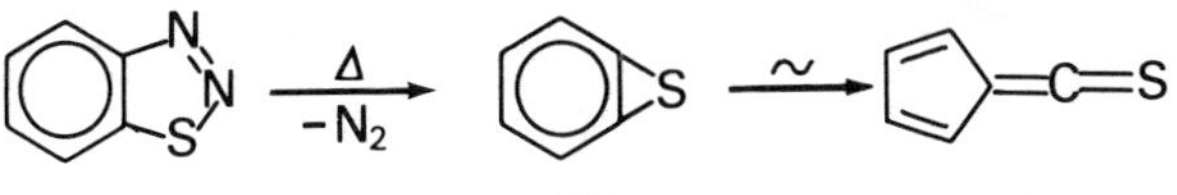

Scheme XXIV

Our knowledge about selenirenes is even less complete. IR spectra recorded after photolysis of matrix-isolated 1,2,3-selenadiazole indicate the formation of selenirenes.[102] On the other hand, thermolysis of bicyclic 1,2,3-selenadiazoles in liquid solution gives cycloalkynes,[103] and gas phase pyrolysis provides preponderantly ring-contracted cyclic selenoketenes.[104] In the case of unsubstituted 1,2,3-selenadiazole or cycloocteno-1,2,3-selenadiazole, gas phase pyrolysis at higher temperatures gives also, besides selenoketenes, acetylene or cyclooctyne, respectively. A microwave spectroscopic study on the pyrolysis products of isotopically labeled 1,2,3-selenadiazole was compatible with the intermediacy of selenirene as a metastable product; however, direct microwave transitions of selenirene could not be observed.[105]

These experiences suggested the investigation of the pyrolysis of 1,2,3-benzothiadiazole (**99**) and both the photolysis and pyrolysis of 1,2,3-benzoselenadiazole (**100**) [giving 6-fulvenethione (**96**) and 6-fulveneselone (**97**), cf. the foregoing section] for the intermediate formation of benzothiirene (**102**) and benzoselenirene (**103**), respectively (Scheme XXV).

Scheme XXV

Thermal cleavage of **99** under mild conditions using the VTPES technique resulted in the appearance of a PE band that could be ascribed neither to the precursor nor to the product, 6-fulvenethione (**96**), although as judged from its location and shape, it could be due to benzothiirene (**102**).[106] The pyrolysis products were then isolated in an argon matrix at 10 K and the IR spectrum was recorded[98]; the IR spectrum exhibited just those absorptions which had been previously ascribed to benzothiirene (**102**) generated in the photolysis of **100** with 254 nm radiation.[98] Obviously, these results strongly suggest that benzothiirene occurs also as a reactive intermediate in the gas phase pyrolysis of 1,2,3-benzothiadiazole (**99**). On the other hand, thermolysis experiments carried out in liquid solution (using isotopically labeled or otherwise marked 1,2,3-benzothiadiazoles) gave contradictory results,[100b,101] suggesting that the reaction depends very sensitively on the conditions applied.

VTPES experiments on 1,2,3-benzoselenadiazole (**100**) under appropriate conditions led to the appearance of an intermediate PE spectrum distinctly different from the PE spectra of the starting material **100** and of the final product, 6-fulveneselone (**97**); the spectrum was ascribed to benzoselenirene (**103**).[106] MIT provided further clear-cut evidence for the novel reactive species.[98] As it was shown, the same species with IR bands at 1443 and 1435 (s), 1156 (m), 1052 (m), 973 (m), 940 (m), 726 (vs), and 655 (s) cm^{-1} was formed, both on irradiation of matrix-isolated **100** and by FVP of **100**. This thermal and photochemical behavior of 1,2,3-benzoselenadiazole (**100**) perfectly parallels the behavior of 1,2,3-benzothiadiazole (**99**). From this and, in particular, from the great similarity of the IR spectra of the intermediates **102** and **103**, there is good reason to assign the structure of benzoselenirene to the reactive intermediate occurring in the photochemical and thermal decomposition of **100**.

Concerning benzoxirene (**104**, not yet detected) a short remark might be added. The MNDO standard reaction enthalpies for the reactions benzoxirene (**104**)→6-fulvenone (**44**) and benzothiirene (**102**)→6-fulvenethione (**96**) amount to −71 and −14 kcal/mol, respectively. There is good reason to assume that the corresponding reaction benzoselenirene (**103**)→6-fulveneselone (**97**) is even less exothermic. Thus these data may provide some basis for understanding why the reactive species **102** and **103** have been characterized, but until now, **104** has not.

104

b. o-Benzyne

The existence of *o*-benzyne (**46**) in the gas phase was proved many years ago, based on self-trapping, to essentially give biphenylene, trapping with external agents, MS, time-resolved MS, time-resolved UV-VIS absorption spectroscopy.[107,108] For a compilation of literature, in particular also for methods for photochemical and thermal generation of **46** using various precursors, the reader is referred to Reference 109. In the early physicochemical studies (UV-VIS spectrum in the gas phase,[107] IR spectrum in an argon matrix,[110,111] rate constant for the dimerization of **46**[108,112]), **46** was always made by photolysis.

In 1978 several attempts were made to subject to gas phase pyrolysis the precursors phthalic anhydride (**105**), indanetrione (**106**), phthaloylperoxide (**107**), o-sulfobenzoic anhydride (**109**) (cf. Scheme XXVI), and in addition

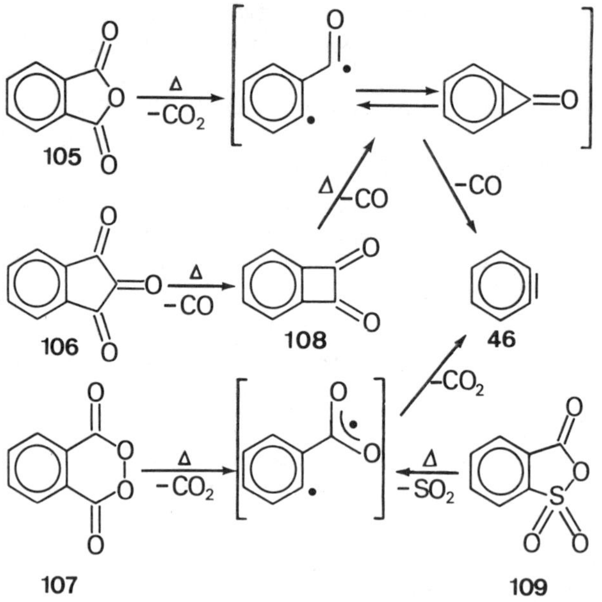

Scheme XXVI

1,2,3-benzothiadiazole-1,1-dioxide (not shown in Scheme XXVI), and to record the PE spectrum of o-benzyne (using the VTPES technique).[113] 1,2,3-Benzothiadiazole-1,1-dioxide exploded on gentle heating in the spectrometer; however, the PE spectra from the four other precursors consistently contained a band system between about 8.5 and 10.5 eV, a region that in all cases is free from photoelectron bands from the respective precursor compounds as well as from bands from biphenylene. The relevant sections of two of the PE spectra obtained, starting from **109** and **106**, are shown in Figure 8-17. This is just the region where three ionizations of o-benzyne are expected to occur, according to LNDO/S PERTCI calculations performed at that time.[113] The calculations were based on the MNDO structure for the $^1A_1$ ground state of the molecule (C≡C bond distance of 125.2 pm) and gave the sequence of ion states $^2A_1(\sigma)$, $^2A_2(\pi)$, and $^2B_1(\pi)$ for the first three ionizations. Accordingly, the first ionization is due to ionization from the

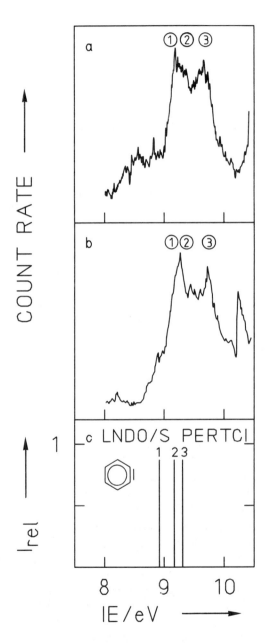

Figure 8-17. Sections of the HeI photoelectron spectra of the pyrolysis products from (*a*) sulfobenzoic anhydride (**109**) and (*b*) indanetrione (**106**). (*c*) LNDO/S PERTCI ionization spectrum of *o*-benzyne (**46**). The ionizations for **46** are assigned as follows: (1) $^2A_1(\sigma)$, (2) $^2A_2(\pi)$, and (3) $^2B_1(\pi)$.

bonding C≡C σ MO, and the next two are due to electron ejection from the two highest occupied π MOs [both ionizations corresponding to the degenerate first {$^2E_{1g}(\pi)$} ionization of benzene]. The calculated VIEs are 8.92(σ), 9.17(π), and 9.31(π) eV, and the measured ones from **109** (**106**) 9.20 (9.23), 9.38 (9.41), and 9.67 (9.70) eV. If it is allowed (on the basis of the calculation alone) to assign the spectral features exhibited in Figure 8-17 to o-benzyne (**46**), the following conclusions can be drawn. The two highest occupied π MOs of **46** are only slightly stabilized relative to the corresponding degenerate pair of π MOs of benzene (9.24 eV). Also splitting of the two π MOs of **46** is only minor (≈0.3 eV). The HOMO of **46** is a σ orbital, occurring about 0.2 eV above the highest occupied π MO (note that the latter discussion is based on Koopmans' theorem which, however, is approximately fulfilled for the respective three ionizations). It is remarkable that the calculated VIEs are shifted by about 0.3 eV to lower ionization energies relative to the measured ones. Since the VIEs of benzene are nearly quantitatively reproduced by the LNDO/S PERTCI procedure,[16b] the shift might indicate that the MNDO geometry is not yet perfect. On the whole, the PE results might indicate that the π electronic structure of o-benzyne is not far from the one of benzene with an additional especially high-lying σ MO, as a consequence of the acetylenelike bending distortion of the C—C≡C—C unit.[114]

The aforementioned results were not disclosed before, for the following reasons: the recorded PE spectra (hence the relevant region between ≈8.5 and 10.5 eV with maximum count rates of 50 c/s for our experimental setup) are of lesser quality. The shape of the relevant bands can vary somewhat from case to case (depending on small changes in temperature or the status of the reactor). However, both findings may be explicable simply as a result of the high reactivity of o-benzyne (**46**) for dimerization[108,112] (essentially equal to the reactivity of a radical in the gas phase for radical–radical reactions). The assignment of bands relies only on calculations. Location of the band maxima of the three ionizations is somewhat uncertain. However, qualitative argumentation (based on o-benzyne electronically being essentially benzene with an additional acetylenelike high-energy σ bond[114]) leads one to expect just three ionizations in the region where the relevant new band system occurs in the VTPE spectra.

The situation has gradually changed now. Recently, a further VTPES study of the precursors **106** and **107** was published.[115] This study is in full agreement with both the experimental findings and experimental problems discussed above. However, the principal agreement between both studies, as well as the recent observation of microwave transitions and derived rotational constants (making use of thermal reaction **108**→**46**),[116] forces one now to disregard previous reservations and to conclude that o-benzyne can be thermally generated in sufficient concentration in the gas phase (in the form of a continuous flow), thus allowing for its direct spectral observation, if wanted, for hours. Nevertheless, more detailed studies are desirable.

In a recent study the singlet–triplet splitting and electron affinity of o-benzyne was determined by negative ion photoelectron spectroscopy.[117] The results are consistent with *o*-benzyne having a singlet ($^1A_1$) ground state [with the first excited triplet [$^3B_2(\sigma\sigma^*)$] state lying 1.637 eV above the ground state] and a triple-bonded structure whose LUMO is primarily acetylenic (a conclusion based on the comparison of the measured electron affinities of *o*-benzyne[117] and a series of cyclic alkynes[118]). The unusual stability of the *o*-benzyne anion (electron affinity of *o*-benzyne: 0.560 eV) is also in harmony with preceding considerations about the low LUMO energy of **46** and the consequences for the reactivity of this system.[119]

LNDO/S PERTCI calculations place the first excited state [$^3B_2(\sigma\sigma^*)$] of *o*-benzyne (**46**) 1.440 eV (vertical value) above the ground state, in reasonable agreement with the measured adiabatic value (see above), and the first excited singlet state [$B_2(\sigma\sigma^*)$] 3.287 eV above the ground state.[120] Experimental work of determining this energy [corresponding to an vertical optical excitation at λ = 377 nm (oscillator strength: 0.042)] is much desired.

c. Furanocyclobutene

The thermodynamic and kinetic properties of the gas phase equilibrium reaction of *o*-xylylene (**22**) and benzocyclobutene (**23**) have been thoroughly investigated[121] (Scheme XXVII). The data (equilibrium constant $K_{800°C} \approx 4$)

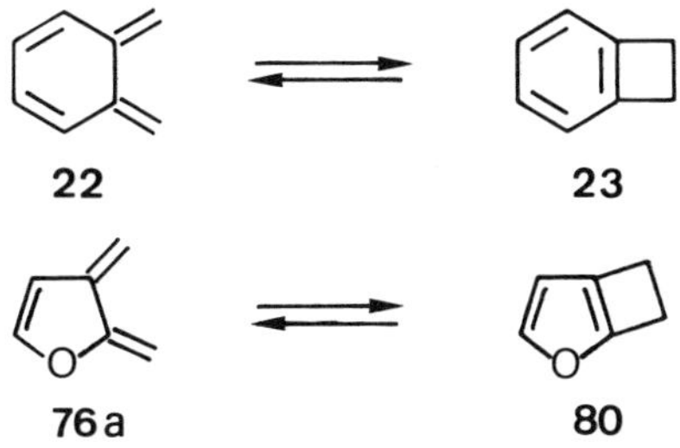

22   **23**

76 a   **80**

Scheme XXVII

suggest that **22** can be observed in equilibrium with **23** at high temperatures, which has meanwhile been verified by heating gaseous **23** to 910°C, subsequent condensation of the reaction mixture with argon in excess at 16 K, and recording the IR spectrum.[76] The data[121] further suggest that **22** could be much more efficiently observed, at rather low temperatures, as an intermediate of the reaction of an appropriate precursor in a thermal flow reactor [rate constant (**22**→**23**) $k_{\rightarrow,225°C} \approx 3.3$ s^{-1}; corresponding half-life $t_{1/2}$ of **22** of ≈0.2 s; reaction time t_R in the reactor in combination with a UV photoelectron spectrometer ≈0.1 s].[122] The latter procedure enabled us to record the PE spectrum of **22** for the first time[18c] (see Subsection B, above).

Before the work of Reference 76 was performed nothing was known about the analogous equilibrium between 2,3-dimethylene-2,3-dihydrofuran (**76a**) and furanocyclobutene (dihydrocyclobutafuran) (**80**) (see Scheme XXVII), apart from the proof that **76a** exists (by MS and NMR spectra)[77,78] (see Subsection B). In particular, **80** was unknown.

Based on measured data[121] for the reaction **22→23** and MNDO data for the reactions **22→23** and **76a→80**, the equilibrium constant for the unknown gas phase reaction **76a→80** was estimated to be about 10^{-13} at 200°C and about 10^{-7} at 800°C.[76] From this result, it could be concluded that it will hardly be possible to observe the thermodynamically unfavorable compound **80** in its thermal equilibrium with **76a** at high temperatures in the same way it was possible for **22** in the equilibrium **22→23**. Moreover, it could be shown that there is little (if any) chance to generate **80** as a reactive intermediate in a (still unknown) gas phase consecutive reaction, **appropriate precursor→80→76a**.[76] For these reasons, furanocyclobutene (**80**), very probably, cannot be generated thermally, which is in accord with hitherto unsuccessful attempts[78,123,124] to obtain **80** thermally.

Therefore, a photochemical route was taken (Scheme XXVIII).[76] The first (thermal) step proceeds quantitatively at about 10^{-2} mbar without interfering

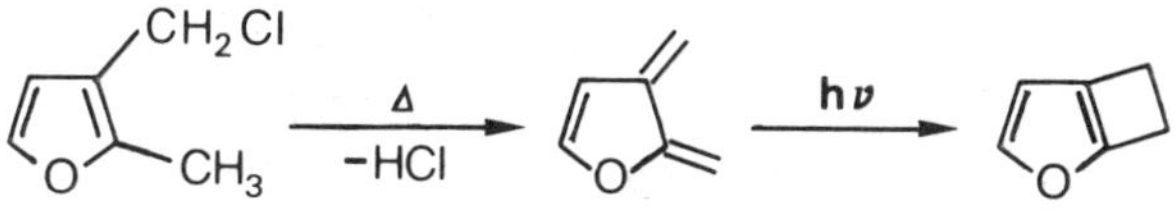

Scheme XXVIII

by-products (which is a distinct advantage over other methods[77,78]). The subsequent ring closure reaction **76a→80** was then accomplished photochemically in the condensed phase (argon matrix) at 16 K. Both the IR and UV-VIS spectra of **76a** and **80** were recorded (for the UV-VIS spectra, see Figure 8-18). The UV-VIS spectrum of **76a** exhibits a vibrationally well-resolved first band at 318 nm and is very similar to the UV-VIS spectrum of **22**[47]; therefore, and for good agreement with the calculated CNDO/S SECI, LNDO/S PERTCI, and HAM/3 excitation energies and oscillator strengths, it is unequivocally proved to be the (hitherto) unknown spectrum of **76a** (see Figure 8-18a).

After irradiation of argon matrix isolated **76a** at $\lambda = 300$ nm for 13 hours, the IR absorption bands, as well as the long-wave ($\lambda = 318$ nm) UV-VIS absorption band, of **76a** disappear completely (for the UV-VIS spectrum obtained after irradiation, see Figure 8-18b). The new spectra unequivocally prove **80** to be the photoproduct: The IR spectrum is in full agreement with calculated data (number, position, and intensity of bands) using the AM1 method. The new UV-VIS absorption band at $\lambda = 223$ nm is well reproduced by theoretical studies, and the spectrum is in excellent agreement with the spectra of other 2,3-alkyl-substituted furans: $\lambda_{max} = 218$ nm in *n*-hexane for 2,3-dimethylfuran and $\lambda = 220.9$ nm in *n*-hexane for 4,5,6,7-tetrahydrobenzofuran.[124b]

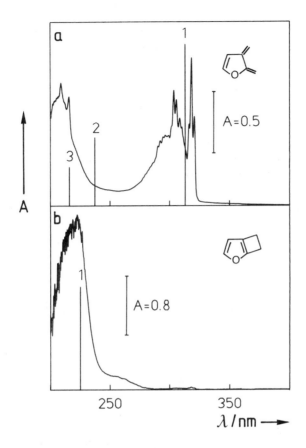

Figure 8-18. (*a*) UV-VIS absorption spectrum of the argon matrix isolated reaction mixture at 16 K after pyrolysis of 2-methyl-3-chloromethylfuran (**79**) in a thermal reactor and HAM/3 electronic excitation spectrum of 2,3-dimethylene-2,3-dihydrofuran (**76a**), A = absorbance. Wavelength, in nanometers (oscillator strength): 1 = 313.2 (0.661), 2 = 237.4 (0.282), and 3 = 216.0 (0.163). (*b*) UV-VIS spectrum of the matrix of (*a*) after UV irradiation at 300 nm and HAM/3 electronic excitation spectrum of furanocyclobutene (dihydrocyclobutafuran, **80**): 1 = 225.3 nm (0.357). (Reproduced with permission from Reference 76.)

The aforementioned results impressively show how a compound that is thermodynamically unfavorable (due to restricted mobility, high ring strain, and lack in stabilization), and thus thermally inaccessible, can conveniently be synthesized from its thermodynamically favorable valence isomer photochemically, under conditions that suppress thermal reisomerization.

E. 1,2-Dithietes and 1,2-Dithiones

Early HMO calculations have predicted that 1,2-dithietes (**110**) are stabilized by electron acceptor substituents and their valence isomers, the 1,2-

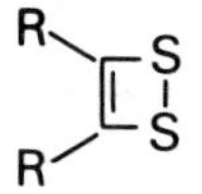

110

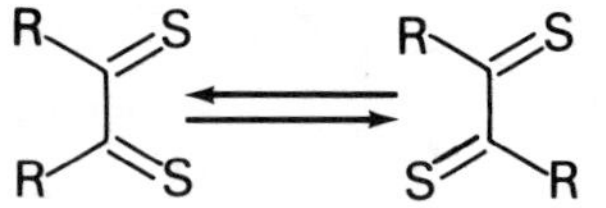

111

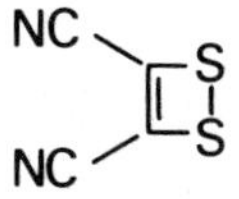

112

dithiones (**111**), by electron donor substituents.[125] Based on these ideas, 3,4-dicyano-1,2-dithiete (**112**) was postulated as a reactive intermediate of the oxidation of disodium dimercaptomaleonitrile.[126] "Chemical experiences" appeared to confirm the aforementioned concept since, for R = dialkylamino or alkylthio, only the dithione form **111**[125c,127] became known, while for R = CF_3, only the dithiete structure **110** is found.[128] The interpretation of the ^{1}H-NMR spectrum of **113** (assuming an equilibrium between **113** and **114**), however, seems to be contradictory to the simple HMO prediction[129] (Scheme XXIX). 3,4-Bis(trifluoromethyl)-1,2-dithiete (**115**) and similar perfluoroalkyl

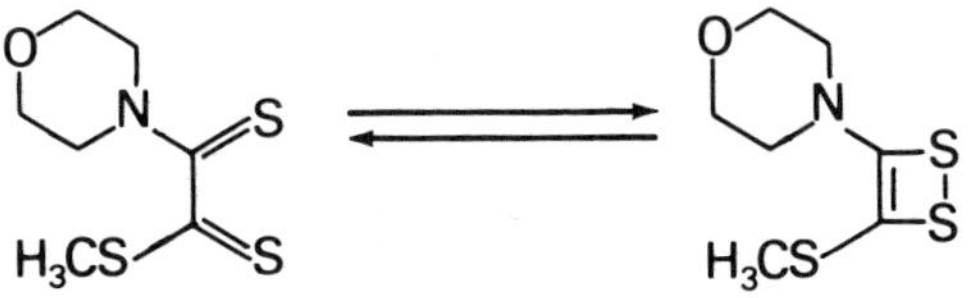

113 114

Scheme XXIX

derivatives,[128] as well as the sterically (kinetically) stabilized compounds **116**, **117**, and **118**,[130,131] could be isolated. The molecular structures of **115** and **118** were determined by gas phase electron diffraction[132] and by single-

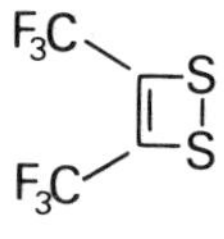

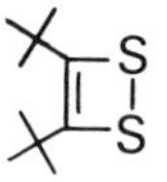

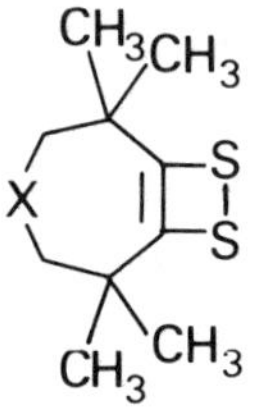

X = CH_2, S, SO_2

115 116 117

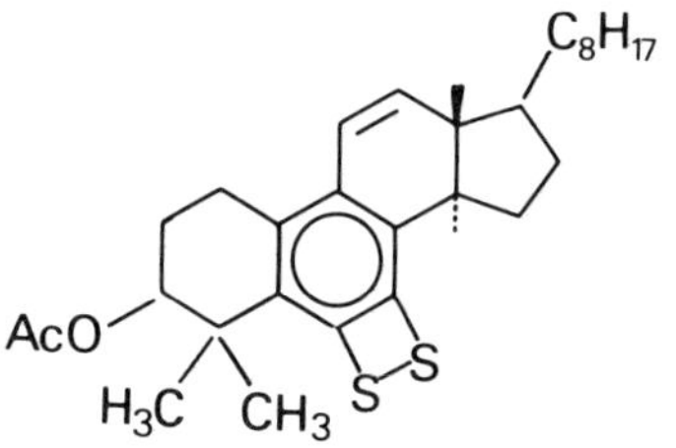

118

crystal X-ray diffraction,[133] respectively. In both cases the dithiete structure was confirmed. As early as 1925, attempts were made to synthesize 1,2-benzodithiete (**69**) by oxidation of 1,2-dimercaptobenzene[134]; later it could be shown that this reaction leads only to dimeric, oligomeric, and polymeric compounds.[135] Only recently, low-temperature UV-VIS spectra recorded after photolysis of 1,3-benzodithiol-2-one (**119**) provided evidence for the formation of unstable 1,2-benzodithiete (**69**)[72] (Scheme XXX). According to

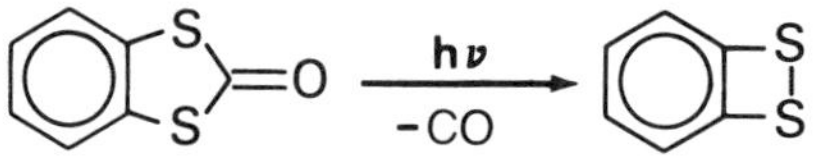

Scheme XXX

the Woodward–Hoffmann rules, the interconversion between the 1,2-dithiete and dithione forms is photochemically allowed, but thermally forbidden.[136,137] Photochemical equilibria between 1,2-dithiete (**121**) or 3-*tert*-butyl-1,2-dithiete (**123**) and their respective dithione forms were found upon irradiation of the dithiolones **120** and **122** in methanol glass at 77 K[138] (Scheme XXXI). Isolation of the reactive 1,2-dithietes and dithiones, how-

Ar = p-$(CH_3)_2N-C_6H_4$

Scheme XXXI

ever, was not possible. The activation energy of the ring-opening reaction **124**→**125** was determined to be only 17.5 kcal/mol, and 22.4 kcal/mol for the reverse reaction (meaning that the dithione form is more stable than the dithiete form by ≈5 kcal/mol).[139] However, the experiments offered no evidence for the relative stabilities of the simpler isomers in the other reactions, since photostationary states were being generated. Thus the question of relative stabilities of 1,2-dithiete (**121**) and dithioglyoxal as well as of simple

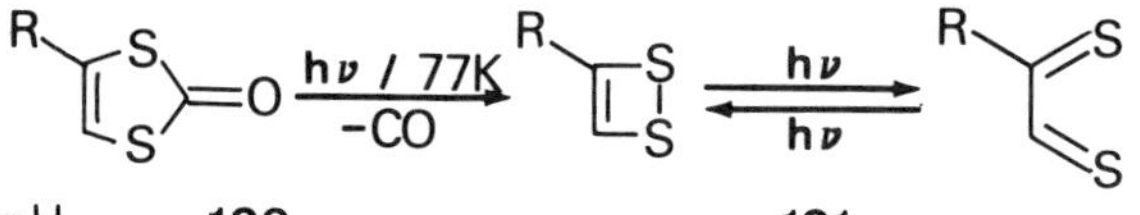

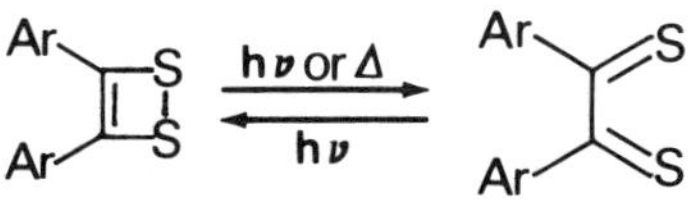

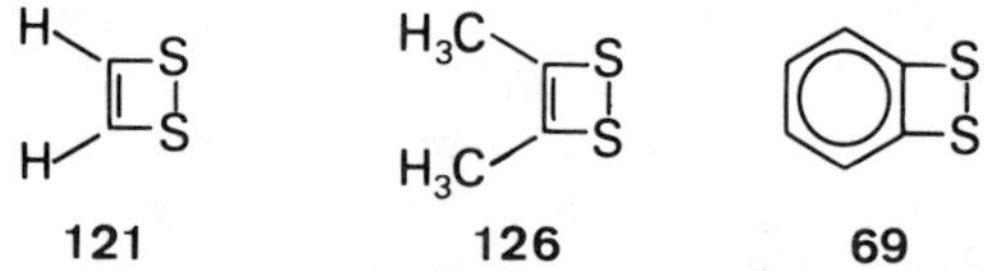

121, **126**, **69**

alkyl-substituted analogues (eg, **126**) was totally open until 1983.[140] Note that EHMO and CNDO/2 calculations were contradictory in their results,[137,141] and ab initio calculations strongly basis-set dependent.[141]

A distinct advantage of VTPES experiments is that they lead (in contrast to photochemical experiments at low temperature) to the thermodynamically most stable product (isomer). Thus VTPES investigation made, for the first time, a decision about the relative stabilities of the dithiete and dithione forms in cases where such decisions have not been possible before.[140]

Thermal extrusion of carbon monoxide from 1,3-benzodithiol-2-one (**119**) (using the VTPES technique) resulted in the appearance of a clear-cut new PE spectrum that could be due to **68** or **69**.[142] CNDO/S PERTCI as well as

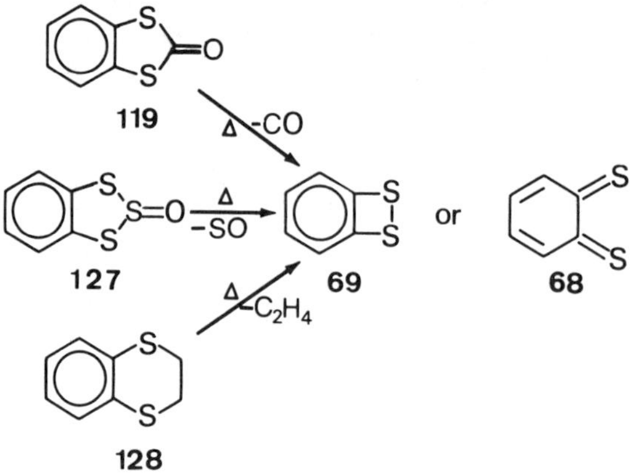

Scheme XXXII

ab initio 4-31G[143] KT (Koopmans' theorem) calculations predict two closely spaced ionizations below 10 eV for **69** and three for **68** (Figure 8-19).[142] The PE spectrum exhibits only two ionizations as expected for **69** (Figure 8-19). This and the finding that the same PE spectrum can be obtained by VTPES of 1,2,3-benzotrithiol-2-oxide (**127**), and benzodithiene (**128**) proves structure **69** for the new product. Further unequivocal proof is provided by the agreement of the IR spectra of thermally (from **119**, **127**, and **128** as precursors) and photolytically (from **119**) generated, matrix-isolated **69**[142] (Figure 8-20). Moreover, note especially the IR absorption at 730 cm^{-1}, along with the series of five weak combination bands in the range of 2000–1700 cm^{-1}

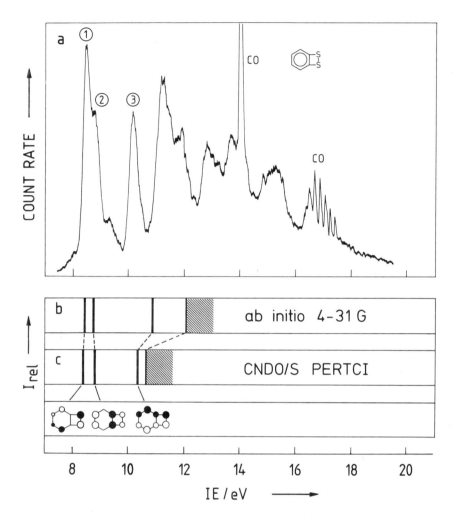

Figure 8-19. (a) HeI photoelectron spectrum of the pyrolysis products [essentially 1,2-benzodithiete (**69**)] from 1,3-benzodithiol-2-one (**119**), and (b) and (c) calculated vertical ionization energies of **69** using the ab initio 4-31G method in combination with Koopmans' theorem and the semiempirical CNDO/S PERTCI method, respectively. The ionizations for **69** are assigned as follows: (1) $^2A_2(\pi)$, (2) $^2B_1(\pi)$, and (3) $^2A_2(\pi)$. (Reproduced with permission from Reference 142.)

(see the insert in Figure 8-20), which are characteristic for 1,2-disubstituted aromatic systems. To further back the results, 4-methyl-1,2-benzodithiete (**129**) was generated in reactions analogous to the reactions shown in Scheme XXXII (Scheme XXXIII).[142] The IR spectrum of argon matrix isolated **129** exhibits a band at 790 cm^{-1}, which is consistent with a strong absorption near 800 cm^{-1}, as expected for a 1,2,4-trisubstituted aromatic ring system.[142]

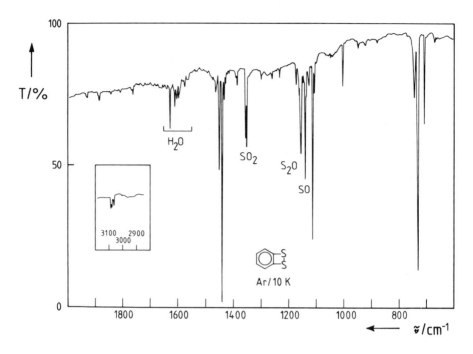

Figure 8-20. IR spectrum of the pyrolysis products [essentially 1,2-benzodithiete (**69**)] from 1,2,3-benzotrithiol-2-oxide (**127**) in an argon matrix at 10 K (inset: 3200–2800 cm^{-1} region). (Reproduced with permission from Reference 142.)

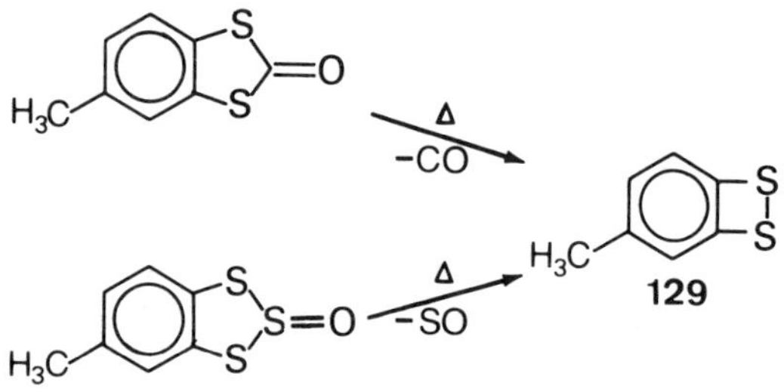

Scheme XXXIII

Finally the IR spectra show that transient 1,2-benzodithiete (**69**) is stable up to 180 K. Between 180 and 200 K, the respective absorptions rapidly disappear.

A VTPES study of 1,3-dithiol-2-one (**120**) gave a new clear-cut PE spectrum.[140] Based on CNDO/S KT, CNDO/S PERTCI, MNDO KT, MNDO PERTCI, and ab initio 4-31G calculation as well as a detailed comparison of the VTPES spectrum with the spectrum of 3,4-bis(trifluoromethyl)-1,2-dithiete (**115**),[140] which is known to possess the dithiete structure, it was clearly established that the new PE spectrum is the spectrum of 1,2-dithiete (**121**) (see Scheme XXXIV). Hence it is confirmed that 1,2-dithiete exists in

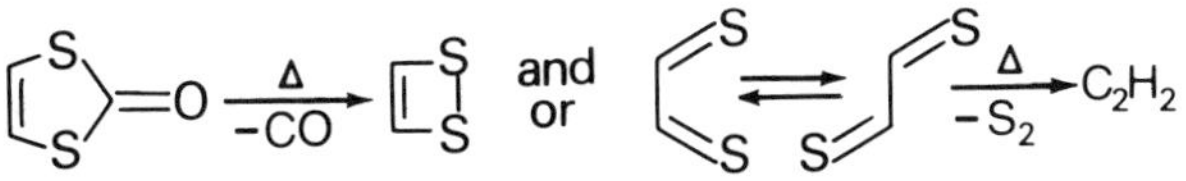

Scheme XXXIV

the gas phase at low pressure, at temperatures as high as about 600°C. A recently published microwave analysis of thermally generated 1,2-dithiete (**121**) using the reaction of Scheme XXXIV[140] confirmed the PE spectroscopic identification.[144] Dithiete **121** possesses a dipole moment of 1.33 debyes and a half-life, at 20 mtorr, of only about 2 seconds.[144] Argon matrix isolated **121** has a strong absorption at 735 cm^{-1} (CH δ_{oop}), which disappears at temperatures as low as 35 K.[140] Thus 1,2-dithiete is an extremely unstable system.

After having established that the thermolysis of 1,3-dithiol-2-one leads to 1,2-dithiete in the VTPES experiment, the route was free for producing other previously unknown, nonisolable derivatives of 1,2-dithiete. For the details of an example, namely a VTPES study of 3,4-dimethyl-1,2-dithiete (**132**) (Scheme XXXV), the reader is referred to Reference 140.

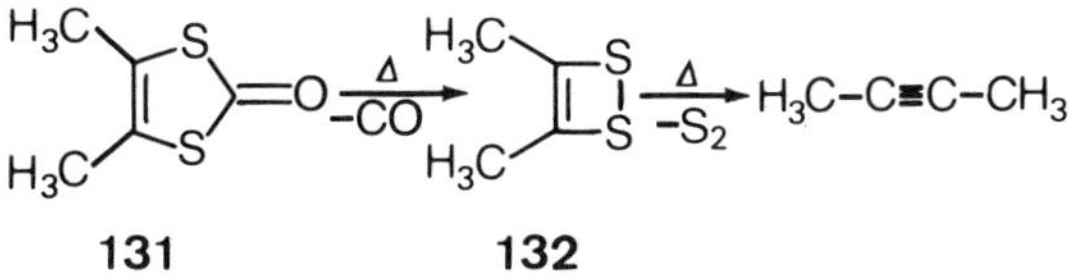

Scheme XXXV

Figure 8-21 shows the correlation diagram for the vertical ionization energies of *cis*-butadiene,[145] 1,2-benzodithiete (**69**),[142] 1,2-dithiete (**121**), 3,4-bis-(trifluoromethyl)-1,2-dithiete (**115**), 3,4-dimethyl-1,2-dithiete (**132**), and 3,4-di-*tert*-butyl-1,2-dithiete (**116**).[146] The diagram makes visible the effects on the ionizations occurring when the *cis*-butadiene and 1,2-dithiete systems are coupled together to 1,2-benzodithiete or when the hydrogen atoms of 1,2-dithiete are substituted by the CF$_3$ (**115**), CH$_3$ (**132**), or *tert*-butyl groups (**116**).[140]

The VTPES and FVP MIT work discussed above clearly demonstrated for the first time that 1,2-dithiete, 3,4-dimethyl-1,2-dithiete, as well as 1,2-benzodithiete are thermodynamically more stable than their ring-opened valence isomers. For a diagram presenting the standard reaction enthalpies of the isomerization reaction 1,2-dithietes → *cis*-1,2-dithiones using the MNDO

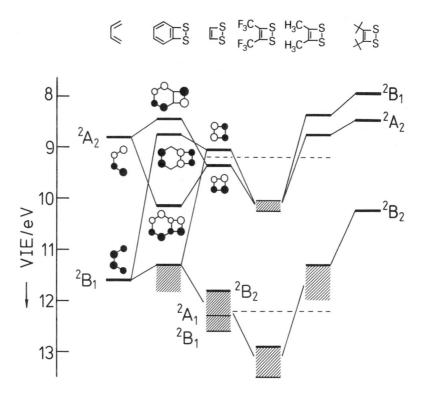

Figure 8-21. Correlation diagram of the vertical ionization energies of *cis*-butadiene, 1,2-benzodithiete (**69**), 1,2-dithiete (**121**), 3,4-bis(trifluoromethyl)-1,2-dithiete (**115**), 3,4-dimethyl-1,2-dithiete (**132**), and 3,4-di-*tert*-butyl-1,2-dithiete (**116**). (Reproduced with permission from Reference 140.)

method, the reader is referred to Reference 140. The results are in full accord with the experimental findings; particularly, the MNDO approach is correct in predicting 1,2-dithiete (**121**) to be more stable than dithioglyoxal (**130**) as well as 3,4-dimethyl-1,2-dithiete (**132**) to be more stable than 1,2-dimethyldithioglyoxal. The energy difference between both isomeric forms is largest in the case of the CF$_3$ grouping. Conversely, the 1,2-dithione structure is favored for NH$_2$ and alkylthio substituents.

Surprisingly, no 1,2-dithiete or 1,2-dithione is formed on photolysis of argon matrix isolated 1,3-dithiol-2-one (**120**); only the secondary product mercaptothioketene (**133**) can be observed[147] (Scheme XXXVI). However,

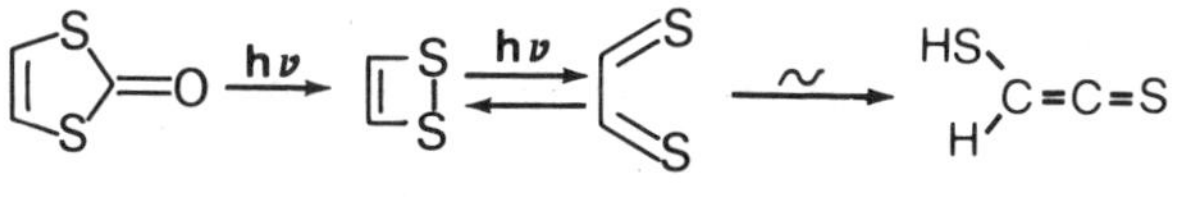

Scheme XXXVI

evidence has been obtained for the photochemical valence isomerizations and wavelength-dependent equilibria shown in Scheme XXXVII.[148] 1,2-Di-

```
     X   S                    X    S              S    X
      \ //    λ>370nm          \  //               \\ /
       C                  ⇌     C         or        C
       ‖                 λ=280nm ‖                   ‖
       C                 λ=340nm C                   C
      / \\                      / \\               // \
     X   S                     X   S              S   X
       116                       134 a              134 b


   F₃C   S       λ>300nm    F₃C   S              S   CF₃
      \ //                     \ //               \\ /
       C                   ⇌    C         ⇌        C
       ‖                        ‖                   ‖
       C                        C                   C
      / \\                     / \\               // \
   F₃C   S                   F₃C   S              S   CF₃
       115                       135 a              135 b
                                         │
                                    -[S ]│ λ>230nm
                                       x │
                                         ↓
                                  F₃C—C≡C—CF₃
```

Scheme XXXVII

tert-butyl-1,2-dithione (**134**) appears to be stable for hours, even at room temperature. A detailed study on the reversible photochemical interconversions between the 1,2-dithiete and 1,2-dithione forms of **115** and **116** isolated in an argon matrix, and based on IR and UV-VIS measurements, will be published elsewhere.[149]

In addition to the 1,2-dithiones discussed above, for example, N,N,N',N'-tetraalkyl dithiooxamides (**136a**), O,O'-dimethyl dithiooxalate (**136b**), and dimethyl tetrathiooxalate (**136c**) should be mentioned as stable,

```
    R    S         R = (alkyl)₂N     a
     \  //
      C                  CH₃O       b
      ‖
      C                  CH₃S       c
     // \
    R    S
      136
```

well-known compounds of 1,2-dithione structure. The structures are in accord with the aforementioned MNDO data. Moreover, dimethyl and diethyl tetrathiooxalates **138** and **140** can be synthesized thermally (by gas phase pyrolysis) and photochemically from the corresponding dithiolone precursors[140] (Scheme XXXVIII). Because of the unstability of **140** above −100°C the aforementioned synthesis is of practical usage only for **138**. Irradiation of **138** and **140** with light of wavelengths exceeding 300 nm did not lead to 1,2-dithietes; instead irreversible changes occurred (in argon matrix at 18 K) with IR absorptions coming up, among other things, in the thioketene region (at 1750 cm⁻¹).[148a]

Therefore it came as a surprise that the corresponding photoisomerization of the cyclic 1,3-dithiol-4,5-dithione (**142**) [made from 2,4,6,8-tetrathiabicy-

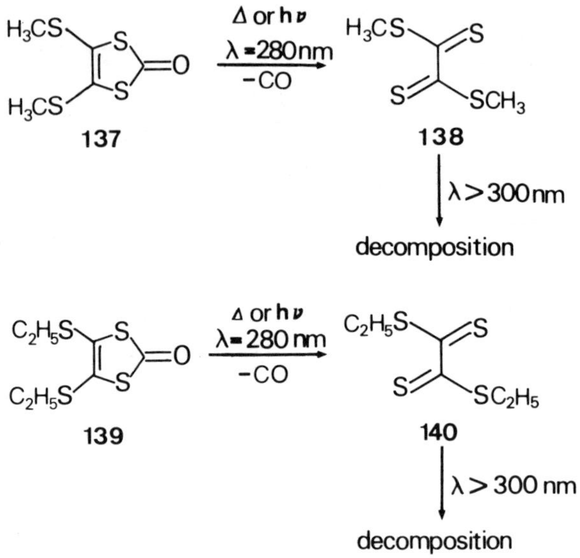

Scheme XXXVIII

clo[3.3.0]oct-1(5)-en-7-one (**141**)] is a very facile, photoreversible reaction[148,149] (Scheme XXXIX). The α-dithione (1,3-dithiol-4,5-dithione) **142**,

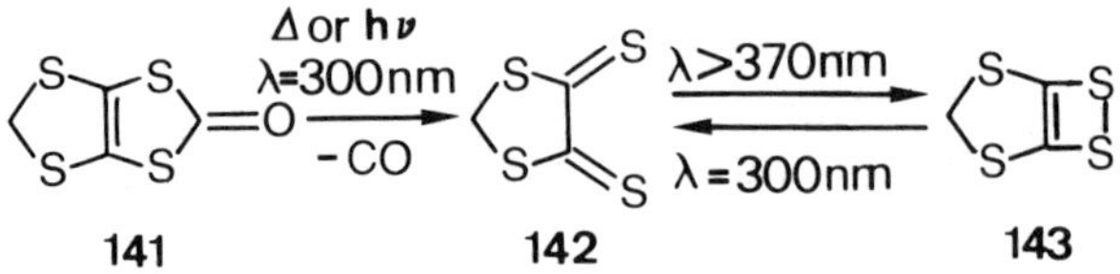

Scheme XXXIX

photoelectron spectroscopically identified for the first time,[141] is formed by FVP or photolysis of the dithiolone precursor **141** isolated in an argon matrix. Compound **142** undergoes a wavelength-dependent, reversible photoisomerization to the colorless, thermodynamically less stable 1,2-dithiete isomer **143**. Note that this is the first case of a tetrathiooxalic acid ester ⇌ 1,2-dithiete isomerization.[149]

F. 1,2-Diselenete and 1,2-Thiaselenete

The structure and properties of the valence isomeric 1,2-dithietes and dithiones have been rather satisfactorily documented (see Subsection E); in contrast, nothing was known about their selenium counterparts, apart from a synthetic route for 3,4-bis(trifluoromethyl)-1,2-diselenete,[150] before the work of Reference 151 was performed. Mixed sulfur–selenium analogues were previously unknown. The syntheses of 1,2-diselenete (**145**) and 1,2-thiaselenete (**147**) have been accomplished by gas phase pyrolysis of 1,3-diselenol-2-one (**144**) and 1,3-thiaselenol-2-one (**146**; using the VTPES technique as well as FVP MIT; Scheme XL).[151]

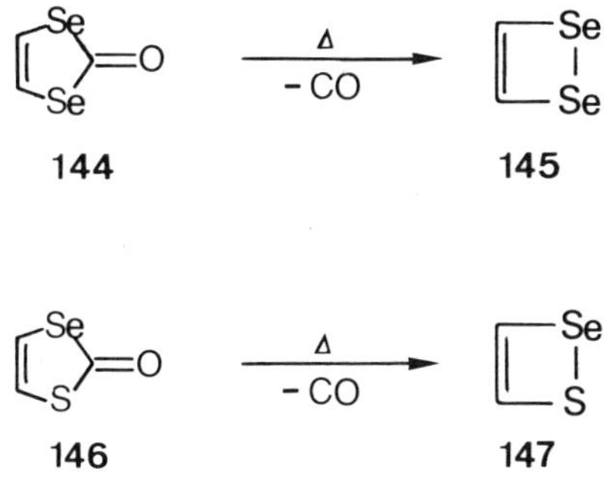

Scheme XL

The PE spectra thus obtained are shown in Figure 8-22, together with the known PE spectrum of 1,2-dithiete (**121**).[140] The spectra of the two new compounds are very similar to the spectrum of 1,2-dithiete. From previous experience this is essentially true for analogous sulfur and selenium heterocycles[153] (because of the small difference in the electronegativities of the two elements); thus the photoelectron spectroscopic findings concomitantly provide unequivocal proof of structure.[151]

These results were confirmed by low-temperature IR spectra in an argon matrix, revealing for **145** and **147** an intense absorption of 734 cm^{-1} (CH δ_{oop} vibration,[151] ie, at practically the same wavenumber, namely 735 cm^{-1}, as 1,2-dithiete[140]). This is consistent with the general rule that the positions of the IR absorption bands, in which the heteroatoms do not participate, remain almost unchanged upon replacing sulfur by selenium.[154]

The pyrolysis reactions described above (Scheme XL) are suitable for the synthesis of other substituted 1,2-diselenetes and 1,2-thiaselenetes, too. This provides the first opportunity to systematically investigate the hitherto unknown chemical and physicochemical properties of these systems (eg, valence isomerism, 6π aromaticity, photochemistry, and electrical conductivity).

G. Silabenzenes and Germabenzenes

First evidence for silaaromatics, namely 1-methylsilabenzene (silatoluene, **148**)[155,156] and 1,4-di-*tert*-butylsilabenzene (**149**),[157] was obtained by either HCl elimination from 1-chloro-1-methyl-1-silacyclohexa-2,4-diene (**150a**)[155]

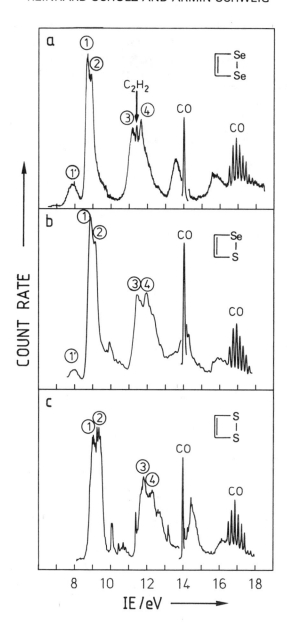

Figure 8-22. HeI photoelectron spectrum of (*a*) 1,2-diselenete (**145**); vertical ionization energy (eV): 1 = 8.71, 2 = 8.82, 3 = 11.18, 4 = 11.64, and 1' = 7.96 (presumably Se$_5$)[152]; (*b*) 1,2-thiaselenete (**147**): 1 = 8.87, 2 = 9.13, 3 = 11.46, 4 = 11.97, and 1' = 8.04 (presumably Se$_5$)[152]; and (*c*) 1,2-dithiete (**121**)[140]; vertical ionization energy (eV) assignment: 1 = 9.05, $^2B_1(\pi)$; 2 = 9.36, $^2A_2(\pi)$; 3 = 11.83, $^2B_2(\sigma)$; and 4 = 12.31, $^2A_1(\sigma)$. (Reproduced with permission from Reference 151.)

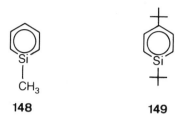

148 **149**

or 1-chloro-1,4-di-*tert*-butyl-1-silacyclohexa-2,4-diene (**150b**)[156] in solution and trapping with bis(trifluoromethyl)acetylene,[155,157] 1,3-dienes,[157] and self-trapping (dimerization)[157] or by pyrolysis of 1-allyl-1-methyl-1-silacyclohexa-2,4-diene (**150c**)[156] in the gas phase in the presence of acetylene or

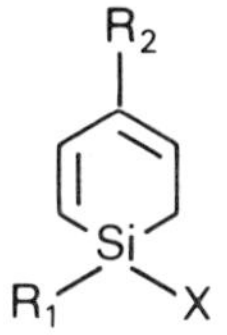

	X	R_1	R_2
a	Cl	CH_3	H
b	Cl	+	+
c	$CH_2-CH=CH_2$	CH_3	H

150   + = t-Butyl

bis(trifluoromethyl)acetylene and identifying the adduct by NMR spectroscopy. In particular, the method of Reference 156 (FVP of 1-X-1-R_1-4-R_2-1-silacyclohexa-2,4-dienes) proved to be the method of choice for generating free silabenzenes for spectroscopic detection (see below).

First spectroscopic evidence (UV-VIS and IR spectra using FVP MIT[158] and PE spectrum[159] using the VTPES technique and in both cases the pyrolytic precursor **150c**[156]) for silatoluene (**148**) was presented in 1980 and, shortly afterward, for silabenzene (**151**), again following the synthetic route of Reference 156 in the FVP MIT[160] and VTPES[161] experiments, supplemented by FVP of 1-acetoxy-1-silacyclohexa-2,4-diene (**154**) in matrix isolation work[160] (Scheme XLI). Subsequent work identified[162] 1-sila-2,5-cyclohexadiene (**153**) as a further pyrolytic precursor (Scheme XLI) and completed the only available substituted silabenzene, silatoluene, by three further substituted silabenzenes, namely 1,4-dimethylsilabenzene (**156**), 1-methyl-4-*tert*-butylsilabenzene (**157**), and 4-*tert*-butylsilabenzene (**158**), ap-

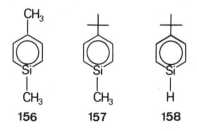

156   **157**   **158**

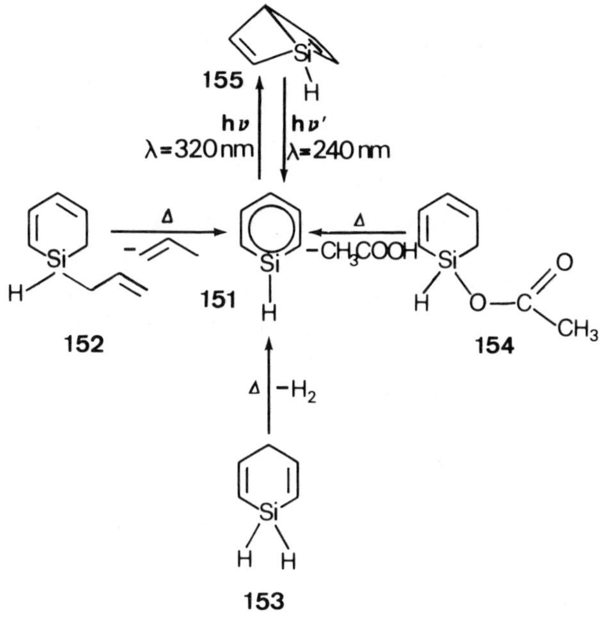

Scheme XLI

plying the VTPES technique to appropriately substituted 1-allyl-1-R_1-4-R_2-1-silacyclohexa-2,4-dienes.[163]

As an example, the PE spectrum of 1,4-dimethylsilabenzene (**156**) is presented (Figure 8-23).[163] In this case the thermal conversion of the precursor to the product **156** was complete at 630°C.

The thermal reactivity of the silabenzenes (essentially known from the aforementioned trapping experiments) is characterized by the ease of nucleophilic attack to the Si⋯C bond (eg, by CH_3OH[158]) and of undergoing Diels–Alder cycloadditions (eg, with $CH\equiv CH$,[156] $F_3CC\equiv CCF_3$,[155-157] and 1,3-dienes,[157] as well as with itself[157,158]). In addition, photochemical transformation of matrix-isolated silabenzene (**151**) to the Dewar isomer (**155**, Scheme XLI) with partial reversion (to a photostationary state) was reported.[164]

Much less is known at present about germaaromatics. Evidence for the intermediacy of 1,4-di-*tert*-butylgermabenzene (**159**) in the dehydrobromination of 1-bromo-1,4-di-*tert*-butyl-1-germacyclohexa-2,4-diene (**160**) in solu-

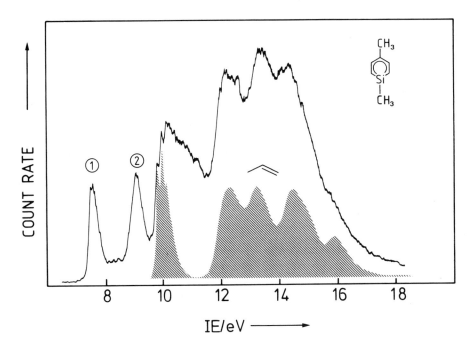

Figure 8-23. HeI photoelectron spectrum of the pyrolysis products [essentially 1,4-dimethylsilabenzene (**156**) and propene (shaded)] from 1,4-dimethyl-1-silacyclohexa-2,4-diene. The ionizations are assigned as follows: (1) $^2B_1(\pi)$ and (2) $^2A_2(\pi)$. (Reproduced with permission from Reference 163.)

tion is based on the characterization of trapping products, as the dimer of **159** (due to self-trapping) or Diels–Alder adducts in the presence of 1,3-dienes.[165]

In 1982, 1-allyl-1-methyl-4-*tert*-butyl-1-germacyclohexa-2,4-diene (**161a**) and 1-allyl-1-methyl-4-ethyl-1-germacyclohexa-2,4-diene (**161b**) (Scheme XLII) were subjected to a VTPES analysis.[163] Since only partial conversions

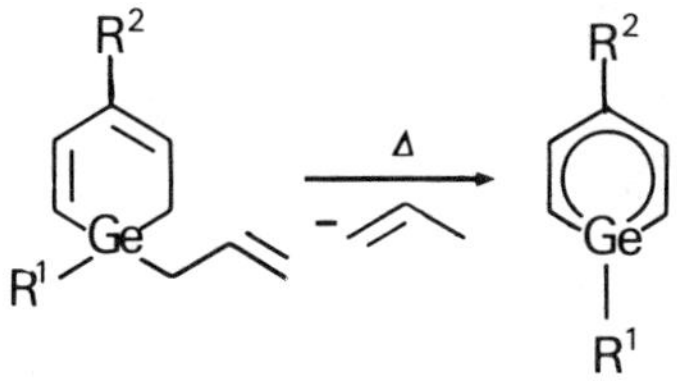

	R¹	R²
a	CH₃	+
b	CH₃	C₂H₅

Scheme XLII

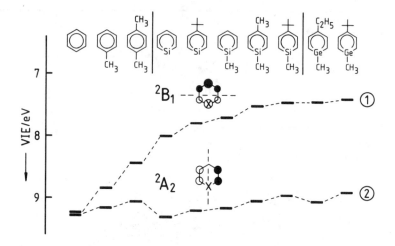

Figure 8-24. Vertical ionization energies for band 1 [$^2B_1(\pi)$] and band 2 [$^2A_2(\pi)$] in the respective photoelectron spectra of benzene, toluene, 1,4-dimethylbenzene, silabenzene (**151**), 4-*tert*-butylsilabenzene (**158**), silatoluene (**148**), 1,4-dimethylsilabenzene (**156**), 1-methyl-4-ethyl-germabenzene (**162b**), and 1-methyl-4-*tert*-butylgermabenzene (**162a**). (Reproduced with permission from Reference 163.)

can be achieved in these reactions, the PE spectra of the germabenzenes 1-methyl-4-*tert*-butylgermabenzene (**162a**) and 1-methyl-4-ethylgermabenzene (**162b**) were obtained from the product and educt spectra by subtraction.

That the germabenzene derivatives **162a** and **162b** were actually formed during pyrolysis of **161a** and **161b** followed from the good agreement of the vertical ionization energies of germabenzene and the correspondingly substituted silabenzene derivatives (Figure 8-24). The agreement is not surprising, since the atomic ionization energies of silicon (8.15 eV) and germanium (7.88 eV) differ only slightly.

From these results it can be expected that the first ionizations of the unsubstituted germabenzene will be at 7.95 [$^2B_1(\pi)$] and 9.25 eV [$^2A_2(\pi)$].

Further VTPES work as well as FVP MIT studies on substituted and unsubstituted germabenzenes are much desired, to achieve deeper insight into the electronic structure, as well as the reactivity pattern, of the hither to rather elusive germabenzenes.

3. OUTLOOK

This chapter is devoted to *chemically* relevant reactive species (intermediates, or transients) of organic chemistry. Only two methods of generation, pyrolysis and photolysis, are used. The methods of isolation of the reactive species (ie, gas phase isolation at low pressure or matrix isolation at low temperatures) as well as the methods of detection used (ie, PE, UV-VIS, and IR spectroscopy) have meanwhile become practically classical techniques; the spectroscopic tools, cryostats, reactors, or lamps, are all commercially available. This and the rich results obtainable in this way (see Section 2) may explain why these methods have become more and more important in modern organic chemistry. It is one of the objects of the present work to demonstrate that the combination of PE with UV-VIS and IR spectroscopy, as well as of gas phase with matrix isolation work, is particularly useful, just as the permanent interaction between experiment and theory is extremely helpful.

Present and future efforts in the field of the spectroscopic detection of reactive species might be directed to the following developments: (a) construction of specially designed high-sensitivity photoelectron spectrometers for studying reactive species (see, eg, Reference 166), (b) construction of more efficient thermal reactors, (c) extension of detection methods of matrix-isolated reactive species (as the measurement of polarized IR spectra, see, eg, Reference 167 or ^{13}C NMR spectra, see eg, Reference 168), (d) extension of theoretical methods for the interpretation of PE (for the calculation of ionization spectra of species including second-row atoms, see Reference 169), IR (as, eg, ab initio calculations of vibrational spectra of Reference 170), and, in particular, UV-VIS spectra of second-row species using the LNDO/S PERTCI approach[169] and their use in the identification of unusual molecules, and (e) direct combination of PE with MIT work by measuring PE spectra of matrix-isolated reactive species (see, eg, Reference 171) or measurements of PE spectra of reactive species generated by gas phase laser flash photolysis methods (see, eg, Reference 172), both methods providing some access (at least in principle) to the PE spectra of photochemically generated species.

The range of reactive species relevant in organic chemistry that can be produced and observed by either gas phase or matrix isolation can be considerably broadened by making use of further well-known generation techniques (not applied in this work) such as discharge methods (microwave or radio frequency discharges) of suitable gas mixtures (precursor, matrix gas, etc), direct photolysis of matrix deposits with high-energy radiation (vacuum UV emission from resonance lamps or a synchrotron radiation source, X-rays, an electron beam, or accelerated H^+ or D^+ ion radiation), and finally, as a useful alternative technique to pyrolysis, photolysis and discharge techniques by bimolecular chemical reaction (using, eg, alkali metal or fluorine atoms together with a suitable reactant). In this way, among other things, such highly reactive species as molecular cations or anions (see, eg, Reference 173) become available for spectroscopic investigation.

Finally, as exemplified by many examples in this chapter (see Section 2), both gas phase and matrix isolation spectroscopic investigations provide direct access (sometimes in combination with suitable calculations) to the structure and energetics of reactive species that, by other chemical or structural methods, would be hardly or not at all accessible. Hence, in the authors' opinion, there is no doubt that the methods applied in the present work or extended along the lines sketched in this section will continue to play an important role in the elucidation of chemically relevant species or, in general, chemical reactions far beyond the 1980s.

ACKNOWLEDGMENTS

We thank the Deutsche Forschungsgemeinschaft and the Fonds der Chemischen Industrie for financial support of the work discussed in this chapter. Norbert Münzel kindly read parts of the manuscript and Dr. Hermann Meyer the entire manuscript; both gave invaluable comments and criticism. We are further indebted to Mrs. Sigrid Bamberger for carefully typing the manuscript, involving great responsibility even if pressed for time, and Harald Specht for the skillful and engaged preparation of the illustrations and all other art work. Thanks are extended to Professor Joel Liebman for his continuing interest and encouragement, without which this chapter would hardly have been finished.

REFERENCES

1. (a) Turner, D. W.; Al-Joboury, M. I. *J. Chem. Phys.* **1962,** *37,* 3007. (b) Al-Joboury, M. I.; Turner, D. W. *J. Chem. Soc.* **1963,** 5141. (c) Turner, D. W.; Baker, C.; Baker, A. D.; Brundle, C. R. "Molecular Photoelectron Spectroscopy." Wiley-Interscience: New York, 1970. (d) Baker, A. D.; Betteridge, D. "Photoelectron Spectroscopy." Pergamon Press: Oxford, 1972. (e) Carlson, T. A. "Photoelectron and Auger Spectroscopy." Plenum Press: New York, 1975. (f) Rabalais, J. W. "Principles of Ultraviolet Photoelectron Spectroscopy." Wiley-Interscience: New York, 1976. (g) Ballard, R. E. "Photoelectron Spectroscopy and Molecular Orbital Theory." Adam Hilger: Bristol, 1978. (h) Gosh, P. K. "Introduction to Photoelectron Spectroscopy." Wiley: New York, 1983. (i) Eland, J. H. D. "Photoelectron Spectroscopy." Butterworths: London, 1984. (j) Kimura, K.; Katsumata, S.; Achiba, Y.; Yamazaki, T.; Iwata, S. "Handbook of HeI Photoelectron Spectra of Fundamental Organic Molecules." Japan Scientific Societies Press: Tokyo; Halsted Press: New York, 1981. (k) Brundle, C.; Baker, A. D., Eds. "Electron Spectroscopy," Vols. 1–5; Academic Press: London, 1977–1984.
2. (a) Hedaya, E. *Acc. Chem. Res.* **1969,** *2,* 367. (b) De Mayo, P. *Endeavour* **1972,** *31,* 135. (c) Hageman, H. J.; Wiersum, U. E. *Chem. Br.* **1973,** *9,* 206. (d) Golden, D. M.; Spokes, G. N.; Benson, S. W. *Angew. Chem.* **1973,** *85,* 602; *Angew. Chem. Int. Ed. Engl.* **1973,** *12,* 534. (e) Wentrup, C. *Chimia* **1977,** *31,* 258. (f) Seybold, G. *Angew. Chem.* **1977,** *89,* 377; *Angew. Chem. Int. Ed. Engl.* **1977,** *16,* 365. (g) Wiersum, U. E. *Aldrichim. Acta* **1981,** *14,* 53. (h) Wiersum, U. E. *Rec. Trav. Chim. Pays-Bas* **1982,** *101,* 317. (i) Wiersum, U. E. *Rec. Trav. Chim. Pays-Bas* **1982,** *101,* 385. (j) Schaden, G. *J. Anal. Appl. Pyrolysis* **1982,** *4,* 83. (k) Wiersum, U. E. *Aldrichim. Acta* **1984,** *17,* 31. (l) Schiess, P.; Rutschmann, S. *Chimia* **1985,** *39,* 213. (m) Karpf, M. *Angew. Chem.* **1986,** *98,* 413; *Angew. Chem. Int. Ed. Engl.* **1986,** *25,* 414. (n) Brown, R. F. C. "Pyrolytic Methods in Organic Chemistry." Academic Press: New York, 1980.
3. (a) Daintith, J.; Maier, J. P.; Sweigart, D. A.; Turner, D. W. In "Electron Spectroscopy," Shirley, D. A., Ed.; North-Holland: Amsterdam, 1972, Chapter III, pp. 289–310. (b) See,

eg, Dyke, J. M.; Jonathan, N.; Morris, A. In "Electron Spectroscopy," Brundle, C. R.; and Baker, A. D., Eds.; Academic Press: London, 1979, Chapter 4, pp. 189–229, and literature quoted therein. (c) Schäfer, W.; Schweig, A. *Z. Naturforsch.* **1975**, *30a*, 1785. (d) Koenig, T.; Balle, T.; Snell, W. *J. Am. Chem. Soc.* **1975**, *97*, 662. (e) Frost, D. C.; McDonald, B.; McDowell, C. A.; Westwood, N. P. C. *J. Electron Spectrosc. Relat. Phenom.* **1977**, *12*, 95. (f) Houle, F. A.; Beauchamp, J. L. *Chem. Phys. Lett.* **1977**, *48*, 457. (g) Solouki, B.; Rosmus, P.; Bock, H.; Maier, G. *Angew. Chem.* **1980**, *92*, 56; *Angew. Chem. Int. Ed. Engl.* **1980**, *19*, 51.
4. (a) Schweig, A.; Vermeer, H.; Weidner, U. *Chem. Phys. Lett.* **1974**, *26*, 229. (b) Müller, C.; Schäfer, W.; Schweig, A.; Thon, N.; Vermeer, H. *J. Am. Chem. Soc.* **1976**, *98*, 5440. (c) Schweig, A.; Thon, N. *Chem. Phys. Lett.* **1976**, *38*, 482. (d) Dechant, P.; Schweig, A.; Thon, N. *J. Electron Spectrosc. Relat. Phenom.* **1977**, *12*, 443. (e) Schweig, A.; Thon, N.; Vermeer, H. *J. Am. Chem. Soc.* **1979**, *101*, 80. (f) Schweig, A.; Thon, N. *J. Electron Spectrosc. Relat. Phenom.* **1980**, *19*, 91.
5. Varma, R.; Hrubesch, L. W. "Chemical Analysis by Microwave Rotational Spectroscopy." Wiley: New York, 1979.
6. (a) Pimentel, G. C. *Spectrochim. Acta* **1958**, *12*, 94. (b) Pimentel, G. C. *Pure Appl. Chem.* **1962**, *4*, 61. (c) Meyer, B. "Low Temperature Spectroscopy." Elsevier: Amsterdam, 1971. (d) Hallam, H. "Vibrational Spectroscopy of Trapped Species." Wiley: New York, 1972. (e) Moskovits, M.; Ozin, G. A. "Cryochemistry." Wiley: New York, 1976.
7. (a) Farmer, J. B.; Lossing, F. P. *Can. J. Chem.* **1955**, *33*, 861. (b) Schissel, P.; McAdoo, D. J.; Hedaya, E.; McNeil, D. W. *J. Chem. Phys.* **1968**, *49*, 5061. (c) Hedaya, E.; Miller, R. D.; McNeil, D. W.; D'Angelo, P. F.; Schissel, P. *J. Am. Chem. Soc.* **1969**, *91*, 1875.
8. Clyne, M. A. A.; Nip, W. S. In "Reactive Intermediates in the Gas Phase," Setser, D. W., Ed.; Academic Press: New York, 1979, Chapter 1, pp. 2–50.
9. Bock, H.; Solouki, B. *Angew. Chem.* **1981**, *93*, 425; *Angew. Chem. Int. Ed. Engl.* **1981**, *20*, 427, and literature quoted therein.
10. Koopmans, T. *Physica* **1934**, *1*, 104.
11. Dewar, M. J. S. "The Molecular Orbital Theory of Organic Chemistry." McGraw-Hill: New York, 1969.
12. (a) Heilbronner, E.; Bock, H. "Das HMO-Modell und seine Anwendung." Verlag Chemie: Weinheim, 1978. (b) Klessinger, M. "Elektronenstruktur organischer Moleküle." Verlag Chemie: Weinheim, 1982.
13. (a) Coulson, C. A.; Longuet-Higgins, H. C. *Proc. R. Soc. London* **1947**, *A192*, 16. (b) Dewar, M. J. S.; Dougherty, R. C. "The PMO Theory of Organic Chemistry." Plenum Press: New York, 1975.
14. (a) See, eg, (a) Cederbaum, L. S.; Domcke, W.; Schirmer, J.; von Niessen, W. *Adv. Chem. Phys.* **1986**, *65*, 115, and literature quoted therein. (b) See also von Niessen, W.; Cederbaum, L. S.; Schirmer, J. *J. Electron Spectrosc. Relat. Phenom.* **1986**, *41*, 235, and literature cited therein.
15. Kreile, J.; Münzel, N.; Schweig, A.; Specht, H. *Chem. Phys. Lett.* **1986**, *124*, 140.
16. (a) Hase, H. L.; Lauer, G.; Schulte, K.-W.; Schweig, A. *Theor. Chim. Acta* **1978**, *48*, 47. (b) Lauer, G.; Schulte, K.-W.; Schweig, A. *J. Am. Chem. Soc.* **1978**, *100*, 4925. (c) Dewar, M. J. S.; Thiel, W. *J. Am. Chem. Soc.* **1977**, *99*, 4899. (d) Schulte, K.-W.; Schweig, A. *Theor. Chim. Acta* **1974**, *33*, 19, and literature quoted therein.
17. Numerous papers in the series of publications "Theory and Application of Photoelectron Spectroscopy." See in particular Schulz, R.; Schweig, A.; Zittlau, W. *J. Mol. Struct. (THEOCHEM)* **1985**, *121*, 115, where a statistical analysis of the performance of semiempirical PERTCI calculations of Koopmans ionizations is presented. Indeed this chapter can be viewed as part 112 of these series; for part 111, see Reference 76.
18. (a) Schulz, R.; Schweig, A.; Zittlau, W. *J. Am. Chem. Soc.* **1983**, *105*, 2980. (b) Schulz, R.; Schweig, A.; Zittlau, W. *Chem. Phys. Lett.* **1984**, *106*, 467. (c) Kreile, J.; Münzel, N.; Schulz, R.; Schweig, A. *Chem. Phys. Lett.* **1984**, *108*, 609. (d) Schweig, A.; Zittlau, W. *Chem. Phys.* **1986**, *103*, 375. (e) Heidenreich, A.; Münzel, N.; Schweig, A. *Z. Naturforsch.* **1986**, *41a*, 1415.

19. See, eg, (a) Lindholm, E.; Åsbrink, L. *J. Electron Spectrosc. Relat. Phenom.* **1980**, *18*, 121. (b) Chong, D. P.; Frost, D. C.; Lau, W. M.; McDowell, G. A. *Chem. Phys. Lett.* **1982**, *90*, 332. (c) Koenig, T.; Klopfenstein, C. E.; Southworth, S.; Hoobler, J. A.; Wielesek, R. A.; Balle, T.; Snell, W.; Imre, D. *J. Am. Chem. Soc.* **1983**, *105*, 2256. (d) Bigelow, R. W. *Chem. Phys. Lett.* **1983**, *100*, 445. (e) Bigelow, R. W. *Chem. Phys.* **1983**, *80*, 45. (f) Spanget-Larsen, J. *Croat. Chim. Acta* **1984**, *57*, 991. (g) Hohlneicher, G.; Distler, D.; Müller, M.; Freund, H.-J. *Chem. Phys. Lett.* **1984**, *111*, 151. (h) Bigelow, R. W. *J. Electron Spectrosc. Relat. Phenom.* **1985**, *35*, 239.
20. (a) Thiel, W.; Schweig, A. *Chem. Phys. Lett.* **1971**, *12*, 49. (b) Thiel, W.; Schweig, A. *Chem. Phys. Lett.* **1972**, *16*, 409. (c) Schweig, A.; Thiel, W. *J. Chem. Phys.* **1974**, *60*, 951. (d) Dechant, P.; Schweig, A.; Thiel, W. *Angew. Chem.* **1973**, *85*, 358; *Angew. Chem. Int. Ed. Engl.* **1973**, *12*, 308. (e) Schweig, A.; Thiel, W. *J. Electron Spectrosc. Relat. Phenom.* **1973**, *2*, 199. (f) Schweig, A.; Thiel, W. *Chem. Phys. Lett.* **1973**, *21*, 541. (g) Schweig, A.; Thiel, W. *Mol. Phys.* **1974**, *27*, 265. (h) Schweig, A.; Thiel, W. *J. Electron Spectrosc. Relat. Phenom.* **1974**, *3*, 27.
21. (a) Downs, A. J.; Peake, S. C. In "Molecular Spectroscopy," Barrow, R. F.; Long, D. A.; and Millen, D. J., Eds.; Specialist Periodical Reports, Vol. 1; The Chemical Society: London, 1973, Chapter 9, pp. 523–558. (b) Chadwick, B. M. In "Molecular Spectroscopy," Barrow, R. F.; Long, D. A.; and Millen, D. J., Eds.; Specialist Periodical Reports, Vol. 3, The Chemical Society: London, 1975, Chapter 4, pp. 281–377. (c) Chadwick, B. M. In "Molecular Spectroscopy," Barrow, R. F.; Long, D. A.; and Sheridan, J., Eds.; Specialist Periodical Reports, Vol. 6; The Chemical Society: London, 1979, Chapter 3, pp. 72–111.
22. (a) Del Bene, J. E.; Jaffé, H. H. *J. Chem. Phys.* **1968**, *48*, 1807, 4050. (b) Ellis, R. L.; Kuehnlenz, G.; Jaffé, H. H. *Theor. Chim. Acta* **1972**, *26*, 131. (c) Kuehnlenz, G.; Jaffé, H. H. *J. Chem. Phys.* **1973**, *58*, 2238. (d) Schulte, K.-W. Doctoral dissertation, Marburg, 1977. (e) Åsbrink, L.; Fridh, C.; Lindholm, E. *Chem. Phys. Lett.* **1977**, *52*, 63, 69, 72. (f) Schweig, A.; Thiel, W. *J. Am. Chem. Soc.* **1981**, *103*, 1425. (g) Lauer, G. Doctoral dissertation, Marburg, 1977.
23. Dewar, M. J. S.; Zoebisch, E. G.; Healy, F. E.; Stewart, J. J. P. *J. Am. Chem. Soc.* **1985**, *107*, 3902.
24. (a) Maier, G. *Angew. Chem.* **1974**, *86*, 491; *Angew. Chem. Int. Ed. Engl.* **1974**, *13*, 425. (b) Bally, T.; Masamune, S. *Tetrahedron* **1980**, *36*, 343.
25. (a) Chapman, O. L.; McIntosh, C. L.; Pacansky, J. *J. Am. Chem. Soc.* **1973**, *95*, 614. (b) Chapman, O. L.; De La Cruz, P.; Roth, R.; Pacansky, P. *J. Am. Chem. Soc.* **1973**, *95*, 1337. (c) Krantz, A.; Lin, C. Y.; Newton, M. D. *J. Am. Chem. Soc.* **1973**, *95*, 2744.
26. (a) Lauer, G.; Müller, C.; Schulte, K.-W.; Schweig, A.; Maier, G.; Alzérreca, A. *Angew. Chem.* **1975**, *87*, 194; *Angew. Chem. Int. Ed. Engl.* **1975**, *14*, 172. (b) Lauer, G.; Müller, C.; Schulte, K.-W.; Schweig, A.; Krebs, A. *Angew. Chem.* **1974**, *86*, 597; *Angew. Chem. Int. Ed. Engl.* **1974**, *13*, 544.
27. (a) Delbaere, L. T. J.; James, M. N. G.; Nakamura, N.; Masamune, S. *J. Am. Chem. Soc.* **1975**, *97*, 1973. (b) Irngartinger, H.; Rodewald, H. *Angew. Chem.* **1974**, *86*, 783; *Angew. Chem. Int. Ed. Engl.* **1974**, *13*, 740.
28. (a) Irngartinger, H.; Riegler, N.; Malsch, K.-D.; Schneider, K.-A.; Maier, G. *Angew. Chem.* **1980**, *92*, 214; *Angew. Chem. Int. Ed. Engl.* **1980**, *19*, 211. (b) Heilbronner, E.; Jones, T. B.; Krebs, A.; Maier, G.; Malsch, K.-D.; Pocklington, J.; Schmelzer, A. *J. Am. Chem. Soc.* **1980**, *102*, 564. (c) Ermer, O.; Heilbronner, E. *Angew. Chem.* **1983**, *95*, 414; *Angew. Chem. Int. Ed. Engl.* **1983**, *22*, 402. (d) Irngartinger, H.; Nixdorf, M. *Angew. Chem.* **1983**, *95*, 415; *Angew. Chem. Int. Ed. Engl.* **1983**, *22*, 403.
29. (a) Kollmar, H.; Staemmler, V. *J. Am. Chem. Soc.* **1978**, *100*, 4304. (b) Masamune, S.; Souto-Bachiller, F. A.; Machiguchi, T.; Bertie, J. E. *J. Am. Chem. Soc.* **1978**, *100*, 4899.
30. Eck, V.; Lauer, G.; Schweig, A.; Thiel, W.; Vermeer, H. *Z. Naturforsch.* **1978**, *33a*, 383.
31. Eck, V.; Müller, C.; Schulz, R.; Schweig, A.; Vermeer, H. *J. Electron Spectrosc. Relat. Phenom.* **1979**, *17*, 67.

32. Hedaya, E.; Miller, R. D.; McNeil, D. W.; D'Angelo, P. F.; Schissel, P. *J. Am. Chem. Soc.* **1969**, *91*, 1875.
33. Maier, G.; Hoppe, M.; Lanz, K.; Reisenauer, H. P. *Tetrahedron Lett.* **1984**, *25*, 5645.
34. Worley, S. D.; Webb, T. R.; Gibson, D. H.; Ong, T.-S. *J. Organomet. Chem.* **1979**, *168*, C16.
35. (a) Maier, G.; Sauer, W. *Angew. Chem.* **1975**, *87*, 675; *Angew. Chem. Int. Ed. Engl.* **1975**, *14*, 648. (b) Maier, G.; Sauer, W. *Angew. Chem.* **1977**, *89*, 51; *Angew. Chem. Int. Ed. Engl.* **1977**, *16*, 49. (c) Maier, G.; Köhler, F. *Angew. Chem.* **1979**, *91*, 327; *Angew. Chem. Int. Ed. Engl.* **1979**, *18*, 308. (d) Wirz, J.; Krebs, A.; Schmalstieg, H.; Angliker, H. *Angew. Chem.* **1981**, *93*, 192; *Angew. Chem. Int. Ed. Engl.* **1981**, *20*, 192.
36. (a) Chapman, O. L.; Chang, C. C.; Rosenquist, N. R. *J. Am. Chem. Soc.* **1976**, *98*, 261. (b) Koenig, T.; Imre, D.; Hoobler, J. A. *J. Am. Chem. Soc.* **1979**, *101*, 6446.
37. Müller, C.; Schweig, A.; Thiel, W.; Grahn, W.; Bergman, R. G.; Vollhardt, K. P. C. *J. Am. Chem. Soc.* **1979**, *101*, 5579.
38. Sustmann, R.; Müller, W.; Roth, W. R.; Wittich, D. *Chem. Ber.* **1985**, *118*, 3939.
39. (a) Oppolzer, W. *Synthesis* **1978**, 793. (b) McCullough, J. J. *Acc. Chem. Res.* **1980**, *13*, 270. (c) Klasinc, L.; McGlynn, S. P. In "Quinones"; Patai, S., Ed.; Wiley-Interscience: New York, 1987.
40. (a) Szwarc, M. *Disc. Faraday Soc.* **1947**, *2*, 46. (b) Szwarc, M. *J. Polym. Sci.* **1951**, *6*, 319. (c) Koenig, T.; Wielesek, R.; Snell, W.; Balle, T. *J. Am. Chem. Soc.* **1975**, *97*, 3225. (d) Koenig, T.; Southworth, S. *J. Am. Chem. Soc.* **1977**, *99*, 2807.
41. Szwarc, M. *Polym. Eng. Sci.* **1976**, *16*, 473.
42. Ha, T. K. *Theor. Chim. Acta* **1984**, *66*, 111.
43. Michl, J. *Top. Curr. Chem.* **1974**, *46*, 1.
44. Thummel, R. P. *Acc. Chem. Res.* **1980**, *13*, 70.
45. Roth, W. R.; Scholz, B. P. *Chem. Ber.* **1981**, *114*, 3741.
46. Flynn, C.; Michl, J. *J. Am. Chem. Soc.* **1973**, *95*, 5802.
47. Tseng, K. L.; Michl, J. *J. Am. Chem. Soc.* **1977**, *99*, 4840.
48. Farr, F. R.; Bauld, N. L. *J. Am. Chem. Soc.* **1970**, *92*, 6695.
49. Bally, T.; Nitsche, S.; Roth, K.; Haselbach, E. *J. Am. Chem. Soc.* **1984**, *106*, 3927.
50. (a) Alder, K.; Fremery, M. *Tetrahedron* **1961**, *14*, 190. (b) Roth, W. R. *Tetrahedron Lett.* **1964**, 1009. (c) Isaacs, N. S. *Can. J. Chem.* **1966**, *44*, 415. (d) Feast, W. J.; Preston, W. E. *J. Chem. Soc. Chem. Commun.* **1974**, 985.
51. Dolbier, W. R.; Matsui, K.; Dewey, H. J.; Horak, D. V.; Michl, J. *J. Am. Chem. Soc.* **1979**, *101*, 2136.
52. Forster, P.; Gschwind, R.; Haselbach, E.; Klemm, U.; Wirz, J. *Nouv. J. Chim.* **1980**, *4*, 365.
53. (a) Watson, P. L.; Warrener, R. N. *Aust. J. Chem.* **1973**, *26*, 1725. (b) Warrener, R. N.; Collin, G. J.; Hutchinson, G. I.; Paddon-Row, M. N. *J. Chem. Soc. Chem. Commun.* **1976**, 373. (c) Warrener, R. N.; Gell, K. J.; Paddon-Row, M. N. *Tetrahedron Lett.* **1977**, 53. (d) Warrener, R. N.; Russel, R. A.; Collin, G. J. *Tetrahedron Lett.* **1978**, 4447. (e) Tanida, H.; Irie, T.; Tori, K. *Bull. Chem. Soc. Japan* **1972**, *45*, 1999. (f) Warrener, R. N.; Paddon-Row, M. N.; Russell, R. A.; Watson, P. L. *Aust. J. Chem.* **1981**, *34*, 397.
54. Hafner, K.; Bauer, W. *Angew. Chem.* **1968**, *80*, 312; *Angew. Chem. Int. Ed. Engl.* **1968**, *7*, 297.
55. Gross, G.; Schulz, R.; Schweig, A.; Wentrup, C. *Angew. Chem.* **1981**, *93*, 1078; *Angew. Chem. Int. Ed. Engl.* **1981**, *20*, 1021.
56. Schulz, R.; Schweig, A.; Wentrup, C.; Winter, H.-W. *Angew. Chem.* **1980**, *92*, 846; *Angew. Chem. Int. Ed. Engl.* **1980**, *19*, 821.
57. McIntosh, O. L.; Chapman, O. L. *J. Chem. Soc. Chem. Commun.* **1971**, 771.
58. Eck, V.; Schweig, A.; Vermeer, H. *Tetrahedron Lett.* **1978**, 2433.
59. Martino, P. C.; Shevlin, P. B. *J. Am. Chem. Soc.* **1980**, *102*, 5429.
60. (a) Grützmacher, H. F.; Hübner, *Justus Liebigs Ann. Chem.* **1971**, *748*, 154. (b) Mamer, O. A.; Rutherford, R. G.; Seidewand, R. J. *Can. J. Chem.* **1974**, *52*, 1983. (c) Ziegler, E.; Stark, H. *Monatsh. Chem.* **1968**, *99*, 1958.

61. (a) Horspool, W. H.; Khandelwal, G. D. *J. Chem. Soc. D* **1970**, 257. (b) Mack, P. O. L.; Pinkey, J. T. *J. Chem. Soc. Chem. Commun.* **1972**, 451. (c) Gray, T. I.; Pelter, A.; Ward, R. S. *Tetrahedron* **1979**, *35*, 2539. (d) Chapman, O. L.; McIntosh, C. L. *J. Am. Chem. Soc.* **1970**, *92*, 7001. (e) Chapman, O. L.; McIntosh, C. L.; Pacansky, J.; Calder, G. V.; Orr, G. *J. Am. Chem. Soc.* **1973**, *95*, 4061. (f) Dvorak, V.; Kolc, J.; Michl, J. *Tetrahedron Lett.* **1972**, 3443.
62. Schulz, R.; Schweig, A. *Tetrahedron Lett.* **1979**, 59.
63. Schulz, R. Doctoral dissertation, Marburg, 1983.
64. Pedersen, A. O.; Lawesson, S. O.; Klemmensen, P. D.; Kolc, J. *Tetrahedron* **1970**, *26*, 1157.
65. Chapman, O. L.; McIntosh, C. L. *J. Am. Chem. Soc.* **1970**, *92*, 7001.
66. Mao, Y.-L.; Bockelheide, V. *Proc. Natl. Acad. Sci. USA* **1980**, *77*, 1732.
67. Brown, R. F. C.; Gardner, D. V.; McOmie, J. F. W.; Solly, R. K. *Aust. J. Chem.* **1967**, *20*, 139.
68. Wentrup, C.; Gross, G. *Angew. Chem.* **1983**, *95*, 552; *Angew. Chem. Int. Ed. Engl.* **1983**, *22*, 543.
69. Schulz, R.; Schweig, A. *Tetrahedron Lett.* **1980**, *21*, 343.
70. Schulz, R.; Schweig, A. Unpublished results.
71. (a) Jacqmin, G.; Nasielski, J.; Billy, G.; Remy, M. *Tetrahedron Lett.* **1973**, 3655. (b) Hartmann, A. G.; Aron, A. J.; Bhattacharya, A. K. *J. Org. Chem.* **1978**, *43*, 3374. (c) Kanakarajan, K.; Meier, H. *J. Org. Chem.* **1983**, *48*, 881.
72. De Mayo, P.; Weedon, A. C.; Wong, G. S. K. *J. Org. Chem.* **1979**, *44*, 1977.
73. De Champlain, P.; Luche, J. L.; Marty, R. A.; de Mayo, P. *Can. J. Chem.* **1976**, *54*, 3749.
74. (a) Hoffmann, R. *J. Chem. Phys.* **1963**, *39*, 1397. (b) Hoffmann, R.; Lipscomb, W. N. *J. Chem. Phys.* **1962**, *36*, 2179, 3489. (c) Hoffmann, R. *J. Chem. Phys.* **1962**, *37*, 2872.
75. Kolshorn, H.; Meier, H. *Z. Naturforsch.* **1974**, *32A*, 780.
76. Münzel, N.; Schweig, A. *Angew. Chem.* **1987**, *99*, 471; *Angew. Chem. Int. Ed. Engl.* **1987**, *26*, 471.
77. Julien, J.; Pechine, J. M.; Perez, F.; Piade, J. J. *Tetrahedron Lett.* **1979**, 3079.
78. Trahanovsky, W. S.; Cassady, T. J.; Woods, T. L. *J. Am. Chem. Soc.* **1981**, *103*, 6691.
79. De Jongh, D. C.; Van Fossen, R. Y.; Bourgeois, C. F. *Tetrahedron Lett.* **1967**, 271.
80. Hageman, H. J.; Wiersum, U. E. *Chem. Br.* **1973**, *9*, 206.
81. De Jongh, D. C.; Brent, D. A. *J. Org. Chem.* **1970**, *35*, 4204.
82. Lage, H. W.; Reisenauer, H. P.; Maier, G. *Tetrahedron Lett.* **1982**, *23*, 3893.
83. Maier, G.; Lage, H. W.; Reisenauer, H. D. *Angew. Chem.* **1981**, *93*, 1010; *Angew. Chem. Int. Ed. Engl.* **1981**, *20*, 976.
84. Baron, P. A.; Brown, R. D. *Chem. Phys.* **1973**, *1*, 444.
85. Koenig, T.; Smith, M.; Snell, W. *J. Am. Chem. Soc.* **1977**, *99*, 6663.
86. Maier, G.; Franz, L. H.; Hartan, H.-G.; Lanz, K.; Reisenauer, H. P. *Chem. Ber.* **1985**, *118*, 3196.
87. Chapman, O. L.; Abelt, L. J. 184th National Meeting of the American Chemical Society, 1982, Abstract No. 137.
88. (a) Müller, C.; Schweig, A.; Vermeer, H. *Angew. Chem.* **1974**, *86*, 275; *Angew. Chem. Int. Ed. Engl.* **1974**, *13*, 273. (b) Schäfer, W.; Schweig, A.; Maier, G.; Sayrac, T. *J. Am. Chem. Soc.* **1974**, *96*, 279. (c) Müller, C.; Schweig, A.; Vermeer, H. *J. Am. Chem. Soc.* **1978**, *100*, 8056.
89. Schulz, R.; Schweig, A. *Angew. Chem.* **1981**, *93*, 603; *Angew. Chem. Int. Ed. Engl.* **1981**, *20*, 570.
90. Schulz, R.; Schweig, A. *Angew. Chem.* **1980**, *92*, 52; *Angew. Chem. Int. Ed. Engl.* **1980**, *19*, 69.
91. (a) Baird, N. C. *Theor. Chim. Acta* **1970**, *16*, 239. (b) Hase, H. L.; Schweig, A. *Tetrahedron* **1973**, *29*, 1759. (c) Müller, C.; Schweig, A.; Vermeer, H. *J. Am. Chem. Soc.* **1975**, *97*, 982. (d) Schäfer, W.; Schweig, A.; Mathey, F. *J. Am. Chem. Soc.* **1976**, *98*, 407. (e) Schäfer, W.; Schweig, A.; Dimroth, K.; Kanter, H. *J. Am. Chem. Soc.* **1976**, *98*, 4410. (f)

Schmidt, H.; Schweig, A. *Z. Naturforsch.* **1976**, *31a*, 215. (g) Hase, H. L.; Müller, C.; Schweig, A. *Tetrahedron* **1978**, *34*, 2983.
92. (a) Baird, M. S.; Dunkin, I. R.; Poliakoff, M. *J. Chem. Soc. Chem. Commun.* **1974**, 904. (b) Bloch, R. *Tetrahedron Lett.* **1978**, 1071. (c) Torres, M.; Clement, A.; Strausz, O. P. *J. Org. Chem.* **1980**, *45*, 2271.
93. Seybold, G.; Heibl, C. *Chem. Ber.* **1977**, *110*, 1225.
94. (a) Schulz, R.; Schweig, A. *Angew. Chem.* **1979**, *91*, 737; *Angew. Chem. Int. Ed. Engl.* **1979**, *18*, 692. (b) Schulz, R.; Schweig, A. *Angew. Chem.* **1984**, *96*, 494; *Angew. Chem. Int. Ed. Engl.* **1984**, *23*, 509.
95. (a) Huisgen, R.; Binsch, G.; König, H. *Chem. Ber.* **1964**, *97*, 2868. (b) Clinging, R.; Dean, F. M.; Mitchell, G. H. *Tetrahedron* **1974**, *30*, 4065.
96. Bock, H.; Hirabayashi, T.; Mohmand, S. *Chem. Ber.* **1981**, *114*, 2595.
97. Torres, M.; Clement, A.; Bertie, J. E.; Gunning, H. E.; Strausz, O. P. *J. Org. Chem.* **1978**, *43*, 2490.
98. Schulz, R.; Schweig, A. *Tetrahedron Lett.* **1984**, *25*, 2327.
99. (a) Krantz, A.; Laureni, J. *J. Am. Chem. Soc.* **1981**, *103*, 486, and literature quoted therein. (b) Torres, M.; Lown, E. M.; Gunning, H. E.; Strausz, O. P. *Pure Appl. Chem.* **1980**, *52*, 1623, and literature quoted therein.
100. (a) Timm, K.; Meier, H. *Chem. Ber.* **1980**, *113*, 2519. (b) Meier, H.; Konnerth, K.; Graw, S.; Echter, T. *Chem. Ber.* **1984**, *117*, 107, and literature quoted therein. (c) Timm, K.; Bühl, H.; Meier, H. *J. Heterocycl. Chem.* **1978**, *15*, 697.
101. (a) Cadogan, J. I. G.; Sharp, J. T.; Trattles, M. J. *J. Chem. Soc. Chem. Commun.* **1974**, 960. (b) Woolridge, T.; Roberts, T. D. *Tetrahedron Lett.* **1977**, 2643. (c) Benati, L.; Montevecchi, P. C.; Zanardi, G. *J. Org. Chem.* **1977**, *42*, 575. (d) White, R. C.; Scoly, J.; Roberts, T. D. *Tetrahedron Lett.* **1979**, 2785.
102. (a) Laureni, J.; Krantz, A.; Hajdu, R. A. *J. Am. Chem. Soc.* **1976**, *98*, 7872. (b) Krantz, A.; Laureni, J. *J. Am. Chem. Soc.* **1977**, *99*, 4842.
103. Meier, H.; Voigt, E. *Tetrahedron* **1972**, *28*, 187.
104. Schulz, R.; Schweig, A. *Z. Naturforsch.* **1984**, *39b*, 1536.
105. Bak, B.; Kristiansen, N. A.; Svanholt, H.; Holm, A.; Rosenkilde, S. *Chem. Phys. Lett.* **1981**, *78*, 301.
106. Bock, H.; Aygen, S.; Solouki, B. *Z. Naturforsch.* **1983**, *38b*, 611.
107. Berry, R. S.; Spokes, G. N.; Stiles, M. *J. Am. Chem. Soc.* **1962**, *84*, 3570.
108. Schafer, M. E.; Berry, R. S. *J. Am. Chem. Soc.* **1965**, *87*, 4497.
109. Hoffmann, R. W. "Dehydrobenzene and Cycloalkynes". Verlag Chemie: Weinheim, and Academic Press: New York, 1967.
110. (a) Chapman, D. L.; Mattes, K.; McIntosh, C. L.; Pacansky, J.; Calder, G. V.; Orr, G. *J. Am. Chem. Soc.* **1973**, *95*, 6134. (b) Chapman, O. L.; Chang, C. C.; Kole, J.; Rosenquist, N. R.; Tomioka, H. *J. Am. Chem. Soc.* **1975**, *97*, 6586.
111. Dunkin, I. R.; MacDonald, J. G. *J. Chem. Soc. Chem. Commun.* **1979**, 772.
112. Porter, G.; Steinfeld, J. I. *J. Chem. Soc. A* **1968**, 877.
113. Schulz, R.; Schweig, A. Unpublished results.
114. Schmidt, H.; Schweig, A.; Krebs, A. *Tetrahedron Lett.* **1974**, 1471.
115. Dewar, M. J. S.; Tien, T.-P. *J. Chem. Soc. Chem. Commun.* **1985**, 1243.
116. Brown, R. D.; Godfrey, P. D.; Rodler, M. *J. Am. Chem. Soc.* **1986**, *108*, 1296.
117. Leopold, D. G.; Miller, A. E. S.; Lineberger, W. C. *J. Am. Chem. Soc.* **1986**, *108*, 1379.
118. Ng, L.; Jordan, V. D.; Krebs, A.; Rueger, W. *J. Am. Chem. Soc.* **1982**, *104*, 7414.
119. (a) Rondan, N. G.; Domelsmith, N. L.; Houk, K. N.; Bowne, A. T.; Levin, R. H. *Tetrahedron Lett.* **1979**, *35*, 3237. (b) Strozier, R. W.; Caramella, P.; Houk, K. N. *J. Am. Chem. Soc.* **1979**, *101*, 1340.
120. Münzel, N.; Schweig, A. Unpublished results.
121. (a) Roth, W. R.; Biermann, M.; Dekker, H.; Jochems, R.; Mosselman, C.; Hermann, H. *Chem. Ber.* **1978**, *111*, 3892. (b) Roth, W. R.; Scholz, B. P. *Chem. Ber.* **1981**, *114*, 3741.
122. Schulz, R.; Schweig, A. *J. Electron Spectrosc. Relat. Phenom.* **1982**, *28*, 33.

123. (a) Chou, C. H.; Trahanovsky, W. S. *J. Am. Chem. Soc.* **1986**, *108*, 4138. (b) Chou, C. H.; Trahanovsky, W. S. *J. Org. Chem.* **1986**, *51*, 4208.
124. (a) Monti, H.; Bertrand, M. *Tetrahedron Lett.* **1969**, 1235. (b) Münzel, N.; Schweig, A. Unpublished results.
125. (a) Bergson, G. *Ark. Kemi* **1962**, *19*, 181. (b) Bergson, G. *Ark. Kemi* **1962**, *19*, 265. (c) Simmons, H. E.; Blomstrom, D. C.; Vest, R. D. *J. Am. Chem. Soc.* **1962**, *84*, 4782.
126. Simmons, H. E.; Blomstrom, D. C.; Vest, R. D. *J. Am. Chem. Soc.* **1962**, *84*, 4772.
127. Hartke, K.; Kissel, T.; Quante, J.; Matusch, R. *Chem. Ber.* **1980**, *113*, 1898.
128. Krespan, C. G. *J. Am. Chem. Soc.* **1961**, *83*, 3434.
129. Wawzonek, S.; Heilmann, S. M. *J. Org. Chem.* **1974**, *39*, 511.
130. Krebs, A.; Kolberg, H.; Höpfner, U.; Kimling, H.; Odenthal, J. *Heterocycles* **1979**, *12*, 1153.
131. Boar, R. B.; Hawkins, D. W.; McGhie, J. F.; Misra, S. C.; Barton, D. H. R.; Ladd, M. F. C.; Povey, D. C. *J. Chem. Soc. Chem. Commun.* **1975**, 756.
132. Hencher, J. L.; Shen, Q.; Tuck, D. G. *J. Am. Chem. Soc.* **1976**, *98*, 899.
133. Barton, D. H. R.; Boar, R. B.; Hawkins, D. W.; McGhie, J. F. *J. Chem. Soc. Perkin Trans. 1* **1977**, 515.
134. Guha, P. C.; Chakladar, M. N. *Q. J. Indian Chem. Soc.* **1925**, *2*, 318.
135. Field, L.; Stephens, W. D.; Lippert, E. L. *J. Org. Chem.* **1961**, *26*, 4782.
136. Cf. Snyder, J. P. In "Organic Sulphur Chemistry," Stirling, C. J. M., Ed.; Butterworths: London, 1975; p. 307.
137. Calzaferri, G.; Gleiter, R. *J. Chem. Soc. Perkin Trans. 2* **1975**, 559.
138. Jacobsen, N.; De Mayo, P.; Weedon, A. C. *Nouv. J. Chim.* **1978**, *2*, 331.
139. Kusters, W.; De Mayo, P. *J. Am. Chem. Soc.* **1974**, *96*, 3502.
140. Schulz, R.; Schweig, A.; Hartke, K.; Köster, J. *J. Am. Chem. Soc.* **1983**, *105*, 4519.
141. (a) Singh, U. C.; Basu, P. K.; Rao, C. N. R. *J. Mol. Struct.* **1982**, *87*, 125. (b) Fabian, J.; Mayer, R.; Carsky, P.; Zahradnik, R. *Z. Chem.* **1985**, *25*, 50.
142. Breitenstein, M.; Schulz, R.; Schweig, A. *J. Org. Chem.* **1982**, *47*, 1979.
143. (a) Ditchfield, R.; Hehre, W. J.; Pople, J. A. *J. Chem. Phys.* **1971**, *54*, 724. (b) Hehre, W. J.; Lathan, W. A. *J. Chem. Phys.* **1972**, *56*, 5255.
144. Rodler, M.; Bauder, A. *Chem. Phys. Lett.* **1985**, *114*, 575.
145. Heilbronner, E.; Gleiter, E.; Hopf, H.; Hornung, V.; de Meijere, A. *Helv. Chim. Acta* **1971**, *54*, 783.
146. Jian-qi, W.; Mohraz, M.; Heilbronner, E.; Krebs, A.; Schütz, K.; Voss, J.; Köpke, B. *Helv. Chim. Acta* **1983**, *66*, 801.
147. Torres, M.; Clement, A.; Strausz, O. P.; Weedon, A. C.; de Mayo, P. *Nouv. J. Chim.* **1982**, *6*, 401.
148. (a) Schulz, R.; Schweig, A. Unpublished results. (b) Diehl, F. Thesis, Marburg, 1987.
149. Diehl, F.; Schulz, R.; Schweig, A. Manuscript in preparation.
150. Davison, A.; Shawl, E. T. *Inorg. Chem.* **1970**, *9*, 1820.
151. Diehl, F.; Schweig, A. *Angew. Chem.* **1987**, *99*, 348; *Angew. Chem. Int. Ed. Engl.* **1987**, *26*, 343.
152. Streets, D. G.; Berkowitz, J. *J. Electron Spectrosc. Relat. Phenom.* **1976**, *9*, 269.
153. (a) Schäfer, W.; Schweig, A.; Gronowitz, S.; Taticchi, A.; Fringuelli, F. *J. Chem. Soc. Chem. Commun.* **1973**, 541. (b) Schweig, A.; Thon, N.; Engler, E. M. *J. Electron Spectrosc. Relat. Phenom.* **1977**, *12*, 335. (c) Spanget-Larsen, J.; Gleiter, R.; Kobayashi, M.; Engler, E. M.; Shu, P.; Cowan, D. O. *J. Am. Chem. Soc.* **1977**, *99*, 2855.
154. Klayman, D. L.; Günther, W. H. H. "Organic Selenium Compounds: Their Chemistry." Wiley-Interscience: New York, 1973.
155. Barton, T. J.; Banasiak, D. S. *J. Am. Chem. Soc.* **1977**, *99*, 5199.
156. Barton, T. J.; Burns, G. T. *J. Am. Chem. Soc.* **1978**, *100*, 5246.
157. Märkl, G.; Hofmeister, P. *Angew. Chem.* **1979**, *91*, 863; *Angew. Chem. Int. Ed. Engl.* **1987**, *18*, 789.

158. Kreil, C. E.; Chapman, O. L.; Burns, G. T.; Barton, T. J. *J. Am. Chem. Soc.* **1980**, *102*, 841.
159. Bock, H.; Bowling, R. A.; Solouki, B. *J. Am. Chem. Soc.* **1980**, *102*, 429.
160. Maier, G.; Mihm, G.; Reisenauer, H. P. *Angew. Chem.* **1980**, *92*, 58; *Angew. Chem. Int. Ed. Engl.* **1980**, *19*, 52.
161. Solouki, B.; Rosmus, P.; Bock, H.; Maier, G. *Angew. Chem.* **1980**, *92*, 56; *Angew. Chem. Int. Ed. Engl.* **1980**, *19*, 51.
162. Maier, G.; Mihm, G.; Reisenauer, H. P. *Chem. Ber.* **1982**, *115*, 801.
163. Märkl, G.; Rudnick, D.; Schulz, R.; Schweig, A. *Angew. Chem.* **1982**, *94*, 211; *Angew. Chem. Int. Ed. Engl.* **1982**, *21*, 221; *Angew. Chem. Suppl.* **1982**, 523.
164. Maier, G.; Mihm, G.; Baumgärtner, R. O. W.; Reisenauer, H. P. *Chem. Ber.* **1984**, *117*, 2337.
165. Märkl, G.; Rudnick, D. *Tetrahedron Lett.* **1980**, *21*, 1405.
166. Morris, A.; Jonathan, N.; Dyke, J. M.; Francis, P. D.; Keddar, N.; Mills, J. D. *Rev. Sci. Instrum.* **1984**, *55*, 172.
167. Arrington, C. A.; Klingensmith, K. A.; West, R.; Michl, J. *J. Am. Chem. Soc.* **1984**, *106*, 525.
168. Yannoni, C. S.; Reisenauer, H. P.; Maier, G. *J. Am. Chem. Soc.* **1983**, *105*, 6181.
169. Heidenreich, A.; Meyer, H.; Schweig, A. Work in progress.
170. Hess, B. A., Jr.; Schaad, L. J.; Čarsky, P.; Zahradnik, R. *Chem. Rev.* **1986**, *86*, 709.
171. Schmeisser, D.; Jacobi, K. *Chem. Phys. Lett.* **1979**, *62*, 51.
172. Imre, D.; Koenig, T. *Chem. Phys. Lett.* **1980**, *73*, 62.
173. Bondybey, V. E.; Miller, T. A. In "Molecular Ions: Spectroscopy, Structure, and Chemistry," Miller, T. A.; and Bondebey, V. E. Eds.; North-Holland: Amsterdam, 1983, pp. 125–173.

Addendum

CHAPTER 8

Extensive test calculations[1] on MNDO and AM1 infrared spectra have established that the wavenumbers are predicted similar to the results from AHF calculations [i.e., generally too high except for the rather proper AM1 values for XH stretching modes (X = C, N, and O)]. The MNDO and AM1 intensities are unreliable.

The infrared spectrum of the reaction products of *11* (Scheme III; Figure 2 of ref. 15) was further analysed by comparison with the infrared spectrum of an authentic sample of *12* in an Ar matrix. The results unequivocally show that *12* was the sole unidentified product.[2] Thus it is clear that anti-tricyclo [4.2.0.0^{2,5}] octa-3,7-triene (*12*) is more stable than the isomeric syn-product and furthermore that the second peak in the photoelectron spectrum at ≈9.9 eV and the shoulder at ≈8.9 eV on the high energy side of the first peak of cyclobutadiene (*1*) (Figure 3 of ref. 15) were indeed due to *12* (as previously supposed).

The LNDO/S PERTCI method in conjunction with the MNDO geometry optimization procedure has proved to be a suitable method for the calculation of electronic spectra of radical cations (ref. 18e). Effects of geometry changes when going from a molecule (or its ionization spectrum) to its ion (or its electronic excitation spectrum) were reasonably accounted for by the combined MNDO and LNDO/S PERTCI approaches. Thus the LNDO/S PERTCI method might be of considerable help in interpreting observed UV/VIS spectra of cations, elucidating interrelations between PE spectra of molecules and UV/VIS spectra of corresponding cations and, last not least in this way, in helping to identify radical cations and cationic products of cations. Preliminary results for [benzocyclobutene]$^+$ and [o-xylylene]$^+$ were communicated (ref. 18e). The results of a detailed spectroscopic and quantumchemical approach to these cations and their radiochemical and photochemical transformations will be published elsewhere.[3]

A detailed study of the UV absorption spectra of benzo[b]thiete (**60**) and thio-o-quinonemethide (**59**) was made.[4] In particular, it was possible to determine the molar absorption coefficients of transient *59* in an Ar matrix. The spectra are fully understood on the base of suitable reference compounds and quantumchemical calculations.

The elimination of HCl in the gas phase using VTPES and FVP MIT has proved to be an excellent general route for generating and detecting heterocyclic o-quinoid and p-quinoid systems as well as heterocyclic radialenes in practically spectroscopic purity (apart from HCl which is not interfering with spectroscopic investigations).[5] In this way, besides free 2,3-dimethylene-2,3-dihydrofuran (*76a*) (ref. 76), 2,5-dimethylene-2,5-dihydrofuran, furanoradialene and, for the first time, 2,3-dimethylene-2,3-dihydrothiophene (*76b*),[6] 2,5-dimethylene-2,5-dihydrothiophene and thiophenoradialene were generated and characterized by photoelectron as well as infrared and UV absorption spectroscopy.[5]

According to MNDO[7] and ab initio[8–10] calculations o-benzyne (*46*) is a system with a C≡C bond (122–126 pm) and five approximately equal (benzene-like) C≑C bonds (around 140 pm). This structure—although not directly confirmed by experiments (microwave spectroscopy or electron diffraction)—was rendered plausible by the infrared spectroscopic observation of the C≡C stretching mode (ref. 110b) and by the comparison of measured and calculated (using this structure) vibrational[7,9–11] and microwave (ref. 116) transitions. The infrared spectrum of *46* was reinvestigated in Ar[11] and N_2[12] matrices. Ab initio calculations (by CI and Green function methods) of the VIEs of *46* appeared.[10] The UV/VIS absorption spectrum of *46* was measured in an Ar matrix and interpreted in accord with other experimental and LNDO/S PERTCI results.[14] Extensive semiempirical calculations including MNDO/S PERTCI (MNDO/for spectroscopy),[15] a new approach to the calculation of ionization and excitation spectra, as well as a critical reconsideration of the experimental conditions, render an alternative interpretation of the UV/VIS spectrum of *46* possible; the full details will be given elsewhere.[16]

The gas phase equilibrium 2,3-dimethylene-2,3-dihydrothiophene (*76b*) ⇄ thiophenocyclobutene (dihydrocyclobutathiophene) was studied. Based on MNDO calculations and experimental data, the equilibrium constant was estimated to be $K_{800\,°C} \approx 10^{-5}$.[6] As expected from qualitative (chemical) ideas, the latter value is in between the corresponding values for the equilibria *22* ⇄ *23* ($K_{800\,°C} \approx 4$) (ref. 76) and *76a* ⇄ *80* ($K_{800\,°C} \approx 10^{-7}$) (ref. 76), but much closer to the latter one. Accordingly, thiophenocyclobutene should be thermally inaccessible (just as *80*) which was clearly confirmed by VTPES and FVP MIT experiments.[6] Thiophenocyclobutene can be produced, however, by irradiating *76b* with light of λ = 352 nm. This photochemical reaction can be reversed upon intense short-time irradiation of λ = 254 nm. The photoelectron spectra and UV/VIS absorption spectra of *22*, *76a* and *76b* as well as the electronic absorption spectra of *23*, *80* and thiophenocyclobutene were considered.[6]

Recent 6-31G* MP4 (Møller-Plesset) energy calculations predicted cis-1,2-dithione (*130a*) to be more stable than 1,2-dithiete (*121*) by ca. 3 kcal/mol and the trans-1,2-dithione (*130b*) to be more stable than its cis-isomer by ca. 4 kcal/mol.[17,18] From 6-31G* ΔSCF results the photoelectron spectra of *121* and the two isomeric forms of *130* were predicted to be rather similar.[17] Based on these energy and spectral calculations, the conclusions of ref. 140 about the relative stabilities of 1,2-dithiete and 1,2-dithiones as derived from a VTPES study were questioned. A recent study, however, based on the comparison of the experimental results of the temperature dependent FVT MIT infrared spectra of the pyrolyses products of *120* and the results from

6-31G* MP2 calculations of the infrared spectra of *121*, *131a* and *130b* unequivocally confirmed that the conclusion of ref. 140, namely that the stable product has the 1,2-dithiete structure, had been correct.[19]

REFERENCES

1. Meyer, H.; Schweig, A. unpublished results.
2. Münzel, N.; Schweig, A.; Specht, H. unpublished results.
3. Münzel, N.; Schweig, A. to be published.
4. Diehl, F.; Kesper, K.; Schweig, A. to be published.
5. Münzel, N.; Kesper, K.; Schweig, A.; Specht, H. submitted for publication.
6. Münzel, N.; Schweig, A. *Chem. Ber.* **1988**, *121*, 791.
7. Dewar, M. J. S.; Ford, G. P.; Rzepa, H. S. *J. Mol. Struct.* **1979**, *51*, 275.
8. (a) Noell, J. O.; Newton, M. D. *J. Am. Chem. Soc.* **1979**, *101*, 51. (b) Bock, C. W.; George, P.; Trachtman, M. *J. Phys. Chem.* **1984**, *88*, 1467.
9. (a) Radom, L.; Nobes, R. H.; Underwood, D. J.; Li, W. K. *Pure Appl. Chem.* **1986**, *58*, 75. (b) Rigby, K.; Hillier, I. H.; Vincent, M. A. *J. Chem. Soc. Perkin Trans. II*, **1987**, 117.
10. Hillier, I. H.; Vincent, M. A.; Guest, M. F.; von Niessen, W. *Chem. Phys. Lett.* **1987**, *134*, 403.
11. Nam, H.-H.; Leroi, G. E. *Spectrodim. Acta* **1985**, *41A*, 67.
12. Brown, R. F. C.; Browne, N. R.; Coulston, K. J.; Danen, L. B.; Eastwood, F. W.; Irvine, M. J.; Pullin, A. D. E. *Tetrahedron Lett.* **1986**, *27*, 1075.
13. Nam, H. H.; Leroi, G. E. *J. Mol. Struct.* **1987**, *151*, 301.
14. Münzel, N.; Schweig, A. *Chem. Phys. Lett.* **1988**, *147*, 192.
15. Meyer, H.; Pietzuch, W.; Schweig, A. unpublished results.
16. Heidenreich, A.; Meyer, H.; Münzel, N.; Schweig, A. manuscript in preparation.
17. Goddard, J. D. *J. Comp. Chem.* **1987**, *8*, 389.
18. Fabian, J.; Hess, B. A. personal communication.
19. Diehl, F.; Fabian, J.; Hess, B. A.; Meyer, H.; Schweig, A. manuscript in preparation.

General Index

A

4-Acetoxy-2-cyclopentenone, pyrolysis of, 322–323
1-Acetoxy-1-silacyclohexa-2,4-diene, pyrolysis of, 351–352
1-Allyl-1-germacyclohexa-2,4-diene, pyrolysis of, 352–354
1-Allyl-1-silacyclohexa-2,4-diene, pyrolysis of, 351–353
AM1 method, for calculation of IR spectra, 294
Amides, lactam linkage, 142–151
Amino acids, electron density analysis, 50
3-Amino-2,4,6-trinitrotoluene (3-amino TNT), sensitivities and detonation of, 258–285
Antibiotics, by bridgehead lactams and, 141–142, 172–175
Sigma-Aromaticity, pros and cons of, 117–123
Atomic energies, 111
    DESTAB (stabilization/destabilization energy), 77
    in small cycloalkanes, 82–87
    STAB (in situ atomic energy), 77–79
    and strain energy, 76
1-azabicyclo[5.2.1]deca-9-10-dione, 8-oxo derivative, 155, 161
1-Azabicyclo[3.1.1]heptan-6-one, 156, 162, 164
1-Azabicyclo[3.2.2]nonan-2-one, 6,7-Benzo, 156, 161, 162, 164, 171
1-Azabicyclo[3.3.1]nonan-2-one, 153, 157, 160–162, 164, 168
    3,4-Benzo derivative, 154
    3-Imino derivatives, 153
    3-Oxa derivative, 155
    5-Phenyl, 153, 161, 171

1-Azabicyclo[3.3.1]nonan-2,6-dione, derivative, 173
1-Azabicyclo[4.2.1]nonan-8,9-dione, 7-oxo derivative, 155, 161
1-Azabicyclo[2.2.2]octan-2-one, 140
    5,6-benzo derivatives, 156, 161
    6,6,7,7-Tetramethyl, 140, 151–152
    6,6,7-Trimethyl, 140
    6,6-Dimethyl, 140, 162, 164
    6,7-Dimethyl, 140
1-Azabicyclo[3.2.1]octan-2-one, 151
    3-Oxo derivative, 155
    derivatives, 151, 162, 164
1-Azabicyclo[3.2.1]octan-7-one (structure-type), 154
1-Azabicyclo[4.1.1]octan-7-one, derivative, 156, 162, 164
1,10-Azabicyclo[6.2.1]undec-9-en-11-one, 155
Azetones, 162–164
Aziridine, strained ring, 102–103

B

Baeyer strain energy, 69, 87, 114, 117, 122
Bases, electron density analysis, 50
Basis sets, 3–4
BBL, *See* Bicyclic bridgehead lactams
Bent bonds, 2, 14, 15, 22, 23, 89, 101
    concave, 125
    convex, 125
    strain energy and, 95–104
Benzene, HOSE model applications, 246–251
Benzocyclobutadiene,
    bond localization in, 299
    formation of, 299
    PE spectrum of, 299
    shakeup ionization of, 299

Benzocyclobutenedione, pyrolysis of, 334
Benzocyclobutene,
  cation UV-VIS spectrum of, 302
  formation of, 301
  isomerization of, 301, 318–319, 337
  stability of, 301–302, 318–319, 337
1,4-Benzodioxene, pyrolysis of, 322–323
1,4-Benzodithiene, pyrolysis of, 342
Benzodithiete,
  formation of, 341–342
  IR spectrum of, 344
  isomerization of, 317–318
  molecular orbitals of, 343, 346
  PE spectrum of, 342–343
  stability of, 317–318, 341–346
1,3-Benzodithiol-2-one,
  photolysis of, 341
  pyrolysis of, 342
Benzopropiolactone,
  formation of, 311
  isomerization of, 311, 318, 321
  stability of, 311, 317–318
o-Benzoquinone,
  formation of, 322–323
  pyrolysis of, 321–323
o-Benzoquinone methide,
  formation of, 310
  ionization energies of, 311
  isomerization of, 310, 318, 321
  stability of, 308–311, 317–318
1,2,3-Benzoselenadiazole, pyrolysis of, 329, 332–333
Benzoselenirene,
  formation of, 332–333
  IR spectrum of, 333
  isomerization of, 332
  PE spectrum of, 332
  stability of, 321, 332–333
1,2,3-Benzothiadiazole, pyrolysis of, 329, 332–333
Benzothiet-2-one (Thiobenzopropiolactone),
  formation of, 312–315
  IR spectrum of, 312, 314
  isomerization of, 312, 317–318, 321
  photolysis of, 312–314
  pyrolysis of, 312–314, 329
Benzothiete,
  formation of, 315
  IR spectrum of, 315
  isomerization of, 315, 317–318
  stability of, 315, 317–318
Benzothiirene,
  formation of, 330, 332
  IR spectrum of, 332, 333
  isomerization of, 321, 332–333
  PE spectrum of, 332–333
  stability of, 332–333
Benzotrithiol-2-oxide, pyrolysis of, 342–344
1,2,3-Benzoxadithiol-2-oxide, pyrolysis of, 324
1,4-Benzoxathiene, pyrolysis of, 324
Benzoxathiete,
  formation of, 315–316
  IR spectrum of, 315
  isomerization of, 315–318, 321
  stability of, 315–318
1,3-Benzoxathiol-2-one, pyrolysis of, 324
Benzoxete,
  isomerization of, 310, 317–318, 321
  stability of, 317–318
Benzoxirene,
  isomerization of, 321, 333
  stability of, 333
o-Benzyne,
  electron affinity of, 336–337
  electronic states of, 337
  formation of, 311, 333–334
  PE spectrum of, 335–336
  reactivity and electronic structure of, 337
  spectroscopic identification, 334
Beryllocyclopropane, 125
Bicyclic bridgehead lactams, 139–175
  1,8-Diazabicyclo[4.2.1]non-7-en-9-one, 155
  1,9-Diazabicyclo[5.2.1]dec-8-10-one derivatives, 155
  1-Azabicyclo[2.2.2]octan-2-one derivatives, 140, 151–152, 161–164
  1-Azabicyclo[3.1.1]heptan-6-one, 156, 162, 164

GENERAL INDEX

1-Azabicyclo[3.2.1]octan-2-one, 151, 155, 162, 164
1-Azabicyclo[3.2.1]octan-7-one, 154–164, 171
1-Azabicyclo[3.3.1]nonan-2-one keto derivatives, 153
1-Azabicyclo[3.3.1]nonan-2, 6-dione, derivative, 173
1-Azabicyclo[4.1.1]octan-7-one, derivative, 156, 162, 164
1-Azabicyclo[4.2.1]nonan-8,9-dione, 7-oxa derivative, 155, 161
1-Azabicyclo[5.2.1]deca-9-10-dione, 8-oxa derivative, 155, 161
1,10-Diazabicyclo[6.2.1]undec-9-en-11-one, 155
β-lactams, 139, 144, 146, 156, 159, 162–166, 169–174
Bredt's rule, 140
α-lactam, 158, 162, 171
β-lactamase, 173–174
strain in, 157–168
twisted, 141–175
Bicyclo[1.1.0]butane, 7, 8, 12, 15, 20, 21
derivatives, 10, 12, 19, 21
Bicyclo[1.1.0]butane-2,4-dicarboxylic acid anhydride, pyrolysis of, 322–323
Bicyclo[1.1.1]pentane, 7, 8, 15
Bicyclo[2.1.0]pentane, 7, 8, 14, 15, 16, 17, 18
Bicyclo[2.1.1]hexane, 7, 8, 16
Bicyclobutane, strained rings in, 102–103
Biologically active molecules, antibiotic development and lactams, 141–142, 172–175
electron density analysis, 41–53
opiates, 41–45
Bis(2,2,2-trinitroethyl) 3,6-dinitraza-1,8-octanedioate, 262
Bis(2,2,2-trinitroethyl) 3-nitrazaglutarate, 262
Bis(2,2,2-trinitroethyl) 4,4,6,6,8,8-hexanitroundecanedioate, 263
Bis(2,2,2-trinitroethyl) 4,4-dinitroheptanedioate, 263
Bis(2,2,2-trinitroethyl) 4-nitraza-1,7-heptanedioate, 262

Bis(2,2,2-trinitroethyl) carbonate, 262
Bis(2,2,2-trinitroethyl) succinate, 263
Bis(2,2,2-trinitroethyl)nitramine, 262
Bis(2,2-dinitrobutyl)nitramine, 261
Bis(2,2-dinitropropyl) oxalate, 263
Bis(2,2-dinitropropyl)nitramine, 261
Bis(1,2)-(trifluoromethyl)dithioglyoxal, formation of, 347
N,N-Bis(2,2-dinitropropyl) 4,4,4-trinitrobutyramide, 263
Bishomocubane, 7, 9, 18
2,3-Bis(methylene)-2,3-dihydrofuran, See 2,3-Dimethylene-2,3-dihydrofuran
Bis(methylene)cyclobutene, bond localization in, 299
3,4-Bis(trifluoromethyl)-1,2-dithiete, 340, 344–347
ionization energies of, 346
isomerization of, 347
molecular orbitals of, 346
N,N'-Bis(2,2,2-trinitroethyl)urea, 263
N,N'-Bis(3,3,3-trinitropropyl)oxamide, 263
"Bond angle strain," 69–71
Bond deviation index, 3–23
transferability of, 18, 22
Bond ellipticity, 102
Bond energy, in situ, 111–114
Bond enthalpies, and strain energy, 76
Bond length,
estimates by x-ray diffractometry, 234–235
HOSE model and, 236–250
strain, 69–71
Bond opposition strain, 69–70
Bond order, 101–103
Bond path, 2, 4, 5, 13, 100, 125
of strained hydrocarbons, 1–23
Bond potential, 8–23
Bonding,
bond length estimates by x-ray diffractometry, 234–235
cyclobutane, 102–104
cyclopropanes, 102–104
definition of chemical bond, 99–100
deviation indices, 1–23

Bonding (*cont.*)
  HOSE model and, 236–250
  metal–metal and metal-ligand bonds, 55–58
  polar effects on C-C bond lability, 179–226
  resonance theory and molecular geometry, 231–253
  strain energy and, 95–104
  strain and in situ bond energies, 111–113
  and types of strain, 69–71
  Westheimer approach to strain energy calculations, 114–117
  See also Carbon-carbon bonds; Electron density analysis
Bonding in cyclopropane, 94, 120
Bonds,
  "bond angle strain", 69–71
  bond eclipsing strain, 69–71
  Coulson-Moffitt bent bond models, 90–95
  See also Carbon-carbon bonds
Bredt's rule, and bicyclic bridgehead lactams, 140
Bridgehead bicyclic lactams, 139, 140
Bridgehead carbons, 12, 15, 17, 19, 20
Bridgehead ketones, IR bands, 162
Bridgehead lactams, 139–175
  antibiotic development and, 141–142, 172–175
  calorimetric data, 168–172
  crystallographic data, 168–172
  linkages, 140–141
  research areas, 140–144
  strain in, 157–168, 172
  syntheses of, 151–156
  UV and PES studies, 156–160
  See also Bicyclic bridgehead lactams
Bridgehead olefins, 140, 141, 163
Buckingham potential, and strain energy, 74–75
Butadiene,
  ionization energies of, 310
  molecular orbitals of, 310
4-*tert*-Butylsilabenzene,
  formation of, 351–352
  ionization energies of, 354
  molecular orbitals of, 354

Butyrolactam (2-Pyrrolidone), 144, 146, 169
  N-Methyl, 162, 164

C

C-C bonds,
  bond dissociation energy, 179–180
  and bond length estimates by x-ray diffractometry, 234–235
  cleavage modes in thermal unimolecular reactions, 180–181
  heterolytic cleavage, 187–216
  polar effects and heterolytic cleavage, 187–216
  polar effects on homolytic cleavage, 181–187
  polar effects on lability, 179–226
  polar effects on pericyclic cleavage, 216–226
  See Carbon-carbon bonds
C-H bond,
  energy, 112
  strengthening, 116–122
Calorimetry, of lactams, 168–172
Canonical structures, HOSE model and, 236–250
Caprolactam, 144, 168–169
Caprylolactam, 149–150, 171
6-Carbonyl-2,4-cyclohexadienethione,
  isomerization of, 312, 317–318
  stability of, 312–315, 317–318
6-Carbonyl-2,4-cyclohexadienone,
  formation of, 311–312
  IR spectrum of, 311–312
  isomerization of, 311, 317–318, 321
  PE spectrum of, 313
  photolysis of, 311
  pyrolysis of, 311–312, 321, 329
  stability of, 311–315, 317–318
Carbon-13 NMR of lactams, 160–161
Carcinogens, electron density analysis, 46–48
Catastrophe theory, 128
Cephalosporins, 141, 172
Charge transfer, HOSE model and, 245–251
Chemical bond, definition of, 99–100
Chemical carcinogens, electron density analysis, 46–48

GENERAL INDEX

Chloranil, electron donor-acceptor (EDA) complexes, 232–234
Chlorocarbonylcyclopentadiene, pyrolysis of, 329–330
CNDO/S calculations, 293, 294
Concentration, Laplacian of electron density, 104–109
Conventional strain energy, 78–79, 79
  in small cycloalkanes, 82–87
  See also Molecular strain; Strain energy
Coulson-Moffitt bent bond models, 90–95
Crystallographic data for amides, 142–144
CSE, *See* Strain; Strain energy
Cubane, 6, 7, 9, 12, 15, 17, 18, 22
Cycloalkanes,
  Coulson-Moffitt bent bond models, 90–95
  Hückel theory and, 117–123
  molecular orbital calculations of strain and, 89–95
  strain energies in, 82–87
  Walsh MOs, 90–95
Cyclobutadiene,
  derivatives of, 295, 298–299
  formation of, 296
  frontier orbital energies of, 298
  molecular structure of, 295
  Non-Koopmans ion state of, 298
  reactivity and electronic structure of, 298
  unsubstituted, 290, 295–298
Cyclobutane, 7–18
  bonding in, 102–104
  Hückel theory and, 117–123
  molecular orbital calculations of strain and, 89–95
  strain energies of, 87–89
Cyclobutane derivatives, 15
Cyclobutene-3,4-dicarboxylic acid anhydride, pyrolysis of, 322–323
Cyclohexane, strain and, 82–86
Cyclopentadienethione, 321, 324–331
  conjugative orbital interaction in, 327–328, 331
  electronic structure of, 324–328, 331
  formation of, 316, 324–327
  IR spectrum of, 326–327
  molecular orbitals of, 331
  molecular structure of, 328
  PE spectrum of, 324–326
  stability of, 327
Cyclopentadienone, 295, 321–323, 327
  conjugative orbital interaction in, 322, 328, 331
  dipole moment of, 322
  electronic structure of, 322, 328, 331
  formation of, 322–323
  ionization energies of, 322, 326, 328, 330–331
  molecular structure of, 328
  PE spectrum of, 328
  spectroscopic identification of, 322
  stability of, 327
Cyclopropane, 4–9, 12–18, 22
  bonding in, 102–104
  Hückel theory and, 117–123
  molecular orbital calculations of strain and, 89–95
  in situ bond energies and, 111–113
  strain calculations from quantum chemistry, 109–111
  strain energies of, 87–89
  Westheimer approach to strain energy calculations, 114–117
Cyclopropane derivatives, 10, 12, 19, 21

D

d orbitals, 54–61
1,1,1,4,6,6,8,11,11,11-Decanitro-4,8-diazaundecane, 262
1,1,1,3,6,6,9,11,11,11-Decanitro-3,9-diazaundecane, 262
1,1,1,5,8,11,14,18,18,18-Decanitro-3,16-dioxa-4,15-dioxo-5,8,11,14-tetrazaoctadecane, 262
1,1,1,3,6,9,12,14,14,14-Decanitro-3,6,9,12-tetrazatetradecane, 262
Deflagration, definition of, 257
Deformation density,
  biologically active molecules, 48–53
  d orbitals, 54–61
  and electron density analysis, 27–31
  organic substances, 38–53
  transition metal complexes, 53–61

DESTAB (stabilization/destabilization energy), 77
    See Atomic energies; Molecular strain; Strain energy
Detonations, 273–285
    definition of, 257
    initiation of, 278–280
    model of, 278
    physical and chemical overview, 273
Dewar-silabenzene, formation of, 352
Dewar-Zimmerman rules, 120
1,4-Di-*tert*-butylgermabenzene, formation of, 352–353
3,4-Di-*tert*-butyl-1,2-dithiete, 340, 345–347
    ionization energies of, 346
    isomerization of, 347
    molecular orbitals of, 346
Diagonal reference states, strain energy and, 82
Diagonal strain energies, 82–86
Dialkyltetrathiooxalates, formation of, 347–348
2,4-Diamino-1,3,5-trinitrobenzene, sensitivities and detonation of, 258–285
1,5-Diazabicyclo[3.3.1]nonan-2-one, derivatives, 152
1,8 Diazabicyclo[4.2.1]non-7-en-9-one, 155
1,9-Diazabicyclo[5.2.1]dec-8-en-10-one, 155
    4-oxo, derivative, 155
1,5-Diazabicyclo[3.2.1]octan-6-one, derivative, 154
6-Diazo-2,4-cyclohexadienone, pyrolysis of, 329
Diazocyclopentadiene,
    conjugative orbital interaction in, 328
    electronic structure of, 328
    ionization energy of, 326, 328
    molecular structure of, 328
    stability of, 328
3,4-Dicyano-1,2-dithiete, stability of, 340
Difference density, 95, 97
    strain energy and, 95–104

2,3-Dihydrobenzo[b]furan-2,3-dione, pyrolysis of, 312
2,3-Dihydrobenzo[b]thiophen-2,3-dione, pyrolysis of, 312
2,2-Dimethyl-2H-indene (Dimethylisoindene),
    cation UV-VIS spectrum of, 305
    Non-Koopmans ionization of, 305
    PE spectrum of, 304–305
2,3-Dimethylene-2,3-dihydrofuran,
    electronic excitation energies of, 319, 338–339
    formation of, 318–319, 338
    ionization energies, 319–320
    isomerization of, 319, 337–338
    Non-Koopmans ionization of, 319–320
    PE spectrum of, 319–320
    stability of, 319, 338
    UV spectrum of, 338–339
3,4-Dimethyl-1,2-dithiete,
    formation of, 345
    ionization energies of, 346
    molecular orbitals of, 346
4,5-Dimethyl-1,3-dithiol-2-one, pyrolysis of, 345
1,4-Dimethylsilabenzene,
    formation of, 351–353
    ionization energies of, 353–354
    molecular orbitals of, 354
    PE spectrum of, 353
2,2-Dinitropropyl-4,4-dinitrovalerate, 263
N,N'-Dinitro-N,N'-*bis*(3,3,3-trinitropropyl)oxamide, 262
3,6-Dinitraza-1,8-octanedinitramine, 261
2,2-Dinitrobutyl 4,4,4-trinitrobutyrate, 263
2,2-Dinitropropane-1,3-diol *bis*(4,4,4-trinitrobutyrate), 263
2,2-Dinitropropyl 4,4,4-trinitrobutyramide, 263
3,3-Dinitro-1,5-pentaneainitramine, 261
N,N'-Dinitro-N,N'-*bis*(3,3-dinitrobutyl)oxamide, 262
N,N'-Dinitro-N,N'-*bis*(3-nitrazabutyl)oxamide, 262

1,4-Dioxa-1,2,3,4-tetrahydronaphthalene, *See*, 1,4-Benzodioxene
1,2-Diselenete, 348–350
  formation of, 349
  IR spectrum of, 349
  PE spectrum of, 348, 350
1,3-Diselenol-2-one, pyrolysis of, 349
1,2-Dithiete, 339–350
  formation of, 341, 345, 346
  ionization energies of, 346, 350
  IR spectrum of, 345
  isomerization of, 340–341, 345–348
  molecular orbitals of, 346
  PE spectrum of, 344–346, 350
  stability of, 339–348
1,2-Dithones, 339–348
1,3-Dithilane-4,5-dithione,
  formation of, 347–348
  isomerization of, 348
1,4-Dithia-1,2,3,4-tetrahydronaphthalene, *See*, 1,4-Benzodithiene
Dithio-*o*-benzoquinone,
  isomerization of, 317–318, 321
  stability of, 317–318
Dithioglyoxal, 340–345
  formation of, 341, 347
  isomerization of, 339–347
  stability of, 339–342
Dithiooxalates, 347
Dithiooxamides, 347
DSE (diagonal strain energies), 83–86
Dunitz-Schomaker strain, 70

E

ESCA, *See* Photoelectron spectroscopy
EDA complexes, HOSE model and calculation of charge transfer, 245–251
Electron density analysis, 95, 98, 125
  of amino acids, 50
  of bases, 50
  biologically active molecules, 41–53
  calculations from quantum chemistry, 109–111
  carcinogens, 46–48
  deformation density, 27–31
  experimental measurements, 26–62
  and Harmonic oscillator stabilization energy (HOSE) model, 245–251
  Laplacian of electron density, 104–109
  multipole refinement models, 32–35
  of neurotransmitters, 48
  of nucleotides, 50
  of opiates, 41–45
  of organic molecules, 38–53
  "outer moments" calculations, 35–36
  of phenothiazines, 48–49
  of strain energy, 95–104
  strain and in situ bond energies, 113–114
  of sugars, 50
  topological analysis, 98–99
Electron donor-acceptor (EDA), molecular geometry and, 232–234
Electrostatic potential, 3–9, 12–23
  experimental measurements, 26–62
  of strained hydrocarbons, 3–23
Enantholactam, 144, 169
Energetic materials, 255–285
  classes of, 256
  conversion steps, 256–258
  definition of, 255–256, 257
  detonations, 273–285
  friction and electrostatic sensitivities, 266–267
  impact sensitivity, 259–263
  initiation of detonation, 278–280
  initiation mechanisms, 267
  molecular properties of related to sensitivity, 267–273
  sensitivity of, 258–273
  shock sensitivity, 263–266
  thermal sensitivity, 259
Energy density, 100, 104
Enthalpies, and strain energy, 76
ESCA, 160
Ethane, bond dissociation energy, 179
E(THEO), electronic consequences of strain energy, 85–86
Ethyl 2,2,2-trinitroethyl carbonate, 263
Ethylene bis(4,14,4-trinitrobutyrate), 263

Ethylenedinitramine(EDNA), 261
Explosion, definition of, 257
Explosives,
  detonations, 273–285
  empirical sensitivity-structure correlations, 269
  nitramines, 269
  nitroaromatics, 267–268
  reactive intermediates generated on initiation, 269–273
  sensitivities of, 258–273
  See also Energetic materials; specific explosives by name

F

F strain, 70
Flash Vacuum Pyrolysis (FVP), 290
Formamide structure, 144–145
6-Fulveneselone, 328–333
  formation of, 329–330
  ionization energies of, 326, 330–331
  IR spectrum of, 330
  molecular orbitals of, 331
  PE spectrum of, 329–330
  stability of, 329–330
6-Fulvenethione, 328–333
  electronic excitation energies of, 330
  formation of, 328–329
  ionization energies of, 326, 330–331
  IR spectrum of, 330
  molecular orbitals of, 331
  PE spectrum of, 329–331
  stability of, 329–333
  UV-VIS spectrum of, 330
6-Fulvenone, 312, 328–331, 333
  conjugative orbital interaction in, 328
  electronic excitation energies of, 330
  electronic structure of, 328
  ionization energies of, 326, 328, 331
  IR spectrum of, 330
  molecular orbitals of, 331
  PE spectrum of, 329–331
  stability of, 330, 333
  UV spectrum of, 330
Fulvene, 320–312

conjugative orbital interaction in, 328
  electronic structure of, 328
  formation of, 308–309, 321
  ionization energies of, 310, 326, 328
  molecular orbitals of, 310
  molecular structure of, 328
Fulvenelike systems, 319–331
  molecular structure of, 328
  orbital interaction in, 322, 326, 328, 331
Fulvenemethylene, See Methylenefulvene
Furanocyclobutadiene, bond localization in, 299
Furanocyclobutene, 319, 337–339
  electronic excitation energies of, 338–339
  formation of, 337–338
  isomerization of, 337–338
  stability of, 338
  UV spectrum of, 338–339
Fused rings, 14–22

G

Germabenzene,
  alkylsubstituted, formation of, 352–353
  ionization energies of, 354
  molecular orbitals of, 354

H

HAM/3 calculations, 294
Harmonic oscillator stabilization energy (HOSE) model, See HOSE model
Heat, and strain energy, 76
Heats of formation,
  CSE in cycloalkanes, 82–86
  energetic materials and, 255–285
  strain energy, and, 76
Heat of polymerization, 142
1,1,1,3,5,5,5-Heptanitropentane, 262
Hetero-o-xylylenes and related compounds, 308–319
Heterofulvenemethylenes, 328–331
Heterolytic cleavage,
  ion pair generation, 202–216

and polar effects on C-C bonds, 187–216
zwitterions and, 187–202
1,1,1,7,7,7-Hexanitroheptanone-2, 263
2,2,4,4,6,6-Hexanitroheptane, 263
2,2,4,7,9,9-Hexanitro-4,7-diazadecane, 261
HMX (1,3,5,7-tetranitro-1,3,5,7-tetraazacyclooctane), 258–285
Homocubane, 7, 9, 12, 18
Homolytic cleavage, and polar effects on C-C bonds, 181–187
Hooke's equation, and strain energy, 68, 102–103
Hooke's law, 68, 74, 87
 for strain, 68
HOSE model,
 applications, 245–251
 dependence on results, 242–245
 formal errors in calculating $HOSE_i$ and C_i, 241–242
 principles of, 236–238
 results compared with other models, 238–241
 and ring geometry for substituted derivatives of Benzene, 246–250
HSE (homodesmotic strain energies), 83–86
Hückel theory, aromaticity and, 117–123
Hybrid orbitals, 89, 111
Hybridization,
 molecular orbital calculations of strain and, 89–95
 in situ bond energies and, 111–113
Hydrocarbons,
 bond path of strained hydrocarbons, 1–23
 electrostatic potential, 3–23
 quantitative assessment of strain energy, 71–79
Hydrogen bonds, electron density analysis and, 50–53

I

I strain, 70
In situ bond energies, strain and, 111–113

Indanetrione, pyrolysis of, 334
2-H-indene, See, Isoindene
Indene, formation from isoindene, 304
2-1H-Indenylmethylacetate, pyrolysis of, 306–307
Infrared spectroscopy, of lactams, 161–168
Intensity of ionization, 291, 293
Isobenzofulvenallene,
 formation of, 308
 ionization energies of, 309–310
 IR spectrum of, 308
 molecular orbitals of, 310
 PE spectrum of, 308–309
 shakeup ionization of, 308
Isobenzofulvene,
 electronic excitation energies of, 308
 formation of, 305–306
 ionization energies of, 306–307, 310
 Non-Koopmans ionization of, 308
 PE spectrum of, 306–307, 310
 stability of, 306
Isoindene, isomerization of, 304

J

Jahn-Teller Distortion, in cyclobutadiene, 298

K

Koopman's Configuration, 291–293
Koopman's Ionization, 292–293
Koopman's Theorem, 291–292

L

Lactams, 139–175
 α-lactam, 164–166, 170–171
 1,2-di-*tert*-butyl, 158, 162, 171
 1,3-diadamantyl, 171
 β-lactamase, 173–174
 β-lactams, 139, 144, 146, 156, 159, 162–166, 169–174
 calorimetric data, 168–172
 caprolactam, 144, 168–169
 cis-peptides, 142–149
 crystallographic data, 168–172
 enantholactam, 144, 169

Lactams (cont.)
  infrared spectroscopy, 161–168
  IR spectra, 157–158, 161–166
  linkage studies, 144–148
  molecular orbitals, 157
  NMR spectroscopy, 160–161
  nonplanar amides and, 149–151
  photoelectron spectroscopy, 157–160
  propiolactam, 144
  strain in, 157–168, 172
  syntheses of bridgehead lactams, 151–156
  trans-peptides, 142–149
  twisted bicyclic bridgehead lactams, 141–175
  ultraviolet studies, 156–159
  UV spectra, 156–157
  valerolactam, 144, 146, 162, 164
  See also Bicyclic bridgehead lactams; Bridgehead lactams
Laplace field, 104
Laplacian concentration of electron density, 104–109
Lenard-Jones potential, and strain energy, 75
Lithiocubane derivatives, 11, 20, 21
LNDO/S calculations, 293–294

M

MATB, 258–285
Matrix Isolation Technique (MIT), 290, 294, et seq.
2-Mercaptobenzoic acid, pyrolysis of, 312–313
Mercaptothioketene, formation of, 346
Metals,
  electron density analysis, 53–61
  metal-metal and metal-ligand bonds, 55–58
2-Methylene-2H-indene, See Isobenzofulvene
6-Methylenecyclohexa-2,4-dienethione, See Thio-o-quinone methide
1-Methyl-4-ethylgermabenzene,
  formation of, 353–354
  ionization energies of, 354
  molecular orbitals of, 354
1-Methylsilabenzene,
  formation of, 349, 351
  ionization energies of, 354
  molecular orbitals of, 354
(2-Methyl-3-furyl)methyl benzoate, pyrolysis of, 318
6-Methylenecyclohexa-2,4-dienone, See o-Benzoquinone methide
Methyl 2,2,2-trinitroethyl carbonate, 262
N-Methyl, 162–168
4-Methyl-1,2-benzodithete,
  formation of, 343–344
  IR spectrum of, 343
2-Methyl-3-chloromethylfuran, pyrolysis of, 319, 338
Methylenebis(2,2,2-trinitroacetamide), 262
Methylenebis(4,4,4-trinitrobutyramide), 263
Methylenedinitramine(MEDINA), 261
Methylenefulvene,
  formation of, 320–321
  ionization energies of, 310, 326, 331
  molecular orbitals of, 310, 331
1-Methyl-4-tert-butylgermabenzene,
  formation of, 353–354
  ionization energies of, 354
  molecular orbitals of, 354
1-Methyl-4-tert-butylsilabenzene,
  formation of, 351–354
  ionization energies of, 354
  molecular orbitals of, 354
Methyl 2,2,2-trinitroethyl nitramine, 261
MNDO calculations, 293–294
Molecular geometry,
  and bond length estimates by x-ray diffractometry, 234–235
  electron donor-acceptor (EDA) complexes, 232–234
  HOSE model and, 236–250
  resonance theory and, 231–253
Molecular orbital calculations,
  of biologically active molecules, 41–53

Coulson-Moffitt bent bond models, 90–95
electron donor-acceptor (EDA) complexes and, 232–234
estimates by x-ray diffractometry, 234–235
Hückel theory and, 117–123
"outer moments" calculations, 35–36
pi complexes and, 124–130
pi complexes and three membered rings, 124–130
of strain, 89–95
strain and in situ bond energies, 111–113
Walsh MO's and, 117–123
Westheimer approach, 114–117
Molecular strain, 1–5, 15–23, 66–132
"bond angle strain", 69–71
bond opposition strain, 69–70
bond parameters of strained rings, 102–104
in bridgehead lactams, 157–168, 172
Buckingham potential, 74–75
chemical consequences of, 79–82
in chemistry, 69
in classical mechanics, 72
classical mechanics concept of strain, 68
Coulson-Moffitt bent bond models, 90–95
in cycloalkanes, 82–87
of cyclobutane, 87–89
of cyclopropane, 87–89
definition of strain energy, 74–79
DSE (diagonal strain energies), 83–86
Dunitz-Schomaker strain, 70
electron density analysis, 104–109, 113–114
electron density approach to, 95–104
electrostatic strain, 70
F strain, 70
HSE (homodesmotic strain energies), 83–86
I strain, 70
Laplacian of electron density, 104–109

Lenard-Jones potential, 75
Pitzer strain, 69–70
quantitative assessment of strain energy, 71–79
quantum chemical calculations, 109–111
in situ bond energies and, 111–113
steric strain, 70
Stoll pressure, 70
"strain-free" reference states, 72–74
superstrain, 70
three membered rings, 124–130
types of, 69–71
Von Baeyer strain, 69–72, 87, 111, 122
Walsh MOs, 89–95
Westheimer approach, 114–117
ZPE (zero-point energy) and calculations of, 83–86
Molecular structure, 128
Molybdenum cycloheptatriene tricarbonyl, molecular structure, 27–30
Multipole refinement models, of electron density analysis, 32–35

N

N-Methyl EDNA, 261
Naphtho[b]cyclopropene, pyrolysis, 308
Neurotransmitters, electron density analysis, 48
3-Nitraza-1,5-pentanedinitramine, 261
Nitramines,
  molecular properties of related to sensitivity, 269
  See also Energetic materials
Nitroaromatics,
  molecular properties of related to sensitivity, 267–268
  See also Energetic materials
Nitroisobutyl 4,4,4-trinitrobutyrate, 263
N-Nitro-N-(3,3,3-trinitropropyl)2,2,2-trinitroethyl carbamate, 262
NMR spectroscopy, of lactams, 160–161

GENERAL INDEX

Non-Koopmans' Configuration, 292–293
Non-Koopmans' Ionization, 292–293
1,1,1,5,7,10,14,14,14-Nonanitro-3,12-dioxa-4,11-dioxo-5,7,10-triazatetradecane, 262
1,1,1,3,6,9,11,11,11-Nonanitro-3,6,9-triazaundecane, 262
Nonplanar amides, and lactams, 149–151
Nucleotides, electron density analysis, 50

O

2,2,4,7,7,10,12,12-Octanitro-4,7-diazatridecane, 261
2,2,5,7,7,9,12,12-Octanitro-4,7-diazatridecane, 261
Octatetraene,
  all *trans*,
    frontier orbitals of, 302, 304
    Non-Koopmans' ion state of, 302–304
Opiates, electron density analysis, 41–45
Organic molecules, electron density analysis, 38–53
Outerpath angle, 101, 103
1-Oxa-4-thia-1,2,3,4-tetrahydronaphthalene, *See* 1,4-Benzoxathiene
1-Oxabenzocyclobutene, *See* Benzoxete
2-Oxabicyclo[2.2.0]hex-5-en-4-one, *See* Photo-alpha-pyrone
2-Oxabicyclo[3.2.0]hepta-1(5), 3-diene, *See* Furanocyclobutene
3-Oxabicyclo[3.2.0]hepta-1,4,6-triene, *See* Furanocyclobutadiene
Oxirane, 125
  strained rings in, 102–103

P

PDA, *See* p-Phenylenediamine (PDA)
Penicillins, 141, 169–170, 172
2,2,4,6,6 Pentanitroheptane, 263
2,2,6,9,9-Pentanitro-4-oxa-5-oxo-6-azadecane, 262
Pentaerythritol tetranitrate, sensitivities and detonation of, 258–285
N,3,3,5,5-Pentanitropiperidine, 261
Peptide distortion, 146–148
Peptides,
  cis-peptides, 142–149
  trans-peptides, 142–149
Pericyclic cleavage, and polar effects on C-C bonds, 216–226
PERTCI calculations, 293–294
PETN (pentaerythritol tetranitrate), 258–285
Phenothiazines, electron density analysis, 48–49
o-Phenylene carbonate, pyrolysis of, 322–323
o-Phenylene sulfate, pyrolysis of, 322–323
o-Phenylene sulfite, pyrolysis of, 322–323
p-phenylenediamine (PDA), electron donor-acceptor (EDA) complexes and, 232–234
Photo-α-pyrone, pyrolysis of, 296
Photoelectron spectra (PE spectra), 290
  calculation of, 293
Photoelectron spectroscopy (PES), 157–160, 290–293
  of lactams, 157–160
Photoionization, 290
Phthalic anhydride, pyrolysis of, 334
Phthaloyl peroxide, pyrolysis of, 334
Pi complex, 124, 128
Pi-aromaticity, 117
Picramide, 258–285
Polar effects,
  on C-C bond lability, 179–226
  on C-C bond heterolytic cleavage, 187–216
  on C-C bond homolytic cleavage, 181–187
  on C-C bond pericyclic cleavage, 216–226
[1.1.1]propellane, 7, 8, 12, 17
Propiolactam structure, 144
Proteases, model substrates for, 175
2-Pyridone, 169
α-Pyrone, formation of, 296

Q

2-Quinuclidone, *See,* 1-Azabicyclo[2.2.2]octan-2-one
Quinomethanes and related compounds, 299-319

R

RDX (1,3,5-Trinitro-1,3,5-triazacyclohexane), 258-285
Reference states,
  diagonal, 82
  strain-free, 72, 82
Resonance structures, HOSE model and, 236-250
Resonance theory, and molecular geometry, 231-253

S

Salicyclic acid, pyrolysis of, 312-313
SE (strain energy), *See* Molecular strain; Strain; Strain energy
Selenirene, formation and spectroscopic detection, 332
Self-consistent field (SCF), 3
Semiempirical valence electron methods,
  electronic excitation energies and intensities calcs, 294
  Ionization energies and intensities calculations, 293
  IR spectra, 294
Shakeup band, 292
Shakeup ionization, 292
Shock wave, definition of, 257
Si compounds, 131, 351-354
1-Sila-2,5-cyclohexadiene, pyrolysis of, 351-352
Silabenzene, 349-354
  formation of, 351-352
  ionization energies of, 354
  molecular orbitals of, 354
  PE spectrum of, 351-354
Spectroscopy,
  4-Acetoxy-2-cyclopentenone, 322-323

1-Acetoxy-1-silacyclohexa-2,4-diene, 351-352
1-Allyl-1-germacyclohexa-2,4-diene, 352-354
1-Allyl-1-silacyclohexa-2,4-diene, 351-353
AM1 method, for calculation of IR spectra, 294
1,2,3-Benzoselenadiazole, 329-333
1,4-Benzodithiene, 342
o-Benzoquinone, 321-323
o-Benzyne, 336-337
Benzocyclobutadiene, 299
Benzocyclobutene, 301
Benzodithiete, 344
Benzothiet-2-one (Thiobenzopropiolactone), 312-315
Benzothiete, 315, 317-318
Benzothiirene, 332-333
Benzotrithiol-2-oxide, 342-344
Benzoxathiete, 315-316
Benzoxete, 317-318
Benzoxirene, 321, 333
Bicyclo[1.1.0]butane-2,4-dicarboxylic acid anhydride, 322-323
1,2-Bis(trifluoromethyl)dithioglyoxal, 347
Bis(methylene)cyclobutene, 299
Butadiene, 310
6-Carbonyl-2,4-cyclohexadienone, 311-315, 317-318
Cyclobutadiene, 295
Cyclopentadienethione, 321-331
Cyclopentadienone, 295, 321-323, 327
1,4-Di-tert-butylgermabenzene, 352-353
3,4-Di-tert-butyl-1,2-dithiete, 340, 345-347
Dialkyltetrathiooxalates, 347-348
Diazocyclopentadiene, 328-330
3,4-Diacyano-1,2,dithiete, 340
2,3-Dihydrobenzo[b]furan-2,3-dione, 312
2,3-Dihydrobenzo[b]thiophen-2,3-dione, 312
2,3-Dimethylene-2,3-dihydrofuran, 319, 338-339

Spectroscopy (*cont.*)
 3,4-Dimethyl-1,2-dithiete, 346
 1,2-Dithiete, 339–350
 1,3-Dithiol-4,5-dithione, 347–348
 1,2-Dithiones, 339–348
 Dithio-*o*-benzoquinone, 317–318
 Dithioglyoxal, 340–345
 Dithiooxalates, 347
 Dithiooxamides, 347
 6-Fulvenone, 312, 328–331, 333
 Fulvene, 320–312
 Furanocyclobutadiene, 299
 Furanocyclobutene, 319, 337–339
 Isobenzofulvenallene, 308–310
 Isobenzofulvene, 305–310
 1-Methyl-4-*tert*-butylsilabenzene, 351–354
 1-Methylsilabenzene, 349, 351
 2-Methyl-3-chloromethylfuran, 319, 338
 4-Methyl-1,2-benzodithete, 343–344
 Methylenefulvene, 320–331
 o-Phenylene sulfite, 322–323
 Phthaloyl peroxide, 334
 o-Xylylene, 290, 298, 300–306
 p-Xylylene, 300–304
Spiropentane, 7, 8, 14, 15, 16
STAB (in situ atomic energy), *See* Atomic energies; Molecular strain; Strain energy
Stabilization energy, HOSE model and, 236–250
Steric energy,
 calculation of, 75–76
 See also Molecular strain; Strain energy
Steric strain, 70
STO-3G, 3
STO-5G, 3
Stoll pressure, 70
Strain, 66
 bond opposition strain, 69–70
 bond parameters of strained rings, 102–104
 in bridgehead lactams, 157–168, 172
 calculations from quantum chemistry, 109–111
 chemical consequences of, 79–82
 in chemistry, 69
 in classical mechanics, 68, 72
 colloquial uses of term, 70–71
 Coulson-Moffitt bent bond models, 90–95
 definition of strain energy, 74–79
 Dunitz-Schomaker strain, 70
 electron density approach to, 95–104
 electronic consequences of strain energy, 85–86
 F and I strain, 70
 Hooke's law, 68
 Laplacian of electron density, 104–109
 limitations of concept, 123–128
 molecular orbital calculations, 89–95
 pi complexes and three membered rings, 124–130
 Pitzer strain, 69–70
 quantitative assessment of strain energy, 71–79
 in situ bond energies and electron density analysis, 113–114
 steric strain, 70
 Stoll pressure, 70
 "strain-free" reference states, 72–74
 three membered rings, 124–130
 types of, 69, 69–71
 unified description of, 104–109
 Von Baeyer strain energy, 71–73, 87, 111, 114, 122
 See also Molecular strain; Strained molecules
Strain in cycloalkanes, 82
Strain energy, 5, 6, 9, 15
 in classical mechanics, 72
 Coulson-Moffitt bent bond models, 90–95
 of cycloalkanes, 87
 cyclobutane, 102–104
 of cyclobutane, 87, 87–89, 180
 of cyclohexane, 85
 cyclopropane, 102–104
 of cyclopropane, 87, 87–89, 119, 130

definition of, 74-79
DSE (diagonal strain energies), 83-86
electron density analysis, 104-109, 113-114
electron density approach to, 95-104
electronic consequences of, 85-86
Hooke's equation, 102-103
HSE (homodesmotic strain energies), 83-86
Hückel theory and, 117-123
Laplacian of electron density, 104-109
molecular orbital calculations, 89-95
quantitative assessment of, 71-79
quantum chemical calculations, 109-111
in situ bond energies and, 111-113
in situ bond energies and electron density analysis, 113-114
in small cycloalkanes, 82-87
"strain-free" reference states, 72-74
three membered rings, 124-130
Walsh MOs and, 117-123
Westheimer approach, 114-117
ZPE (zero-point energy) and calculations of, 83-86
See also Molecular strain; Strain
Strain limitations, 123
Strained hydrocarbons,
bond path and electrostatic potential, 1-23
See also Hydrocarbons
Strained molecules, 66-132, 295-299, 332-339
bond deviation index, 4-5
electrostatic potential, 5-6
hydrocarbons, 3-23
quantitative assessment of strain energy, 71-79
structures of, 3-4
types of strain, 69-71
See also Molecular strain
Stress, 68, 71
See Molecular strain; Strain

Sugars, electron density analysis, 50
o-Sulfobenzoic anhydride, pyrolysis of, 334
Surface delocalization, 103, 108, 119, 131
Surface orbital, 92, 94, 108

T

TATB, 258-285
TCNQ geometry, HOSE model and, 245-251
TENA, 258-285
TET, 258-285
1,1,1,3-Tetranitrobutane, 262
2,2,5-Tetramethylhexan-3,4-dithione, formation of, 347
2,4,6,7-Tetrathiabicyclo[3.2.0]hept-1(5)ene, formation of, 347-348
3,3,4,4-Tetranitrohexane, 263
3,3,9,9-Tetranitro-1,5,7,11-tetraoxaspiro[5,5]undecane, 263
4,5,6,7-Tetrahydrobenzofuran, pyrolysis of, 318
Tetrahedrane, 7, 8, 16, 19
Tetrahedrane derivatives, 16
Tetrakis(2,2,2-trinitroethyl) orthocarbonate, 263
1,3,5,7-Tetranitro-1,3,5,7-tetrazacyclooctane (HMX), 258-285
Tetranitroaniline, sensitivities and detonation of, 258-285
Tetrathiooxalic acid esters,
formation of, 347-348
stability of, 347
Tetryl, 261
1,2-Thiaselenete,
formation of, 348-349
IR spectrum of, 349
PE spectrum of, 349, 350
1-Thiabenzocyclobutene, *See* Benzothiete
1,3-Thiaselenol-2-one, pyrolysis of, 349
Thio-o-benzoquinone,
electronic excitation energies of, 316
formation of, 315-316, 321

Thio-*o*-benzoquinone (*cont.*)
 ionization energies of, 316
 isomerization of, 315–318, 321
 stability of, 315–318
Thio-*o*-quinone methide,
 electronic excitation energies, 315
 formation of, 315
 IR spectrum of, 315
 isomerization of, 315, 317–318, 321
 stability of, 315, 317–318
Thiobenzopropiolactone, *See* Benzothiete-2-one
Thiosalicyclic acid, *See* 2-Mercaptobenzoic acid
Three membered rings, pi complexes and, 124–130
TNT, 258–285
Transpeptidase, 173–174
Transition metal complexes,
 electron density analysis, 53–61
 metal-metal and metal-ligand bonds, 55–58
1,1,1,3,6,6,8,10,10,13,15,15,15-Tridecanitro-3,8,13-triazapentadecane, 262
2,2,2-Trinitroethyl 2,4,6,6-tetranitro-2,4-diazaheptanoate, 262
2,2,2-Trinitroethyl 2,5,5-trinitro-2-azahexanoate, 264
2,2,2-Trinitroethyl 2,5-dinitrazahexanoate, 262
2,2,2-Trinitroethyl 3,3,3-trinitropropyl nitramine, 262
2,2,2-Trinitroethyl 3,3-dinitrobutyl nitamine, 262
2,2,2-Trinitroethyl 4-nitrazavalerate, 262
2,2,2-Trinitroethyl *N*-(2,2,2-trinitroethyl)nitraminoacetate, 262
3,5,5-Trinitro-3-azahexyl nitrate, 262
3-[*N*-2,2,2-Trinitroethylnitramino]propyl nitrate, 262
2,2,2-Trinitroethyl 4,4,4-trinitrobutyrate, 263
2,2,2-Trinitroethyl 4,4-dinitrovalerate, 263
2,2,2-Trinitroethyl carbamate, 262
4,4,4-Trinitrobutyric anhydride, 263
5,5,5-Trinitro-2-pentanone 2, 263

N-(2,2,2-Trinitroethyl)-3,3,5,5-tetranitropiperidine, 263
2,2,2-Trinitroethyl, 4,4-dinitrohexanoate, 263
1,3,5-Trinitrobenzene, 258–285
1,3,5-Trinitroso-1,3,5-triazacyclohexane, 258–285
1,3,5-Trinitro-1,3,5-triazacyclohexane (RDX), 258–285
2,4,6-Trinitrotoluene, 258–285
Trinitro-RDX, 258–285
N-(2,2,2-Trinitroethyl)nitraminoethyl nitrate, 262
N-Methyl-*N*'-trinitroethyl EDNA, 262
N,*N*'-bis(2,2,2-Trinitroethyl)-MEDINA, 262
Triprismane, 7, 8, 17, 18
Triprismane derivatives, 10, 13, 19–21
Tris(2,2,2-trinitroethyl) orthoformate, 263
Twisted bicyclic bridgehead lactams, 141–175

U

Ultraviolet photoelectron spectroscopy, 290–355
 of lactams, 156–159
 See also Spectroscopy

V

Valerolactam(2-Piperidone), 144, 146
 N-Methyl, 162, 164
Variable Temperature Photoelectron Spectroscopy (VTPES), 290
Vertical ionization energy (VIE), 291
 calculation of, 293
2-Vinylidene-2H-indene, *see* Isobenzofulvenallene
Von Baeyer strain energy, 69–73, 87, 111, 122

W

Walsh MOs,
 Hückel theory and, 117–123
 and strain energy, 89–95
Walsh orbital, 90, 120

Westheimer approach, to strain energy calculations, 114–117
Westheimer equation, 74, 114
Winkler-Dunitz parameters, 144–145, 147–149

X

X-ray crystallographic studies, 25–28
  biologically active molecules, 48–53
  d orbitals, 54–61
  electron density analysis, 27–61
  linkage studies of lactams, 142–149
  organic substances, 38–53
  strain energy and, 95–104
  transition metal complexes, 53–61
X-ray diffractometry, and bond length estimates, 234–235
Xylylenes and related compounds, 299–319

o-Xylylene, 290, 298, 300–306
  formation of, 301–302
  frontier orbitals of, 302, 304
  ionization energies of, 302–304
  isomerization of, 301, 317–318, 337
  Non-Koopmans' ionization of, 302–304
  PE spectrum of, 302–304
  stability of, 301, 317–318, 337
p-Xylylene, 300, 304
  formation of, 300
  Non-Koopmans' (shakeup) ionization of, 300
  polymerization of, 300

Z

Zero flux surface, 98
ZPE (zero-point energy), in strain calculations, 83–86